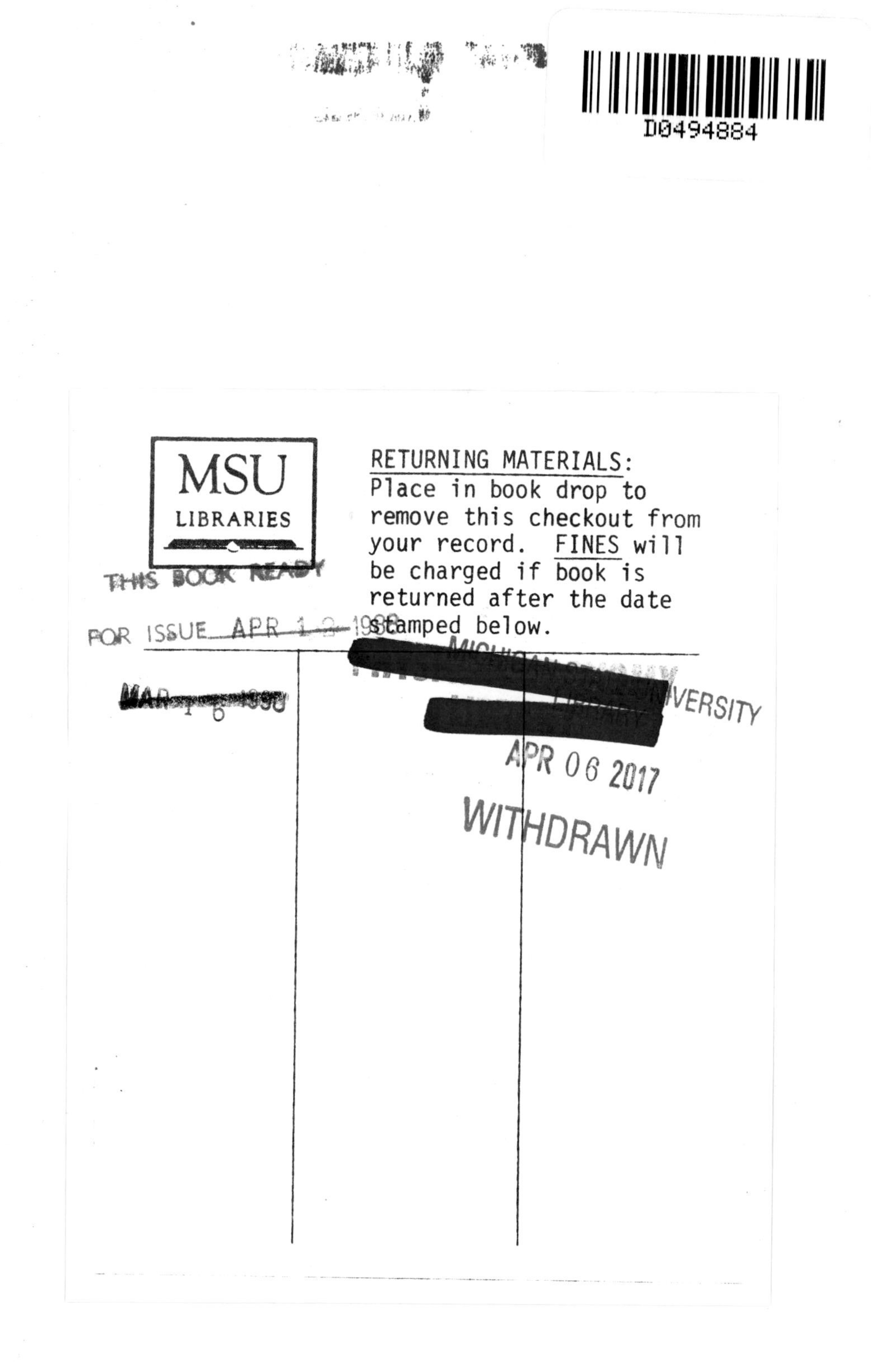

NEW DIRECTIONS IN SPECTROPHOTOMETRY

A meeting held in Las Vegas, Nevada
March 28 -30, 1988

Edited by

A. G. Davis Philip
Van Vleck Observatory, Union College and
Institute for Space Observations

D. S. Hayes
Institute for Space Observations and
Fairborn Observatory

Saul J. Adelman
The Citadel and
Institute for Space Observations

Van Vleck Observatory Contribution No. 6

L. Davis Press
Schenectady, N. Y.
1988

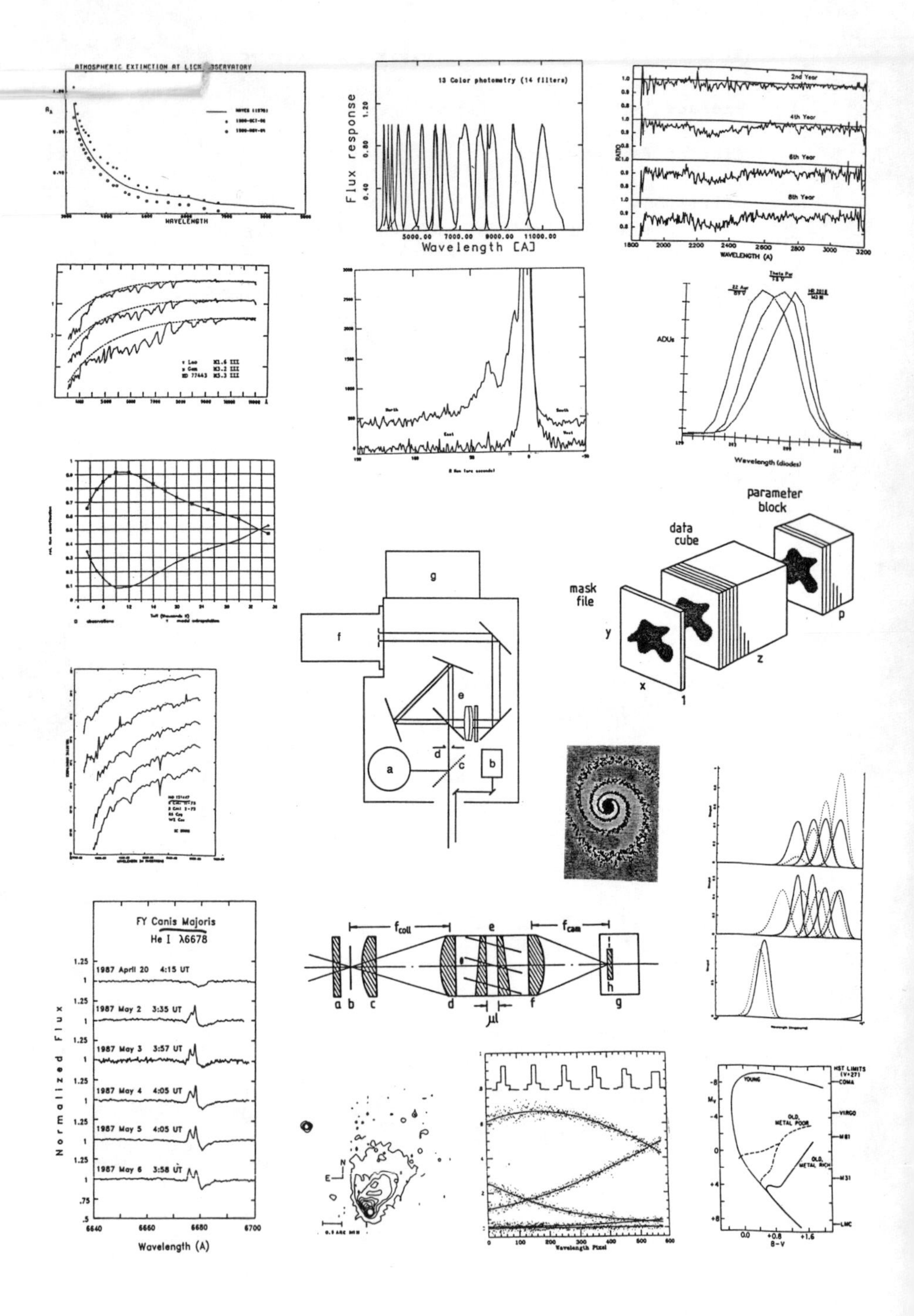

ATMOSPHERIC EXTINCTION AT LICK OBSERVATORY
WAVELENGTH
13 Color photometry (14 filters)
Flux response
Wavelength [A]
2nd Year
4th Year
6th Year
8th Year
RATIO
WAVELENGTH (A)
ADUs
Wavelength (diodes)
parameter block
data cube
mask file
g
f
e
d
c
a
b
y
x
z
p
FY Canis Majoris
He I λ6678
1987 April 20 4:15 UT
1987 May 2 3:35 UT
1987 May 3 3:57 UT
1987 May 4 4:05 UT
1987 May 5 4:05 UT
1987 May 6 3:58 UT
Normalized Flux
Wavelength (A)
f_{coll}
f_{cam}
μl
a b c d e f g h
Wavelength Pixel
YOUNG
OLD, METAL POOR
OLD, METAL RICH
HST LIMITS (V=27)
COMA
VIRGO
M81
M31
LMC
B−V

TABLE OF CONTENTS

Section I - Day One

Section III - Day Three

PREFACE

The meeting, "New Directions in Spectrophotometry" was held March 28-30, 1988 at the Alexis Park Hotel in Las Vegas, Nevada. The decision to hold such a meeting was made by Saul J. Adelman and Donald S. Hayes after extensive discussions with colleagues interested in photometric and spectrophotometric problems. They felt that a meeting on spectrophotometry was long overdue and hoped that it would stimulate research and new instruments especially those using array type detectors which had the potential to achieve or surpass the accuracies and precisions of rotating grating scanners.

To our knowledge "New Directions in Spectrophotometry" was the first meeting held to discuss this important astrophysical technique. It allowed investigators to assess the current status of this method of examining astronomical energy distributions and to present a number of astrophysical problems which can be profitably investigated by spectrophotometry. These discussions formed the basis for a consideration of the current and future instrumental needs. In discussing this topic the IUE, small automatic telescopes, and array detector experiences played a major role. Forty five astronomers from six countries attended the three day meeting.

The Scientific Organizing Committee was co-chaired by Saul J. Adelman and Donald S. Hayes. Other members of the SOC were Ralph Bolhin, Rusell M. Genet, Hollis R. Johnson, Rolf Kudritzki, A. G. Davis Philip, Diane Pyper Smith, and Benjamin Taylor. Session chairmen were:

Session	1.	Saul J. Adelman
	2.	Benjamin J. Taylor
	3.	Hollis Johnson
	4.	Ralph Bohlin
	5.	D. S. Hayes
	6.	J. B. Oke
	7.	Diane Pyper Smith
	8.	Russell Genet
	9.	David L. Crawford
	10.	Robert Kurucz
	11.	A. G. Davis Philip
Workshop		D. S. Hayes and S. J. Adelman

"New Directions in Spectrophotometry" was hosted by the University of Nevada, Las Vegas and Fairborn Observatory. The reception at the UNLV Museum of Natural History was made possible by the assistance of Dr. John Unrue, Vice President for Academic Affairs and Provost. Local organizing was coordinated by Diane Pyper Smith.

Carol J. Adelman was instrumental in helping with the mailings and in preparing the abstract book. Many of the incident expenses for organizing this meeting were paid by grants from The Citadel Development Foundation to SJA. We appreciate the cooperation of Gwen Smith, Daniel Long and their staff of the Alexis Park Hotel.

Authors of papers at the meeting submitted their contributions over BITNET or on floppy disks. Having the papers in electronic form made the editing process much easier and quicker. The files, received in ASCII form, were transferred to the word processor "Final Word" and printed out on a Laser printer. These versions were mailed back to authors for final checking. The authors are thanked for their help in this cooperative effort in producing the proceedings volume. Mary Bongiovanni organized the typing of the discussions following each paper. Most of the discussion was proofread during the meeting and only the last few sections had to be done after the meeting.

June, 1988

Schenectady

The Editors

LIST OF PARTICIPANTS

France

Francois Querci — Observatory of Midi-Pyrenees

Italy

Maria L. Malagnini-Sicuranza — University of Trieste

Sweden

Lars O. Loden — Astronomical Observatory, Uppsala

Switzerland

Lukas Labhardt — Astronomical Institute, Univ. of Basel

USA

Saul J. Adelman	The Citadel
Thomas B. Ake III	CSC/Space Telescope Science Institute
Roger A. Bell	University of Maryland
J. Bland	Institute for Astronomy, Univ. of Hawaii
Ralph C. Bohlin	Space Telescope Science Institute
Louis J. Boyd	Fairborn Observatory
John W. Briggs	Los Gatos, California
Carla Cacciari	Space Telescope Science Institute
Roy Campbell	Southwestern Adventist College
Tod Colegrove	University of Nevada, Reno
David L. Crawford	Kitt Peak National Observatory
Paul B. Etzel	San Diego State University
Russell M. Genet	Fairborn Observatory
Carol A. Grady	CSC/IUE Observatory
Edwin J. Grayzeck	University of Nevada, Las Vegas
Donald S. Hayes	Fairborn Observatory
Keith D. Horne	Space Telescope Science Institute
Hollis R. Johnson	Indiana University
Mike Joner	Brigham Young University
William C. Keel	University of Alabama
Robert L. Kurucz	Smithsonian Astrophysical Observatory
Stephen J. Little	Bently College
Kelly McDonald	University of Colorado
Albert Merville	MIRA
John B. Oke	California Institute of Technology
Klaus Olesch	Spring Grove, Illinois
Geraldine J. Peters	University of Southern California
A. G. Davis Philip	Van Vleck Observatory and Union College
William V. Schempp	Photometrics, Ltd.
Alfred B. Schultz	University of Nevada, Las Vegas

Diane Pyper Smith	University of Nevada, Las Vegas
Benjamin J. Taylor	Brigham Young University
Terry J. Teays	IUE Observatory
David A. Turnshek	Space Telescope Science Institute
Arthur R. Upgren	Van Vleck Observatory
Wayne H. Warren, Jr.	National Space Science Data Center
Nathaniel M. White	Lowell Observatory
Ramon L. Williamson II	Space Telescope Science Institute
Robert F. Wing	Ohio State University

W. Germany

Karl-Heinz Mantel	University Observatory, Munich
Joachim Puls	University Observatory, Munich

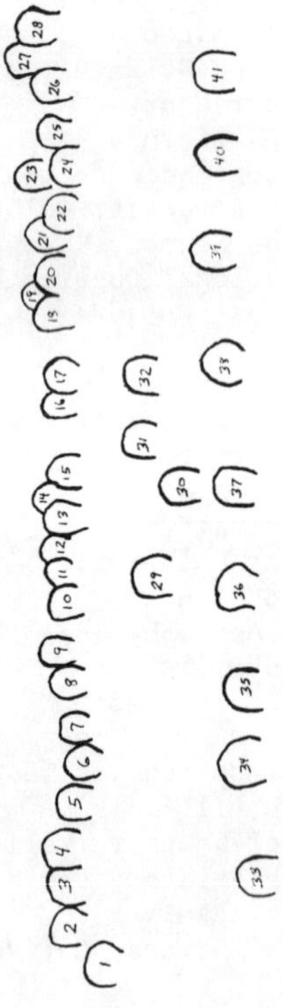

1. G. Peters
2. N. White
3. T. Colgrove
4. D. Philip
5. R. Campbell
6. D. Pyper Smith
7. K. Olesch
8. R. Wing
9. R. Genet
10. B. Oke
11. L. Loden
12. W. Keel
13. H. Johnson
14. J. Briggs
15. A. Upgren
16. R. Williamson
17. B. Taylor
18. A. Merville
19. L. Labhardt
20. K. Mantel
21. J. Puls
22. W. Weaver
23. W. Warren
24. C. Cacciari
25. T. Teays
26. K. McDonald
27. M. Joner
28. R. Bell
29. L. Boyd
30. T. Ake
31. D. Turnshek
32. K. Horne
33. D. Hayes
34. R. Kurucz
35. R. Bohlin
36. S. Little
37. F. Querci
38. C. Grady
39. E. Grayzeck
40. D. Crawford
41. S. Adelman

REVIEW PAPERS

Session Chairmen

Day One

Session 1.	Saul J. Adelman
Session 2.	Benjamin J. Taylor
Session 3.	Hollis R. Johnson
Session 4.	Ralph C. Bohlin

Day Two

Session 5.	D. S. Hayes
Session 6.	John B. Oke
Session 7.	Diane Pyper Smith
Session 8.	Russell M. Genet

Day Three

Session 9.	David L. Crawford
Session 10.	Robert L. Kurucz
Session 11.	A. G. Davis Philip

THE HISTORY AND LEGACY OF PHOTOMULTIPLIER SCANNERS

Benjamin J. Taylor

Brigham Young University

This review will be a set of snapshots - not a panorama. Given its thirty-five minute spoken precursor's length, no other course can be feasible. Within a year or two, I hope to expand it into a full-scale history of photomultiplier scanners. My literature search for this history has reached 1968; this limit will influence my choice of sources to some degree. I'll pursue two trains of historical thought and touch on the following questions.

1) How have photomultiplier scanners been perceived?

2) How have they been used, and how may their use pattern help us now?

3) How has their accuracy compared to that of area-detector devices?

4) What can we learn from a comparative history of primary and secondary calibrations?

5) What quality of scanner data does the literature bequeath to us? What can we learn from that quality, and what effects should it have on present planning?

At the end of World War II, a substantial gap opened in instrumental practice. The 1P21 became available - but at first, astronomers did not fully exploit it. To describe the problem, I've plotted a schematic diagram (see Fig. 1) with resolution depicted on the ordinate and quantum efficiency on the abscissa. One sees how spectroscopy was being left out of sensitive-detector usage.

The gap was not to be closed by simply putting a photomultiplier in the focal plane of a spectrograph. The reason was that such an instrument had to be "sequential" - that is, it had to measure resolution elements one after another. Seeing fluctuations could affect these measurements differentially - and seriously. This effect was important as long as the spectrograph slit was at least as small as the stellar image. (See, for example, Hiltner and Code 1950).

A. G. Davis Philip, D. S. Hayes and S. J. Adelman, (eds.)
New Directions in Spectrophotometry 3 - 15

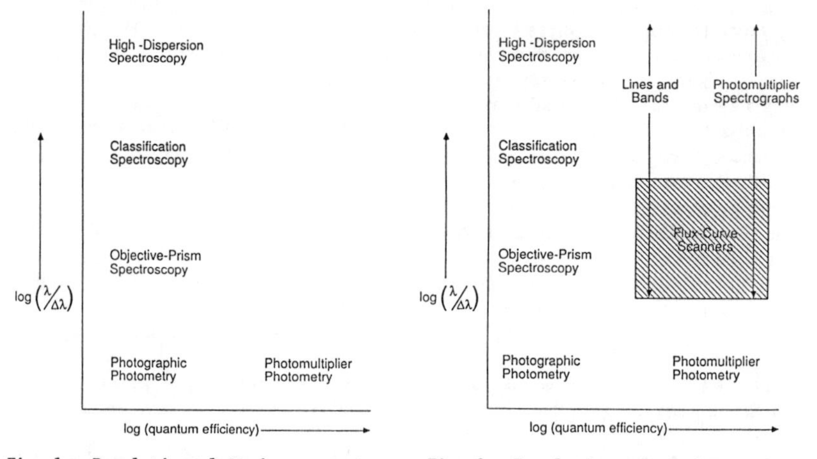

Fig. 1. Resolution plotted vs. quantum efficiency for instruments in use just after World War II.

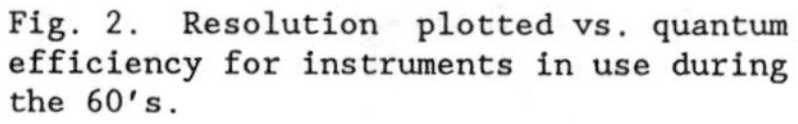

Fig. 2. Resolution plotted vs. quantum efficiency for instruments in use during the 60's.

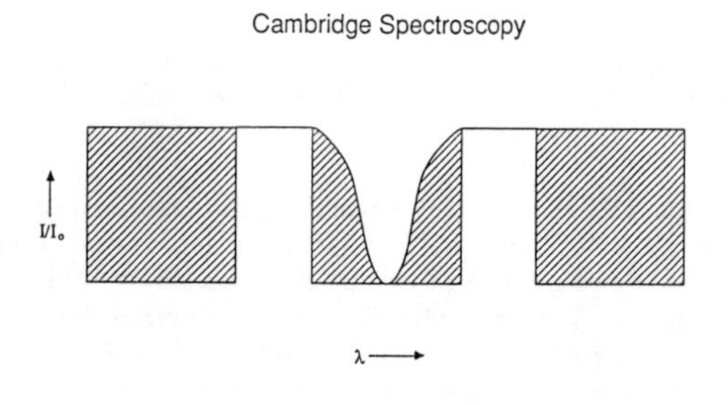

Fig. 3. The "scanner Cambridge spectroscopy" procedure for measuring absorption features.

In response to this problem, the 1950's developed two kinds of instruments. In one, the spectrograph slit was kept narrow, so resolution was not sacrificed. Seeing fluctuations were dealt with by comparing the outputs from "scan" and "monitor" tubes (Code and Liller 1962). I shall call these instruments "photomultiplier spectrographs."

In the other kind of instrument, the entrance slit was replaced by an aperture through which essentially all the starlight passed. The size of the stellar image limited the resulting resolution to a few Ångstroms at best (Oke 1965b). I shall call these instruments "flux-curve scanners" or simply "scanners."

Scanners and photomultipler spectrographs differed quite noticeably. For one thing, photomultiplier-spectrograph data spanned too narrow a wavelength range to require correction for atmospheric extinction. Scanner data did require such correction, and this gave scanner work a resemblance to filter photometry. This resemblance was fortified by the need to transform most scanner data to a standard system. It was increased when continuous scanning gave way to pulse counting and discrete-wavelength measurements in the 1960's. At that point, one effectively used scanners to do N-passband, narrow-band filter photometry. No similar statement could be made about photomultiplier spectrographs.

Another difference between the instruments rose from choice of problem. Flux-curve work, unlike spectroscopy, has an inescapable upper limit to resolution. One finds out nothing new about large-scale wavelength trends by narrowing an exit slot below a certain generous size. Instead, one simply throws away signal - often needlessly. In addition, small exit slots risk data bias by localized wavelength features. For these reasons, the resolution limit I've mentioned above was acceptable in scanners. No similar limit applied to photomultiplier spectroscopy, so the two kinds of instrument occupied different areas in the resolution-sensitivity plane (see Fig. 2).

If those areas had been disjoint, the historian's task had been simpler. In fact, they overlapped; scanner resolution could be used for absorption features as long as they were strong enough. One measured a "feature band" and one or two "side bands," then calculated an index which depicted the light lost to the "feature band" (see Fig. 3). A similar procedure could be used for emission lines. Photomultiplier spectroscopy of this sort was done prominently by the Cambridge observers (see, for example, Griffin and Redman 1960). Three-passband "scanner Cambridge spectroscopy" was first done by Oke on QSO emission lines (Oke 1965a).

In practice, the scanner was a vigorously prosecuted success. The literature for 1967 may be used to support this point. I count 46 papers from that year which either presented new scanner data or discussed such data extensively. It is, therefore, surprising that, at that very time, some astronomers were dismissing the scanner as a failure. In very condensed form, they reasoned thus: "A scanner is a failed spectrograph."

Behind this reasoning were two fundamental misperceptions. One failed to distinguish scanners from photomultiplier spectrographs, and so aimed criticisms at one which were entirely appropriate only for the other. Confusion between the two instruments was by no means peculiar to the scanner's critics. Only Code and Liller's (1962) review seems to have made the distinction at all, and no one seems to have stressed it. It would have helped if scanners had been thought of as photometers - as they fairly can be - but they seem to have been described almost always in spectroscopic terms. Scanner Cambridge spectroscopy, if known to critics of scanners, would have helped to increase the confusion.

I and others encountered the other misperception early in the 1970's. A fit name for it is "high-dispersion imperialism." Defenders of this attitude focus exclusively on lines and bands, and maintain that essentially nothing can be learned about them at scanner resolution. By deciding that such resolution is unfit for use, they tacitly decide also that flux curves are unfit for study.

For any who plan flux-curve devices, this kind of thinking is a legitimate concern. Let's see how this concern might be confronted. Suppose we want to do this and also to list projects for a flux-curve device, and that we've searched the literature for a catalog of scanner projects done in the past. We proceed by extracting two lists of feasible projects from this catalog. The projects in one list are fit for a research proposal; those in the other list are not.

Tables I through III show some samples of what we may expect to find. For convenience, I've compiled these tables from the literature for 1966-8. I've also assumed that we're considering an automated telescope or other small telescope, so I've excluded galaxy projects from the tables.

Inevitably a list of topics from twenty years ago will contain some which are outdated now. I've included four of these - with two involving Cambridge spectroscopy - in Table I. "Emission-line objects" are usually planetary nebulae or H II regions; the paper I've cited in this case is from an extensive series on these objects. Potassium flares turned out, on strong circumstantial grounds, to be due to light from matches which had accidentally been introduced into coudé optical trains (Wing, Peimbert, and Spinrad 1967). In both cases, one would now choose an area detector and reserve the freedom to use resolutions which are inconveniently high for flux curves. Oke (1965) prefigured this idea when he remarked on the difficulty of using a scanner to split the λ 3727 O II lines.

Table II also includes Cambridge spectroscopy - but now as part of a feasible project. This difference in judgment stems from two

TABLE I

Topics which are no longer suitable

Topic	Kind of Measurement	Do now with
Emission-line objects (cf. Aller, Czyzak and Kaler, 1968)	Cambridge spectroscopy	Area detector (say IDS)
Potassium flares (Wing Piembert and Spinrad 1967)	Cambridge spectroscopy	Area Detector (say IDS)
Instability-strip variables (cf. Bessell 1967)	Flux curves	Strömgren photometry
K dwarfs (Whiteoak 1967)	Flux curves	VRI photometry (if color ordering only)

TABLE II

Do it but don't advertise it beforehand

Topic	Kind of Measurement	Do now with
M giants (Wing 1967)	Flux curves and Cambridge spectroscopy	-----

TABLE III

Old Standbys

Flux curves for unusual stars		Normal stars:
(partial 1966/88 list:		Secondary standards
Ap/Am stars	H-poor stars	Comparison with model atmospheres
FGK subdwarfs	Metal-poor K III	Radii and effective temperatures
Lambda Boo stars	Horizontal-branch	
W-R stars	Stars)	
Reddening curves		Synthetic photometry

important differences of circumstance. For one thing, this is a "mixed-mode" project for Cambridge spectroscopy and flux curves combined. Wing, Spinrad, and Kuhi (1967) illustrated the power of mixed-mode data by using them to diagnose the states of infrared stars. For problems of this class, one needs only integrated band or feature strengths and not detailed wavelength structure. This leads to a second difference of circumstance: within wide limits, one does not damage such measurements by sacrificing resolution. Since the same is true of flux curves, this kind of problem permits wide exit slots and generous amounts of admitted light.

Mixed-mode work has since become familiar in practice. Nevertheless, it might be risky to mention such observing in a proposal. Suppose one's proposal falls into the hands of a high-dispersion imperialist who is out of touch with the history of observing practice. Such a person could focus on the Cambridge spectroscopy, ignore the unfamiliar flux-curve question, say "This could be done better at high dispersion," and blue-pencil the project.

To combat this, one could mention only projects such as those in Table III. All these projects focus on flux curves; Cambridge spectroscopy does not appear. One project - synthetic photometry - uses flux curves in modeling filter photometry. This proves to be one of the earliest uses to which flux curves have been put (cf. Arp 1961).

Consider a high-dispersion imperialist's reaction to this list. If he is to argue against these projects in any detail, he must resort to assertions which are hard to defend the moment he has to make them explicit. For instance, his principles rule out filter photometry for the same reasons they rule out flux-curve work. He must say this if he is to explain why he objects to synthetic photometry. If he instead objects to comparisons with model atmospheres, he must claim that theoreticians should henceforth ignore flux curves and compare their results only with line and band data. Such statements would provoke vociferous cries of protest at a meeting like this one. One can hope that a referee would think of this and be bluffed out of making those statements to begin with.

This exercise requires some footnote comment. For one thing, some who read it will doubtless think the whole issue to be pointlessly paranoid. Such people clearly have not been reading the same sort of referee's reports I have. My experience here is unequivocal: obtuse, myopic referee sentiments are a force to be reckoned with in astronomy. Referees are not bound to maintain opinions which are answerable to either hard reality or good sense. They can and do single out well-established practices in both the literature and the field, and object to them as if no one had ever seen them before. In addition, they may reject papers and research proposals on suspiciously thin pretexts. It is therefore wise to be careful - especially with a research proposal, since a misfortune in choice of referees here is a misfortune from which one has no ready appeal.

Another comment concerns the reasoning I've used so far about flux-curve devices. For sake of simplicity, I've neglected to mention relative accuracy - even though this issue should be of the first importance in practice. Here, some more history is needed.

In my experience, scanner critics clearly expected that the scanner would be obsoleted by area-detector instruments. The reality proved to be quite different. Once the IDS (Robinson and Wampler 1972) was in service, observers required faint secondary standard stars to calibrate IDS flux curves. These standard stars were not fundamentally calibrated by any area detector. Instead, the burgeoning literature of IDS and similar flux curves came to rest foursquare on a foundation of photomultiplier measurements. Here was an ironic denouement - and a pointed tacit assessment of relative accuracies. (See Hayes 1985 for a review of work on faint secondary standards.)

The issue of relative accuracy is clouded by a shortage of hard evidence. I note a certain reticence by a number of contemporary authors about the exact sizes of their flux-curve errors. Nevertheless, the issue must be addressed for an obvious reason: a fit candidate to fully succeed the photomultiplier must match the photomultiplier's accuracy. This follows because some flux-curve projects have stringent accuracy requirements; synthetic photometry and comparisons with model atmospheres are examples. Searching for more such projects in the catalog I've mentioned above would doubtless be useful.

A decade ago, Lockwood, White, and Tüg (1978) published an account of their labors in constructing an absolute stellar calibration. They made it clear (if it had not been so before) that anyone who does this has volunteered for a severe and exacting task. Hayes (1985) would later list its three components: choice of flux standard, the logistics of light-path geometry, and photometry of star and source. Such a complex effort offers numerous lodgments for displays of the imbecile force of Murphy's Law. No matter how skilled the experimenters are, there is every reason for a skeptic to fear that the results will be wrong.

In principle, calibrating secondary standards should be a less strenuous effort. More can be made similar or identical, so less should go wrong. The key differences in such work are those between the light paths of the primary and secondary standards. One can contrive to eliminate those differences (as Bahner did; see Bahner 1963) or else pay reasonably careful attention to atmospheric extinction. Either way, there is good prospect for success.

In the 1960's, an assessment like this could have led by broad upland paths to a reasonable prediction. One could have guessed that primary calibrations would settle only reluctantly and gradually into a tolerable lack of disagreement. One could have guessed further that long before this happened, secondary standards would be well in hand. Both of these predictions would have been wrong--the second one bewilderingly so. We need to look at the literature history to see what happened instead.

In retrospect, the 1960's could almost be called the "no-calibration" era. The decade began with an absolute calibration for which systematic errors as large as a tenth of a magnitude were suspected (Code 1960). Oke (1964) then adopted a model-atmosphere flux curve as an interim substitute, and this was modified three years later (Bessell 1967). During the first half of the decade, secondary standards were drawn from the "Whitford-Code" and "first-epoch Oke" lists (Code 1960, Oke 1960). Later, the "second-epoch Oke list" (Oke 1964) was cited in almost all papers.

The "no-calibration era" was followed by a "two-calibration era." The Hayes (1970) and Oke-Schild (1970) efforts were reported within nine months of each other. There were now two "scanner establishments" to choose from, each having its own primary calibration and its own flux curves for secondary standards. The differences between the primary calibrations were duly noted; moreover, Stickland (1970) foreshadowed serious later trouble by pointing out that the secondary standards differed as well. As a rule, astronomers responded to all this by choosing one "establishment" or the other without comment.

The "two-calibration era" was ended by two papers. Both were watersheds - one for secondary standards, the other for primary calibrations. The latter was Hayes and Latham's (1975) reworking of the Lick and Palomar calibrations with fabricated extinction coefficients. The sequel to this paper has been an era of such positive results as a skeptic could scarcely have expected. The fabricated Hayes-Latham Palomar coefficients turned out to agree with coefficients actually measured at the site (Hayes and Philip 1983). Moreover, no less than five subsequent calibrations agree well with that of Hayes and Latham (Hayes 1985).

The other watershed paper was that of Breger (1976a). This paper's analysis was sound and well-conceived - but it nonetheless began an era of open confusion about secondary standards. In somewhat altered form, that era is with us yet. A more sobering contrast to the primary calibrations could not well be imagined.

Breger had assembled a catalog of flux curves (Breger 1976b). He was, therefore, in a position to make quality tests. He did this by using synthetic photometry to compare the flux curves to Strömgren data (see Breger's Fig. 1). His results had two salient features: there often was about twice the scatter one might have hoped for, and there were clearly right-ascension trends in a number of data series.

In the near-infrared, affairs were worse yet. I found this out for myself a few years later (see Taylor 1979). Over a 4000 Å baseline, near-infrared flux curves could disagree by as much as 0.15 mag. Breger sought to calibrate "consensus" standards from his assembled database, but the state of the IR data usually prevented him from doing this beyond 6100 Ångstroms.

These results were an indictment of scanner work. To see this, one must make a point of regarding the scanner as a photometer. Bright stars yield ample amounts of signal to a scanner - just as they do in filter photometry. Sequential-scanner measurements do take more time than filter photometry, so one can't observe as many stars per night. There is no obvious reason, however, why the difference between (say) 40 and 120 stars per night should be critical. It was reasonable to expect scanner work to measure up to first-line filter photometry - and it is reasonable to ask why it failed to do so.

The wavelength limit of the Breger standards is especially troubling. To see why, let us suppose we have imported a speaker from another time line for this meeting. Let us further suppose that in his time line, the problems are with UBV photometry instead. The speaker might say at a certain point, "After sixteen years of diligent effort, we've finally guaranteed that you can standardize (U-B) accurately. (B-V) is still a mess, though, so there you'll have to struggle along for a few more years yet." At this point, the speaker is blown off the stage by a roar of derisive laughter. I trust I need not belabor the point.

It was bad enough that the wavelength limit was so unaesthetic. It was worse that the march of events had contrived to make it such a nuisance in practice. The limit was clearly traceable to the 1P21, which had often been used with scanners in the 1960's. When Breger's work was published, however, tubes with more extended red responses were used routinely - and that was the trouble. No matter what choice of photosurface one made - S20, GaAs, or S1 - there was an "awkward region" which was never less than 2000 Ångstroms wide, and where the Breger standards could not be used.

In the years following Breger's work, a number of devices were applied to the "awkward region." Breger himself had referred readers to the original measurements, and at least one set of authors produced a careful discussion by following this advice (Lane and Lester 1980). Others used Vega alone, or else selected one particular list of standards. (See Taylor 1984 for a brief review.)

One device deserves mention because it has led to a potential difficulty. In retrospect, it is not surprising that someone should simply have resorted to one of the existing "establishments." Stone and Baldwin (1983) and Baldwin and Stone (1984) used the Hayes (1970) lists to calibrate flux curves for faint southern standards. There was evidence, however, for systematic error in some of the Hayes data (Breger 1976a, Taylor 1979). I ultimately set aside the suspect data because they did not agree with a "consensus extension" of the Breger standards (Taylor 1984). There is, therefore, good reason to reexamine the Stone-Baldwin results.

So far, I've said little about faint secondary standards and nothing about the Soviet scanner "establishment." When Hayes (1985) reviewed these topics, he found himself to be largely balked by a lack of overlap between data sets. One could read his discussion, remember Breger's work, and murmur to oneself:

"If there's no overlap, that's bad, because then I can't find disagreement."

"If there is overlap, that's worse, because then I _can_ find disagreement."

Admittedly this is a little too pessimistic - but still this history inspires some questions.

To begin with, what had compromised the quality of the measurements? Breger ascribed the right-ascension errors to choice of secondary standards. Stickland (1971) discusses this effect on his data, and it surely is part of the answer. Such a problem would have been worst for early southern-hemisphere observers; it's significant that their data tend to suffer more from systematics than the average.

A need for a second explanation emerges if one looks closely at Breger's plots. Kubiak (1973), Stickland (1971), and Wolff, Kuhi, and Hayes (1968) all appear to have adopted the Hayes (1970) standards. Nonetheless, there are differences between the residuals for the three sets of data. There are also differences between these residuals and those for the Hayes data themselves. An explanation which comes readily to mind is treatment of extinction - especially the practice of using mean extinction coefficients. (This has been suggested by Taylor 1979).

Next: what could have been done differently? This topic resembles antic referee behavior: people may feel uncomfortable if someone starts talking about it in public. Unless I forestall them, I expect to hear excuses like this: "Well, you know how hard it was to use one of those old sequential clunkers." Such sentiments neglect a tart maxim: the most annoying limits on any experiment are usually between the experimenter's ears - with the author not being excepted. It is there that responsibility may fitly be placed for putting things right.

Photometry has had a history of discovering its problems through experience. This has been true of VRI work, for instance (see Bessell 1979), and I doubt it would have been fair to expect anything else of scanner photometry. Once disagreements are known, however, our response is a different story. A ubiquitous, unspoken folkway of the discipline says: "Make a reasonable-looking choice, apply it, and get on with the business at hand." Correcting the disagreements is often left to a minority who have a taste for such problems. Such tendencies appeared in this problem. It shouldn't be hard to improve on them.

First: if observers must cope with disagreeing standards, they should make it possible for others to correct their results later. This is scarcely an original suggestion; the literature contains some first-rate examples of this procedure (see, for example, Kubiak 1973 and Lane and Lester 1980). Not all observers follow it, however, and when they do not, the labor of checking their work is greatly increased.

Second: users of disagreeing standards should measure a number of them and make a point of reporting the results. This suggestion dovetails comfortably with the one I've just described. Again, it isn't original; something similar has been done in Strömgren photometry for some years (see, for example, Lindemann and Hauck 1973). If several observers do this, they can bequeath to others a powerful data base bearing on the problem.

Third: more astronomers might undertake extensive work on standards, and the discipline should make a point of encouraging those who do this. I am especially concerned about that last point, because I suspect that "toolmakers" sometimes find that their efforts are disparaged by their colleagues for the sin of not being flashy. Solving three arcane mysteries of the Universe before breakfast is not the only valuable astronomy one can do. We can have better sense than to hinder ourselves through shortsighted value judgments.

I'll close by asking some questions about the data this history has bequeathed to us. First, how consistent are the diverse available standards? One might hope that many of them form a "consensus network," with both direct comparison and synthetic photometry displaying their consistency. If synthetic Cousins VRI photometry were to do this, this would be especially encouraging because of the quality of that system. If some standards disagreed with such a "consensus network," we could be reasonably confident about where the error lies. There is a first suggestion of a "consensus network" at short wavelengths in the account given by Hayes (1985). Clearly, however, a lot of work remained to be done at that time.

Second: what else can be done about standards besides checking their consistency? Some years ago, I mentioned additional bright standards for some parts of the sky, additional measurements in some awkward wavelength regions, and other possibilities (Taylor 1984). Since then, I have made some slow progress toward realizing some of them. It would be worthwhile to update that list and see how close some of its items are to being completed.

Third: while compiling data for Taylor (1984), I gained the impression that better allowance for extinction has generally improved the quality of flux curves since the Hayes-Latham paper was published. Can this be confirmed? If so, this would counter a common tacit assumption that the accuracy limits Breger found couldn't easily be bettered.

Fourth: how many existing data measure up to contemporary standards, and how many should be abandoned? For questions like these, the discipline can have another self-defeating value judgment. The pejorative phrase "old stuff" can waft through the corridors of meetings like the aroma of burned insulation. The danger in this attitude is that some problems can needlessly be done twice - or more often yet.

This question is especially important because of its bearing on judgments about telescope time. One could "saturate" a flux-curve device with repeated measurements of variable stars, for instance - but such a policy may not be worth considering if there are major gaps in our database for nonvariable, "normal" stars. Some attention has been given to this matter (Hayes 1986), but more is needed.

Since the example of Thucydides, historians have hoped that review of the past might benefit the future. Hopefully, I've displayed some ways in which this can be done in flux-curve work.

REFERENCES

Arp, H. C. 1961 Astrophys. J. **133**, 874.
Bahner, K. 1963 Astrophys. J. **138**, 1314.
Baldwin, J. A. and Stone, R. P. S. 1984 Monthly Notices Roy. Astron. Soc. **206**, 341.
Bessell, M. S. 1967 Astrophys. J. Letters **149**, L 67.
Bessell, M. S. 1979 Publ. Astron. Soc. Pacific **91**, 589.
Breger, M. 1976a Astrophys. J. Suppl. **32**, 1.
Breger, M. 1976b Astrophys. J. Suppl. **32**, 7.
Code, A. D. 1960 in Stellar Atmospheres, J. L. Greenstein, ed. The University of Chicago Press, Chicago, p. 50.
Code, A. D. and Liller, W. C. 1962 in Astronomical Techniques, W. A. Hiltner, ed., The University of Chicago Press, Chicago, p. 281.
Griffin, R. F. and Redman, R. O. 1960 Mon. Notices Roy. Astron. Soc. **120**, 287.
Hayes, D. S. 1970 Astrophys. J. **159**, 165.
Hayes, D. S. 1985 in IAU Sumposium No. 111, Calibration of Fundamental Stellar Quantities, D. S. Hayes, L. E. Pasinetti, and A. G. Davis Philip, eds., Reidel, Dordrecht, p. 225.
Hayes, D. S. 1986 in Highlights of Astronomy, Vol. 7, J. -P. Swings ed., Reidel, Dordrecht, p. 819.
Hayes, D. S. and Latham, D. W. 1975 Astrophys. J. **197**, 593.
Hayes, D. S., and Philip, A. G. D. 1983 Astrophys. J. Suppl., **53**, 759.
Hiltner, W. A., and Code, A. D. 1950 Journ. Opt. Soc. America **40**, 149.

Kubiak, M. 1973 Acta Astron. **23**, 23.
Lane, M. C. and Lester, J. B. 1980 Astrophys. J., **238**, 210.
Lindemann, E. and Hauck, B. 1973 Astron. Astrophys. Suppl., **11**, 119.
Lockwood, G. W., White, N. M. and Tüg, H. 1978 Sky and Telescope, **56**, 286.
Oke, J. B. 1960 Astrophys. J. **131**, 358.
Oke, J. B. 1964 Astrophys. J. **140**, 689.
Oke, J. B. 1965a Astrophys. J. **142**, 810.
Oke, J. B. 1965b Ann. Rev. Astron. Astrophys., **3**, 23.
Oke, J. B. and Schild, R. E. 1970 Astrophys. J. **161**, 1015.
Robinson, L. B. and Wampler, E. J. 1972 Publ. Astron. Soc. Pacific, **84**, 161.
Stickland, D. J. 1970 Observatory **90**, 206.
Stickland, D. J. 1971 Monthly Notices Roy. Astron. Soc. **153**, 501.
Stone, R. P. S. and Baldwin, J. A. 1983 Monthly Notices Roy. Astron. Soc. **204**, 347.
Taylor, B. J. 1979 Astron. J. **84**, 96.
Taylor, B. J. 1984 Astrophys. J. Suppl. **54**, 259.
Wing, R. F. 1967 Ph. D. thesis, University of California, Berkeley.
Wing, R. F., Spinrad, H. and Kuhi, L. V. 1967 Astrophys. J., **147**, 117.
Wolff, S. C., Kuhi, L. V. and Hayes, D. S. 1968 Astrophys. J., **152**, 871.

DISCUSSION

WING: In your historical survey, I wonder if you came across a scanner at Lick that preceded your arrival there and the introduction of the movable-grating scanner by J. Wampler. I'm referring to a "photomultiplier spectrograph" scanner that was available at the Crossley telescope around 1963-65, when G. Preston mounted a photomultiplier on the old Wright spectrograph, a prism instrument used a half a century earlier for slitless spectroscopy of planetary nebulae (Reference in the Lick Obs. Bull.). I don't think anything was published from this scanner, but it played an historical role by leading me and other students into the field of spectrophotometry and starting us thinking in terms of designing photometric systems according to what's in the spectrum.

TAYLOR: I didn't find anything; evidently nothing from the Preston scanner was ever published. I'd certainly like to include it, though.

ANALYSIS OF CATALOGS OF SPECTROPHOTOMETRIC DATA

Lukas Labhardt

Astronomical Institute, University of Basel

In recent years several compilations of observed stellar energy distributions have become available (cf. Table I of Cacciari 1988). These catalogs are important for a wide range of astronomical applications such as synthesis of spectra of stellar systems, verification of theoretical stellar energy distributions, (automated) classification of digital spectra, evaluation and calibration of photometric systems.

Depending on the motivations behind the original observations, the various compilations of spectrophotometric data differ from each other with regard to a) range of spectral types, luminosity classes and metallicities, b) wavelength coverage, c) spectral resolution and sampling and d) photometric quality. Since there are only a few or no stars in common, a direct comparison between the energy distributions included in different catalogs is of limited value or simply not possible. Kiehling (1987) reports of average absolute differences of 0.01 - 0.1 mag when comparing his spectrophotometric data of southern and equatorial late-type stars with the corresponding energy distributions measured by other authors. A thorough check of the photometric quality of spectrophotometric data sets by means of synthetic photometry is often hampered by the lack of homogeneous filter photometric observations for all stars included in the individual catalogs.

Following up the review by Hayes (1986) an analysis of the following catalogs of continuous energy distributions was undertaken: Gunn and Stryker (1983), Jacoby, Hunter and Christian (1984), and Pickles (1985). Each of these spectrophotometric libraries includes a large number of Population I stars, covers a large wavelength range and a wide range in spectral types, and is available in machine readable form. Our comparison of synthetic colors and homogeneous UBV colors taken from Nicolet (1978) gives standard errors of 0.03 - 0.05 mag, both for the catalog of Gunn and Stryker (cf. also Labhardt and Buser 1985) and the library of Jacoby et al. Pickles has published energy distributions for mean spectral types only, i.e. in a form not suited for such a detailed comparison, but his own comparison of colors synthesized from spectra of individual stars with observed BVRI

A. G. Davis Philip, D. S. Hayes and S. J. Adelman, (eds.)
New Directions in Spectrophotometry 17 - 19

photometry gives similar standard errors.

To derive intrinsic energy distributions of the catalog stars, an interstellar reddening law is used with color excesses that are determined by comparing the observed colors with the mean intrinsic colors (FitzGerald 1970) corresponding to the best available spectral type. Applying different interstellar reddening laws (Whitford 1958, Schild 1977, Seaton 1979, Howarth 1983) results in different amounts of absorption corrections; the synthesized intrinsic UBV colors of a heavily reddened catalog star differ therefore by up to 0.06 - 0.1 mag.

A detailed presentation of this work is in preparation (Labhardt 1988). Support from the Swiss National Science Foundation is gratefully acknowledged.

REFERENCES

Cacciari, C. 1988 in New Directions in Spectrophotometry, A. G. D. Philip, D. S. Hayes and S. J. Adelman, eds., L. Davis Press, Schenectady, p. 159.
FitzGerald, M. P. 1970 Astron. Astrophys. **4**, 234.
Gunn, J. E. and Stryker, L. L. 1983 Astrophys. J. Suppl. **52**, 121.
Hayes, D. S. 1986 in Highlights of Astronomy, Vol.7, J. -P. Swings, ed., Reidel, Dordrecht, p. 819.
Howarth, I. D. 1983 Monthly Notices Roy. Astron. Soc. **203**, 301.
Jacoby, G. H., Hunter, D. A. and Christian, C. A. 1984 Astrophys. J. Suppl. **56**, 257.
Kiehling, R. 1987 Astron. Astrophys. Suppl. **69**, 465.
Labhardt, L. and Buser, R. 1985 in IAU Symposium No. 111, Calibration of Fundamental Stellar Quantities, D. S. Hayes, L. E. Pasinetti, A. G. D. Philip, eds., Reidel, Dordrecht, p. 519.
Nicolet, B. 1978 Astron. Astrophys. Suppl. **34**, 1.
Pickles, A. J. 1985 Astrophys. J. Suppl. **59**, 33.
Schild, R. E. 1977 Astron. J. **82**, 337.
Seaton, M. J. 1979 Monthly Notices Roy. Astron. Soc. **187**, 73P.
Whitford, A. E. 1958 Astron. J. **63**, 201.

DISCUSSION

HAYES: This is very important work, addressing the comparison of spectrophotometric data with filter-photometric data using spectrophotometry. The results, standard errors of 0.03 - 0.05 mag., show that the quality and homogeneity of the data are not adequate for this comparison to be really useful. This points out the great need for new catalogs of spectrophotometric data covering a range of spectral types, luminosity classes and metallicities. Such catalogs would also need to be observed in the major filter-photometric systems with great care. When the catalogs exist, the great power of synthetic photometry for evaluating and calibrating photometric systems would be realized.

STELLAR WINDS AND THE SPECTROPHOTOMETRY OF HOT STARS

J. Puls, A. Pauldrach, R. P. Kudritzki,
R. Gabler and A. Wagner

University Observatory, Munich

ABSTRACT: Stellar Winds are an ubiquitous feature of hot luminous stars. These winds can reach terminal velocities up to 4000 km/s and show mass loss rates from 1×10^{-9} to 1×10^{-4} solar masses per year. Therefore they are of great astrophysical importance. On the one hand they change the star's evolution significantly and transport nuclear processed material into the interstellar medium; on the other hand they modify the star's energy distribution in nearly all ranges and have to be taken into account for an accurate spectral diagnosis.

The most promising attempt to explain these winds is the theory of radiatively driven winds. As was shown by Lucy and Solomon (1970), the radiation pressure in even a few strong uv-resonance lines is capable of accelerating and maintaining stellar winds of hot stars. The first approach to a self-consistent treatment was given by Castor, Abbot and Klein (1975), which, however, suffered from over-simplifications and was of only qualitative character.

In the first part of this paper, the present stage of knowledge as developed by the Munich group (R.P. Kudritzki, A. Pauldrach, J. Puls, R. Gabler and A. Wagner) in the last three years is presented. After a significant improvement of the theory it is now possible to treat the stellar wind phenomenon in a nearly complete self-consistent way, taking into account the full NLTE-problem of about 4000 coupled levels (elements hydrogen to zinc in the first six ionization stages), where the radiative transport is calculated with respect to the complete line-overlap case. The resulting wind structure and energy spectrum comes out to be close to observations and is dependent only on the four input parameters T_{eff}, log g, R-star and chemical composition.

In the second part of the paper, the influence of stellar winds on the energy spectrum and spectrophotometry of hot stars is discussed, e.g.:

- the observed IR-excess is shown to be the result of the wind-structure.

A. G. Davis Philip, D. S. Hayes and S. J. Adelman, (eds.)
New Directions in Spectrophotometry 21 - 24

- Line-blocking line- and wind blanketing will change the photospheric temperature structure and the uv-part of the spectrum.

It is concluded that the use of hydrostatic model atmospheres (even in NLTE) together with today's observational precision is insufficient to determine stellar parameters with the theoretical possible accuracy. Future instruments like HST and ROSAT should allow us to test the predictions of the radiatively driven wind theory and to obtain information from the faintest object.

REFERENCES

Castor, J. I., Abbott, D. C. and Klein, R. 1975 Astrophys. J. **195**, 157.
Lucy, L. B. and Solomon, P. 1970 Astrophys. J. **159**, 879.

DISCUSSION

GENET: How computationally intensive are your models?

PULS: Our model calculations are very computer intensive, as for the self-consistent solution the rate equations, transfer equations and hydrodynamical equations have to be iterated. Depending how far the model is away from the initial guess of the hydrodynamical structure, we need typically an amount of time equivalent to one to ten hours per model on a Cray.

HORNE: Is the accuracy of the energy distributions from the unified stellar atmospheres good enough for use in absolute calibrations by constructing models for standard stars?

PULS: For the optical range, the unified stellar atmospheres are as good as standard NLTE calculations. For the UV part (wind blanketing, line blanketing) and the IR (excess) however, we have detailed calculations and comparison with observations only for our standard star ζ Pup (beginning with detailed comparisons for CSPN's). As long as we have no complete grid of models compared to observations the use of the unified stellar atmospheres for calibration purposes should be delayed.

WEAVER: How do you compare the three wind calculation techniques your group used in the 1987 paper, and which are you using now?

PULS: The different approaches were designed to investigate different kinds of simplifications and approximations. At the moment, we calculate absolute self-consistent wind models, treating the multi-line-effects in the Sobolev approximation.

WEAVER: What are the advantages and disadvantages of using the Abbott and Lucy Monte Carlo approach to calculating wind models?

PULS: The Monte-Carlo method is valuable in the following case: when the occupation numbers are known (self-consistent solution of radiative transfer and rate-equations), the Monte-Carlo procedure gives comprehensive results both for the line force and the spectrum synthesis. When you calculate the mean intensity needed for the rate equations via Monte-Carlo, however, you must assume complete resonance scattering, as to consider also the collisional processes would require an enormous amount of computer time. The rate equations themselves, however, depend significantly on the net-radiative bracketts, which are identically zero for complete resonance scattering, so that in this way the rate equations would be completely determined by the collisional processes. This intrinsic contradiction is avoided by using the Sobolev approximation in the radiative transfer. When the correct occupation numbers are known, then the difference between complete resonance scattering and the presence of small collisional terms is marginal and the results are essentially the same for both methods.

GRADY: Do your models reproduce the high velocity discrete absorption components which are common in the resonance profiles of C IV, Si IV, and N V in OB stars? I'm particularly interested in one of the stars included in your mass loss rate calculations, λ Cep.

PULS: At the present stage of technique, we are calculating only stationary winds, which of course do not reproduce the observed high velocity components. To explain these features, a lot of theory has to be done, although there are plenty of suggestions. (For example, shells of dense matter moving throughout the wind.) Also, the problem of instabilities has to be investigated (compare the work of Lucy, Rybicki and Owocki, etc.) before a final solution of this topic can be given.

ADELMAN: How sensitive are your results to uncertainties in the atomic data?

PULS: The line force in total and in this way the hydro-dynamical structure is rather insensitive to very accurate atomic data, as it is the sum over thousands of contributing lines. For a detailed comparison of the emergent profiles, however, precise atomic data are very important. Together with leading atomic physics experts in our institute (K. Butler) and in cooperation with the London-Boulder-Meudon opacity group, we are at the moment improving cross sections and completing the atomic structure of the required ion species. Especially for the lower ionization stages, which came out a bit too small by number in our present calculations, we expect a significant improvement.

PETERS: I am intrigued by your comment on the O VI calculations. Have you compared your theoretical predictions with available observations? Copernicus data on O VI are confusing and suggest that parameters other than the usual ones may be important. We have the case of τ Sco (BOV) which has strong O VI but other early B stars (even BO.5 - Bl varieties) do not display such prominent O VI wind lines.

PULS: For ζ Pup, our standard star, the results concerning O VI are comprehensive. For a paper in preparation, we have calculated the O VI profiles for a number of O-stars, where the profiles came out to be close to observations. Concerning B-stars, we are just at the beginning of our self-consistent calculations and at present I can make no comment concerning this topic.

YET ANOTHER PROGRESS REPORT: NEW MODEL ATMOSPHERES SOON

Robert L. Kurucz

Harvard-Smithsonian Center for Astrophysics

ABSTRACT: I am working hard to improve model atmospheres because existing models have numerical errors, an unphysical treatment of convection, an inadequate or non-existant treatment of statistical equilibrium, an arbitrarily chosen microturbulent velocity, an arbitrarily chosen helium abundance, and a greatly underestimated line opacity for iron group elements. To solve some of these problems I have computed a new line list for the iron group elements using a large grant of time at the San Diego Supercomputer Center. Now I have started to compute opacities for the temperature range 2000 K to 200,000 K, for abundances ranging from 0.0001 solar to 10 times solar, for microturbulent velocities 0, 1, 2, 4, and 8 km/s. Then I will compute corresponding grids of models, fluxes, colors, and spectra.

[As this paper is substantially the same as "Atmospheres for Population II Stars", pp. 129-136 in IAU Colloquium No. 95, The Second Conference on Faint Blue Stars, A. G. Davis Philip, D. S. Hayes, and J. W. Liebert, eds., L. Davis Press, Schenectady, 1987, it is not printed here.]

DISCUSSION

JOHNSON: What value will you use for the mixing length in treating convection in the cooler models?

KURUCZ: I'll use 1.0 - I like round numbers. But I will be able to compute whole grids at any mixing length you like. The opacities take computer time, models do not.

GENET: Can you contrast the accuracy of model results with observational results?

KURUCZ: If you recall the plot for Vega, the stated errors are 2 and 3% for the spectrophotometry. I think the models must be similar. These are systematic errors. The scatter is much smaller. I would like to see both models and observations better than 1%.

WEAVER: What is the lowest T_{eff} model you will calculate?

A. G. Davis Philip, D. S. Hayes and S. J. Adelman, (eds.)
New Directions in Spectrophotometry 25 - 26

KURUCZ: I have only diatomic molecules, so only K stars.

JONER: With the addition of the molecular lines and the opacity changes incorporated in the new models, will you still use Vega as the zero point for the photometric grids?

KURUCZ: If the photometric system is perfectly known only one star should be necessary. In the real world, using several stars produces an empirical correction to the theoretical photometric system, to make it more like the actual system. To test the models (and photometry) I would use Vega. To make useful calibrations I would use several stars.

JONER: Will you now be able to predict an accurate solar color?

KURUCZ: When my models can reproduce the solar spectrum they should easily be able to predict the colors. I would use a multiple star calibration.

BOHLIN: For the comparison of your models with UV observations of hot stars, what models would you have the most confidence in?

KURUCZ: None. I am always surprised that there is such good agreement. Maybe there are canceling errors - not enough line opacity, but too much microturbulence.

WHITE: With what solar flux observations did you compare your model?

KURUCZ: First Labs and Neckel. Now Neckel and Labs.

WHITE: Lockwood, Tüg, and myself have made a direct Vega-Sun calibration and we find a discrepancy similar to that on your diagram. That is, blueward of the peak flux or redward of the peak flux can be made to fit with a flux zero shift, but not both at the same time. The point is that the observations may not be doing justice to the model rather than the other way around. Are we at a stage of precision where both the models and the observations need equal improvements?

KURUCZ: Yes.

PETERS: For two early B stars that I am familiar with, ι Her and τ Sco the flux shortward of 1100 Å is observed (from Voyager UVS) to be more intense than your 1979 models predict. There is apparently not a deficiency of opacity in the models but rather too much of the flux is absorbed. Do you have an explanation for this?

KURUCZ: Increasing the line opacity raises the flux anywhere there are no lines. Perhaps most of the additional lines are in the Fe III region around 1500 Å and there happens to be only a small increase at 1100 Å.

COMPOSITE SPECTRA AND BINARIES: THE NEED FOR NEW SPECTROPHOTOMETRY

Thomas B. Ake

Space Telescope Science Institute

ABSTRACT: I would like to draw attention to a group of objects that tend to be rather neglected but include some promising systems for deriving fundamental stellar parameters, the composite stars. In particular, I will review the status of recent observations of a subset of these, the middle to late-type giants and supergiants with hot companions. Observations with IUE indicate that there are a number of bright objects that would profit from a systematic spectrophotometric survey and monitoring over a wavelength range from the atmospheric cutoff in the UV to the near infrared. As an example, the first ζ Aur stars with G-type primaries are being discovered with IUE that could have easily been found from the ground. In these systems, where a hot secondary undergoes an atmospheric eclipse when passing behind the outer envelope of a supergiant primary, observations in the optical UV would clearly show the eclipse. Combined with spectral classifications, near reddening-free line indices, and measures at longer wavelengths, fundamental parameters such as the sizes and temperatures of the components are derivable. Data on other objects also will be discussed.

1. INTRODUCTION

There is a class of peculiar stars whose oddity arises only because of the company they keep: the objects of composite spectral type. Classically these objects are defined to be those stars which appear to be single, but whose spectra exhibit features of two different, though normal, spectral classes. Some of these are found to be spectroscopic binaries, but others remain true spectrum binaries where no velocity variations are seen in the lines. Because of the constraint that the components are normal, the more complex and exotic systems, such as cataclysmic variables, symbiotics, and other objects where one of the stars is near the end of its evolution, are excluded from this group. Essentially we are considering visual binaries with components too close to resolve but separated enough that there is little or no interaction between the components.

Generalizing this definition, we can include objects that are not obvious spectrum binaries but that are photometrically composite, i.e., their energy distributions cannot be matched by single normal stars.

A. G. Davis Philip, D. S. Hayes and S. J. Adelman, (eds.)
New Directions in Spectrophotometry 27 - 36

These systems have components of widely different temperature that can sensibly be resolved when observed over a broad wavelength range. In most cases, there must be a crossover wavelength where the objects contribute equal light and thus would appear spectroscopically composite, but this region may not be readily accessible to the observer. In some cases, this point occurs where the normal absorption line spectrum is too sparse or otherwise unsuitable for recognizing a peculiar object. To appear composite, the components of the system must satisfy certain conditions between temperature and luminosity class and apparent brightness difference in the wavelength region in which they are observed (Hynek 1951). If the components are too similar in spectral class or the magnitude difference is so large that the flux contribution from the fainter component is negligible, the multiplicity of the system is unperceived.

Since we are dealing with unresolved visual binaries, we can assume that some of their general characteristics apply to composites as well. For instance, it has been known for some time that visual binaries can be divided into three classes of systems: systems where a main-sequence primary has a later, fainter type main-sequence companion; systems where the components are similar in both color and magnitude; and middle- to late-type post-main-sequence primaries with earlier type companions. This distinction is easily understood in terms of the HR diagram and stellar evolution, and thus indicates what kinds of composites to expect.

For the purposes of this discussion, I would like to concentrate on the last of the classes, cool giants and supergiants with hot companions. The value of binary star studies lies in the derivation of fundamental stellar parameters, such as luminosities, masses and radii, and the paucity of such determinations for middle-type supergiants, for example, is due to the small number of well-studied binaries. Part of this problem arises because some of the systems are going unrecognized. Rather than summarizing the major results in this field, I hope to whet your appetite with some recent IUE results and demonstrate what can be gained by further careful spectrophotometric observations. In particular, newly-discovered eclipsing systems now being found with IUE could have been found from the ground, and indeed in one recent case, probably has been.

2. RECOGNIZING COMPOSITES

To find which objects deserve further study, we need some criteria for selection. Composite stars can be found by a variety of techniques. The premier of these is through spectroscopic programs, such as objective prism surveys. At the resolution employed in these, systems with discordant types are recognizable, but classifications to only 1/2 or 1 spectral class can be assigned. Lately, the Michigan survey in the south and the Case survey in the north are increasing the pool of composite candidates (see Bidelman 1985 for the latest list). Naturally at higher spectral resolution the more difficult cases can be identified, especially stars of more equal type or large magnitude difference. Corbally (1987) has summarized the warning signs at classification dispersion that one is dealing with a composite.

A second method for finding these systems is through radial velocity surveys. If intrinsic variability, such as pulsation, can be ruled out, objects with variable velocity are prime candidates for closer examination simply because one knows they are multiple. Velocity spectrometer measurements have been available long enough now and are of such precision that the longer-period spectroscopic binaries, which the giant and supergiant composites tend to be, are being found. One important case, HR 6902 (Griffin and Griffin 1986), will be discussed in section 4.

Finally, there are the photometric techniques where one discovers stars in which the colors do not match the spectral type or simply are just plain odd. These include objects that stand out on color-color plots or those found by cross-referencing photometric surveys in widely different wavelength regions. As an example of the first case, Madore (1977) suggested 38 Cepheids had B-type companions based on open-loop trajectories in their (U-B), (B-V) plots, indicating the presence of a constant light component. Böhm-Vitense and Proffitt (1985) and others have confirmed with IUE that many, though not all, of Madore's objects did have hot companions. While the color-color method was not fool-proof, the important point is that a group of possible composite objects was identified as deserving further study. As an example of the second case, Parsons (1981) identified seven supergiants with hot companions by matching the ANS and TD-1 catalogs with ground-based photometry, and was able to derive photometric temperature types, reddenings and magnitude differences based on the total energy distribution.

3. DECONVOLVING THE COMPONENTS

Once a composite is identified, what can be done to decide what stars are actually involved? For the spectroscopist, clues from relative line strengths and the gradient of the photographic density are used. For difficult composites, direct comparisons with synthetic exposures provide added insight, particularly when using trailed exposures of close binaries whose individual types and magnitudes can also be reasonably separated, but in the end the result is somewhat subjective and depends on the cleverness of the observer to account for all the clues (Bidelman 1984).

For systems with components of decidedly different color and large differential brightness, broad or intermediate band photometry over a wide wavelength range can be used. Here the solution is somewhat more analytical, if for no other reason than the number of data points is smaller, but the result can be even more uncertain than by spectroscopic techniques if reddening is not adequately known. Reddening lines in color-color diagrams have slopes that are annoyingly similar to the run of colors with spectral type, and an extra unknown flux distribution added to a normal star, when including the cosmic dispersion of colors with type and measurement errors, is not always obvious. Observations in the satellite UV are of help here since the 2200 Å bump yields an independently-determined reddening value, although one must then contend with which reddening curve is appropriate for the system.

The most comprehensive studies are those that combine both techniques and use information from the line spectrum and colors together to resolve the components. Spectra can narrow the temperature and luminosity classes of the components, or at least do reasonably well for the primary in large Δm systems. Photometry then is fit by adding energy distributions of normal objects, adjusted for various reddening values, until the spectra and photometry provide consistent results.

Such a method is utilized explicitly by spectrophotometric methods. Two of the more recent results are those of Beavers (1982) and Schmidtke (1983, 1984), both of whom used photoelectric scanners to derive spectral indices of absorption features in normal stars and then deconvolved composite systems using least squares methodologies. By the proper choice of absorption features and continuum points, the indices are relatively reddening-free, and once the components are identified, the continuum observations can be compared to standard colors to derive the magnitude differences. The large number of indices that can be computed from spectrophotometric scans allows one to derive unique parameters for most spectral types and guard against anomalies, such as Ca II emission, which can confuse the analyses if a small number of points are observed. Beavers sampled points at 49 wavelengths in 10 Å bins from 3500-4400 Å, while Schmidtke chose a longer baseline (3750-10400 Å) in 20 or 40 Å bandpasses at 48 wavelengths. Although of higher resolution, Beavers' system was useful only for components with $\Delta m < 1.8$ mag. and of similar spectral class, but Schmidtke could separate a variety of components with Δm up to ~3 mag. due to the longer wavelength baseline. This magnitude difference is in the range of that for the supergiant binaries.

4. G-TYPE ZETA-AURIGAE SYSTEMS

Returning to the idea of finding binaries by extending the wavelength range of the observations, Parsons and I have been using IUE profitably in studying binary candidates identified either by peculiar color or radial-velocity variation. Typically for these systems, the point at which the components contribute equal light is in the 2600-2800 Å region, so using the SWP camera one can observe the secondary essentially uncontaminated. Coupled with optical photometry, knowledge of the spectral class of the primary, and indications of the reddening from the 2200 Å feature, rather complete deconvolution is possible. We have undertaken these studies with the intent of improving the luminosity calibrations for the primaries, but we have been finding some rather interesting objects.

In Table I, I present systems referred to in Parsons and Ake (1987) for which we have classified the secondary based on IUE standards. These types then are assigned on a consistent system. The visual-magnitude difference and reddening are determined as described by Parsons (1981). While some of the objects were known composites optically, others are IUE discoveries. Note that many of the objects are spectroscopic binaries with well-determined orbital elements.

Further observations of some of these has lead to the discovery of the first ζ Aur systems with G-type primaries. In the ζ Aur stars, an atmospheric eclipse occurs as the hotter secondary passes behind the outer envelope of the cooler primary star. The classical systems are K supergiants with B-type companions, although sometimes the M-type VV Cep stars and the peculiar F supergiant ϵ Aur are included (Wright 1970). ϵ Aur is a unique object that does not fit in the group because the atmospheric eclipse occurs when the secondary is in front of the primary, but the VV Cep stars are an extension of the phenomenon to cooler primaries. Because of the color difference between these types of components, they are not difficult to recognize in the photographic region even though the magnitude differences are 2 - 3 mag. in the visual. To become a member of the group, a composite of this type must have an inclination favorable for eclipses.

When proceeding to types earlier than K, however, the effect of the color difference becomes less efficient in offsetting the Δm, and these stars are more difficult to recognize. Perhaps because of this difficulty, most observers tend to shy away from the earlier-type composites, but with careful study new eclipsing systems are being identified. There are now known to be two or perhaps three G-type systems of the class, 22 Vul (Ake, Parsons and Kondo 1985), HR 2554 (Ake and Parsons 1987), and most likely HR 6902 (Griffin and Griffin 1986). If we relax the criterion to include G giants, τ Per (Ake et al 1986) can be included as a fourth member.

Would it have been possible to find these objects from ground spectrophotometry? For 22 Vul (G 3 Ib + B 9 V), Keenan and Pitts (1980) had noted that Ca H and K were weak and Parsons (1981) determined that the secondary must be a B 7 - 9 star nearly three mag fainter, based on the TD-1 fluxes. Optical photometry of the eclipse (Parsons, Ake and Hopkins 1985) found the depths to be 0.05 mag in V, 0.14 mag in B and 0.35 mag in U. For HR 2554 (G 6 II + A 0 V), no spectroscopic anomalies have been reported, although its (U-B) color is too blue for ts type. Unfortunately the eclipses (every 6.4 months) have not been well-placed for optical observers, but the situation should improve in the next few years. Just completed IUE observations of the Nov. 1987 eclipse suggest that the system is only grazing. HR 6902 (G 9 II + B 8 V) was noted as composite in the HD catalog, but an improved orbit, spectral classifications and a possible eclipse (based on a spectrometric subtraction technique) have been established from

TABLE I

Composite Systems Observed With IUE

Name	HD No.	Spectral Type	Period (d)	UV Type	dV	E(B-V)
Tau Per	17878+	G 5 III + A 3	1515.868	A 2	1.9	0.04
HR 958	19926+	K 5 IIIep +A	115.0?	A 5	2.5	0.04
HR 1129	23089+	(G0 III + A 3)	212 ?	B 8.5	1.9	0.16
HR 1206	24497	F 8 V / A 2:+K:	v ?	A 2	1.4	0.04
Mu Per	26630	G 0 Ib	283.30	B 9.5	3.5:	0.14:
52 Per	26673+	G 5 Ib + A2	1577.7	B 9.8	2.1	0.12
58 Per	29094+	G 8 II + B9:	10518.	B 8	2.2	0.32
	31244+	K 3 II-III +B5		B 9.3	2.2	0.00
Zeta Aur	32068	K 4 II + B7 V	972.16	B 6-7	2.0	0.12
	34807	G 8 III + A :		A 0	0.2	0.16
HR 2024	39118	G 8 III + A 0	v	B 8.5	2.0	0.04:
HR 2554	50337	G 6 II + A :	195.24	A 0	3.0	0.04
HR 2786	57146	G 2 Ib	v	B 9.3	3.7	0.04
HR 2859	59067AB	G 8 I-II + B	v ?	B 2 V	1.0	0.15
HR 2902	60414+	M 2 Iab + B	9752.	~B 7 p	~2.0	0.16:
HR 3080	64440	K 1-2 II + A	2554.0	A 0	2.5	0.10
Eps Car	71129+	K 3: III + B 2:		B 2 Vp	2.6	0.00
HR 3386	72737+	K 0 III + A3		B 9.3	1.0	0.04
HR 3459	74395	G 1 Ib		B 9	3.3	0.00
HR 3591	77258	G 8-K1 III +A	74.15	A 7	0.8	0.00
	101007	M 3- Ib + B		B 3:p	3.5:	0.26
HR 4492	101379+	G 5-8 III+A 0-1	v ?	A 0	1.3	0.04
HR 4511	101947	G 0 Ia		B 1 Ib	3.2	0.26
12 Com	107700	G 0 III-IV +A 3	396.49	A 2	0.7	0.02:
DL Vir	120901+	(A 2 V +F 9 III)	v	A 2:	~2.4	0.20:
HR 5308	124147	K 5 III + B-A		B 8	2.8	0.12:
HR 5637	134270	G 1-3 Ib-II		B 7	2.7	0.18
HR 5667	135345+	G 5 Ia + B		B 7	1.8	0.00
HR 6384	155341	M 1-2 II-III+A	v	B p	~4.0	~0.2
HR 6560	159870	(G 5 III + A 7)	178.55:	A 7	1.0	0.04:
V777 Sgr	161387	K 5 Ib + A	936.07 L	B p	3.8:	0.16:
nu Her	164136	F 2 II		B 9.5	2.5:	0.00:
HR 7031	172991+	K 3 II + B 7	v	B 8	2.0	0.04
Beta Sct	173764	G 4 II	832.5	B 9.5	3.3	0.20
Nu-1 Sgr	174974	K 2 I:	v	B 9.5	5.7	0.12

Name	HD No.	Spectral Type	Period (d)	UV Type	dV	E(B-V)
113 Her	175492+	gG 4 + A 5	245.295	A 0	1.4	0.12:
Psi Sgr	179950	G 8: III + A 8		A 8	0.4	0.00
HR 7428	184398+	K 2 II-IIIe +A	108.57	A 2 e	2.9:	0.04:
SU Cyg	186688	F 0-G 1		B 6p	~3.3	
Delta Sgr	187076+	M 2 Ib-II + A	3725.	B 9.5	2.4	0.04:
	187299	G 5 Iab-b + B	1901.0	B 9:	3.3:	0.48
	187321+	(G 5 II + B 9)		B 8	1.9	0.08:
31 Cyg	192577	K 2 II + B 3 V	3784.3	B 4	2.3	0.06
22 Vul	192713	G 3p Ib-II	249.11	B 9	3.3	0.04
32 Cyg	192909	K 3 Ib + B	1147.15	B 6:	3.6:	0.04
35 Cyg	193370	F 5 Ib	2452.6	B 7	4.5	0.28
	193469	K 5 Ib		B 9:	4.9:	0.24
Beta Cap	193495+	G 8 II + B 8:	1374.126	B 7-8	1.8	0.06
HR 7795	194069	G 5 II + A:		A 5	3.2	0.04
47 Cyg	196093+	K 2 Ib + B	v	B 4	3.5	0.00:
HR 7895	196753+	K 0 II-III +A 3	v	B 7	1.8	0.12
49 Cyg	197177	G 8 IIb	v ?	A 0	2.6	0.00:
Xi Cyg	200905	K 4.5 Ib-II	v	A 1	5.6	0.04:
Alp Equ	202447+	G 0 II + A 3	98.81	A 4	0.6	0.04
HR 8147	203156	F 3 II	v	B 7+	2.4	0.06
HR 8242	205114+	G 2 Ib + A-B	1322 ?	B 8	2.3	0.22
5 Lac	213310+	M 0 Iab + B	42 years	B 6p	3.4	0.10
Eta Peg	215182	G 8 II: + A:	817.43	A 2	2.8	0.00
	215318+	(G 1 II + A 2)		A 2	1.4:	0.04:
HR 8752	217476	G 0 Ia v	v ?	B 0.5IV	4.9	0.64
Psi And	223047	G 5 Ib		B 9p	3.3	0.02

Spectral Type: "Best" available MK classification; in parentheses if derived by other means.

Period : Period determined from radial velocity; "v" means probable orbital variations in radial velocity.

UV type : Classification of IUE low-dispersion far-UV spectrum relative to IUE standard spectra.

ΔV : Magnitude difference corresponding to hot/cool flux ratio in V, estimated from energy distribution.

E(B-V) : Excess in (B-V) color due to interstellar reddening, as determined by fit to energy distribution.

the ground alone. Griffin and Griffin find that Δm = 1.9 mag by their subtraction method; using their types, published UBV colors and assuming no reddening, I find Δm = 2.3 mag and predicted eclipse depths of 0.87, 0.33 and 0.12 mag in U, B, V. I suspect these values are too large, but without data over a larger wavelength range, better estimates cannot be made. I am unaware whether the last expected eclipse in August 1987 was observed to verify the event. For τ Per (G5 III + A2V), an astrometric orbit is available, and because conjunction occurs near periastron for this eccentric system, only a two day eclipse out of a four year period is seen! The last eclipse was not well covered either by IUE or from the ground, and in U, B, V the depths are at least 0.16, 0.12 and 0.06 mag. One of the IUE observations was taken closer to centrality and suggests a depth in V of at least 0.13 mag Further observations are planned for the upcoming 16 Jan 1989 eclipse.

While not extreme, in each of these systems the eclipses are certainly detectable optically with careful work (and good weather!). From the Δm values derived for them, a spectrophotometric system such as Schmidtke's could have identified them as composites, and thus candidates for radial velocity studies. In these cases the situation was reversed - velocity variables were selected to search for composite systems. Note also from the standpoint of deriving radii from the eclipse data, broad-band observations for the ζ Aur stars must be interpreted with caution since the observed duration of ingress and egress phases is affected by the added atmospheric absorption far from geometrical eclipse. Such effects can be better sorted out with smaller bandpasses.

5. CONCLUDING REMARKS

While IUE provides valuable data on the astrophysical conditions in the eclipsing systems, as a survey instrument it is a much less efficient way of discovering eclipses themselves. How many other cool + hot binaries can be found? Somewhere between 20 and 30% of supergiants are reported to be binary, and with the tendency to form components of equal masses, the secondaries should not be much later in type than the O- and B-type progenitors of the primaries. Massa and Endal (1987) estimate there are perhaps 240 non-variable supergiant binaries in the HD catalog with periods less than 15 years and velocity amplitudes for the primary > 4 km/sec. Thus there appears to be a considerable number of discoveries yet to be made.

While I have focussed on the ζ Aur-like systems, I want to emphasize that the Michigan and Case surveys can be expected to yield some surprising systems. I mention one last object, HR 6384 (Table I), classified as M2 II-III + A by Houk and Cowley. The IUE spectrum of this star (Ake and Parsons 1985) has characteristic features of a ζ Aur system in partial eclipse. It has since been found to be both a photometric (Walker, Marino and Herdman 1986) and radial velocity variable (Andersen et al. 1985), but the period is only about 80 days. For a system with a period this short and yet the magnitude difference so large, the secondary is most likely permanently obscured by circumsystem material and the system is likely to be interacting. Follow-up observations of objects in the early results papers from these surveys should not be forgotten.

ACKNOWLEDGEMENTS

Much of the work described here has been undertaken with my colleague Sid Parsons, without whom much of the eclipsing system discoveries would not have been made. I would like to express my gratitude to the IUE Project in handling the special scheduling demands required for some of these observations, and gratefully acknowledge the support of NASA contracts NAS5-25774 and NAS5-28749.

REFERENCES

Ake, T. B., Fekel, F. C. Jr., Hall, D. S., Barksdale, W. S., Fried, R. E., Hopkins, J. L., Landis, H. J. and Louth, H. 1986 Inform. Bull. Var. Stars No. 2847.
Ake, T. B. and Parsons, S. B. 1985 Inform. Bull. Var. Stars No. 2686.
Ake, T. B. and Parsons, S. B. 1987 Inform. Bull. Var. Stars No. 3002.
Ake, T. B., Parsons, S. B. and Kondo, Y. 1985 Astrophys. J. **298**, 772.
Andersen, J., Nordstrom, B., Ardeberg, A., Benz, W., Imbert, M., Lindren, H., Martin, N., Maurice, E., Mayor, M. and Prevot, L. 1985 Astron. Astrophys. Supp. **59**, 15.
Beavers, W. I. 1982 Astrophys. J. Supp. **49**, 273.
Bidelman, W. P. 1984 in The MK Process and Stellar Classification, R. F. Garrison, ed., David DUnlop Obs., Toronto, p. 45.
Bidelman, W. P. 1985 Astron. J. **90**, 341.
Böhm-Vitense, E. and Proffit, C. 1985 Astrophys. J. **296**, 175.
Corbally, C. J. 1987 Astrophys. J. Supp. **63**, 365.
Griffin, R. and Griffin, R. 1986 J. Astrophys. Astr. **7**, 195.
Hynek, J. A. 1951 in Astrophysics, A Topical Symposium, J. A. Hynek, ed., McGraw-Hill, New York, p. 448.
Keenan, P. C. and Pitts, R. E. 1980 Astrophys. J. Supp. **42**, 541.
Madore, B. F. 1977 Monthly Notices Roy. Astron. Soc. **178**, 305.
Massa, D. and Endal, A. S. 1987 Astron. J. **93**, 579.
Parsons, S. B. 1981 Astrophys J. **247**, 560.
Parsons, S. B. and Ake, T. B. 1987 Bull. Amer. Astron. Soc., **19**, 708.
Parsons, S. B., Ake, T. B. and Hopkins, J. L. 1985 Publ. Astron. Soc. Pacific **97**, 725.
Schmidtke, P. C. 1983 in IAU Coll. 62, Current Techniques in Double and Multiple Star Research, Lowell Obs. Bull., Vol. IX, No. 1, R. S. Harrington and O. G. Franz, eds., Lowell Obs., Flagstaff, p. 228.
Schmidtke, P. C. 1984 Bull. Amer. Astron. Soc. **16**, 912.

Walker, W. S. G., Marino, B. F. and Herdman, G. 1985 Inform. Bull. Var. Stars, No. 2775.
Wright, K. O. 1970 in Vistas in Astronomy, A. Beer, ed., Pergamon Press, London, p. 147.

DISCUSSION

ETZEL: Has anyone tried to use least-squares decomposition on composite binaries synthesized from known (single) standards?

AKE: Beavers and Cook (Ap. J. Suppl. 44, 489, 1980) describe two tests of their decomposition procedure. First, they try to classify some of their single standard stars using their mean indices. Secondly, they create artificial composites to see if the fitting technique properly resolves them. In both cases they can assign types to about 1-2 subclasses.

KURUCZ: Can you estimate how many stars are identical binaries that you could not detect?

AKE: The canonical percentage is that 20-30% of stars are binary (e.g. for the Cepheids) and due to the preference to form equal mass components this is probably the best estimate of the fraction.

WING: I would like to add a comment about the work of Paul Schmidtke (Ph.D. thesis, OSU, 1981) that you mentioned. The observations were obtained with the HCO scanners at Kitt Peak and Cerro Tololo, and his set of wavelengths was modeled fairly closely on the work of Spinrad and Taylor (Ap. J. 1969, 1971) at Lick. The Lick study was aimed at determining the stellar content of external galaxies from their integrated spectra; Schmidtke undertook the much more manageable problem of separating the components of composite binaries. He found that the key to success was to cover a wide range of wavelength, and to include spectral features from both the hot component (Balmer and Paschen lines) and the cool component (metallic lines, molecular bands) near both ends of the wavelength range.

NEW INSIGHTS INTO THE CAUSE(S) FOR THE MASS LOSS IN Be STARS

Geraldine J. Peters

Space Sciences Center, University of Southern California

ABSTRACT: Spectroscopic and photometric observations of Be stars in the optical and ultraviolet regions have recently confirmed interesting long-term (months-years) and short-term (≈ one day) variations which can potentially give us some insight into the cause(s) for the mass loss in these stars. The current consensus is that rapid rotation alone cannot be responsible for the mass loss and that other mechanisms such as nonradial pulsations, magnetic fields, and binary mass transfer might be important. Selected observations that show the nature of the episodic mass loss and long-term variations are presented and some ways in which intensive monitoring of a *limited number of objects* with an Automatic Spectrophotometric Telescope could advance our understanding of these incredibly active stars are enumerated.

1. INTRODUCTION

The new detectors on ground-based telescopes that yield high signal/noise (e.g. CCD's and Reticons) along with the UV observations from space-borne telescopes have revealed hitherto unknown activity in early type stars. Evidence for nonradial pulsations has been found in main sequence band OB stars as well as in O - F supergiants (cf. reviews by Smith 1988, Baade 1987a,b). One of the more significant discoveries provided by the UV spectra of early type stars was that these objects typically display *variable* stellar winds (cf. Underhill and Doazan 1982). But of all the classes of early type stars nowhere are the nonradial pulsations and the mass loss more spectacular and variable than in the Be stars. The envelopes of Be stars have long been known to vary, usually unpredictably, but the activity observed in ground-based spectrograms was routinely interpreted within the framework of the "classical" Struve (1931) model. Although from the earliest studies it was apparent that as a class Be stars were rapid rotators, the philosophy of the Struve model was based upon the idea of *critical rotation*. The general consensus now is that Be stars do not rotate at their critical values (Slettebak 1987) and, in fact, $\omega/\omega_{cr} \leq 0.85$ (Stalio, Polidan, and Peters 1987). A number of other models are now being considered to explain the Be phenomenon. There is

A. G. Davis Philip, D. S. Hayes and S. J. Adelman, (eds.)
New Directions in Spectrophotometry 37 - 47

considerable interest currently in the possibility that nonradial pulsations, perhaps in conjunction with rapid rotation, are the most important factors, but other possible causes for the mass loss include magnetic fields, binary mass transfer, and several wind models (cf. Proc. from IAU Colloquium No. 92, Slettebak and Snow 1987, for details).

To gain some insight into the cause(s) for the mass loss in Be stars, one strategy that has gained popularity lately is to perform multiwavelength studies of the activity in specific objects. This approach differs from earlier techniques which involved observing a large number of stars and performing statistical studies. These earlier methods tended to produce the ubiquitous fragmentary data bases that only served to confuse the interpretation. We are discovering that Be stars display interesting short-term and long-term variability which was missed in earlier studies in which the star was observed sporadically in only the optical spectral region. Examples of recent successful multiwavelength campaigns include studies of the Be stars 59 Cyg (Doazan et al. 1985), θ CrB (Doazan et al. 1986), and ω Ori (Sonneborn, et al. 1988).

Activity cycles have been studied in late type stars now for several decades (Baliunas and Vaughan 1985). But in the case of the cool stars, the interpretation of the observations is aided by our experience with the Sun and the extensive knowledge of solar physics that exists. For the early type stars, however, we have no paradigm. We have never seen a hot star close-up and do not know what to expect. Therefore, when a certain type of activity is discovered in a particular object, it is essential that we follow up on the observations, observe this star intensively and acquire data with sufficient time resolution.

In this paper selected observations that demonstrate the nature of some of the short-term and long-term variations in Be stars that have recently been found are presented. Various ways in which intensive monitoring of a *limited number of Be stars* with an Automatic Spectrophotometric Telescope could advance our understanding of the activity in these stars are enumerated throughout.

2. SHORT-TERM ACTIVITY

Spectroscopic and photometric variations which occur on a time scale of minutes-few days probably reflect photospheric activity such as nonradial pulsations or episodic/transient ejection of material. Observations of nonradial pulsations, as well as transient activity, are discussed extensively by Smith (1988). Here I wish to consider a type of short-term spectral variation that I call *abrupt* activity. This is a sudden change in the state of the circumstellar environment that persists for days-months afterwards and differs fundamentally from the type of transient activity we associate with ejection of matter in the form of a "puff".

In 1985, using a CCD detector with the Coudé Feed Telescope at Kitt Peak National Observatory (KPNO), I observed the rapid development of Hα emission in the Be star μ Cen in just *two* days (Peters 1986). In this event, the emission phase lasted for about a month, but the changes that took place on or near the photosphere from which we believe the ejection originated happened on a time scale of a day or less.

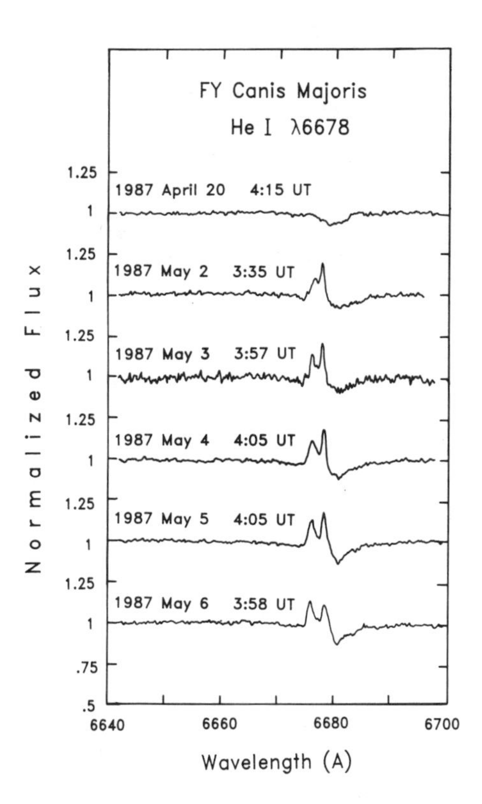

Fig. 1. Striking changes in He I 6678 in the Be star FY CMa during 1987 April. They provide an example of *abrupt* activity (from Peters 1988b).

Using the same CCD instrumentation at KPNO, I recently observed an even more striking abrupt change in the circumstellar material in FY CMa = HR 2855 (Peters 1988b). In this case, in less than *twelve* days (between 1987 April 20 - May 2), the violet-shifted peak of the double

Hα emission feature increased in strength by 30%, while He I 6678 changed from being a simple absorption feature to a structured, inverse P Cygni line with a peak emission intensity of 1.2 I_{cont}! These observations are shown in Fig. 1. When the star was next observed in late August 1987 (Fig. 2), its red spectrum appeared to be fundamentally unchanged from its appearance in May. Therefore, the optical spectrum was either invariant for 110 days, or if the activity is periodic, the star was observed at the same phase.

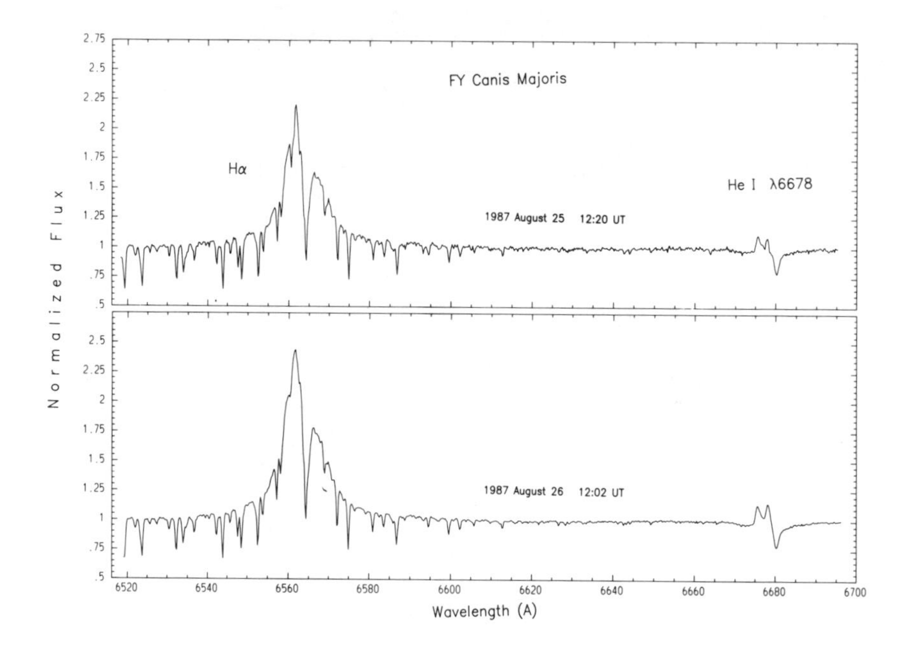

Fig. 2. Hα and He I 6678 observed in FY CMa in 1987 August show very little change from their profiles in May. Since the star was observed through 7 - 9 air masses (about 2° above the actual horizon just before the Sun rose), telluric water vapor lines are prominent. Any program aimed at automatic monitoring of Hα will have to include a method for removing water vapor absorption.

To what extent are abrupt changes such as the ones seen in FY CMa commonplace in Be stars? To a lesser degree, a similar event was observed in 59 Cyg during the same observing run! These observations are shown in Fig. 3. On 1987 April 18, Hα showed a peak intensity of 1.65 I_{cont} and He I 6678 contained emission that just filled the absorption feature. An observation on May 2 revealed that Hα had increased slightly in strength and that the helium line had developed a weak, violet-shifted emission component that was marginally detectable with the CCD instrumentation. By May 6, Hα showed a further increase in strength as well as a profile change and the He I line could now clearly be seen at an intensity of 2% above the continuum. Observations during late August 1987 revealed very little change in the spectrum from May 6, but the helium line was now seen at a strength of 1.06 I_{cont}. In this case it took but four days to produce a quasi-permanent change in the circumstellar envelope.

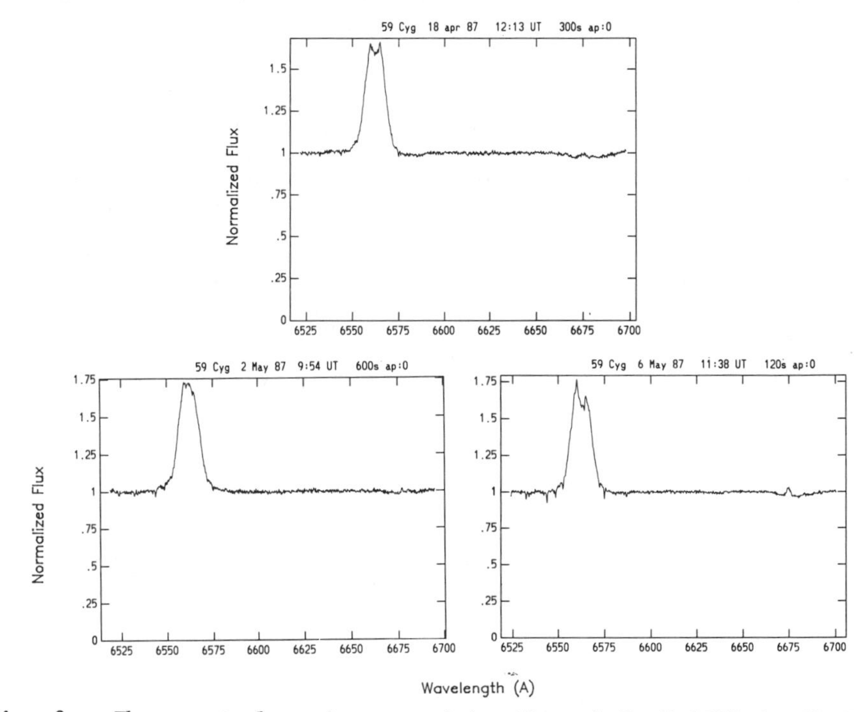

Fig. 3. The spectral region containing Hα and He I 6678 in 59 Cyg observed with the TI 3 CCD and the Coudé Feed Telescope at KPNO during 1987 April - May. Note the abrupt change in the strength and profile of the Hα emission feature and the development of an emission component (with a profile similar to Hα's) in He I 6678. The spectrum resembled the one seen on 1987 May 6 when next observed in late August of 1987.

Abrupt changes such as the ones I have described above could presumably by detected with the aid of an automatic spectrophotometric telescope. It is clear that frequent monitoring of selected stars which have a history of activity is needed as the time scale for the transformation in the envelope is apparently only a few days. But there is one important problem with which we will have to deal: the presence of the pervasive telluric water vapor absorption around Hα. These lines can prominently be seen in Fig. 2. The strength of these features is a function of air mass and the ambient humidity during the observation. Therefore, an observational program aimed at studying temporal changes in Hα will have to include observation of a nearby early type star that is devoid of emission and preferably rapidly rotating.

3. LONG-TERM ACTIVITY

Spectroscopic variability on a time scale of years-decades has been known to exist in Be stars for over a half century (cf. Underhill and Doazan 1982). In addition, long-term photometric trends have been confirmed within the past thirty years. The variations can be cyclic, secular trends, or unpredictable, striking events. Notable spectroscopic changes include long-term oscillations in the V/R ratio of the Balmer emission lines, the peak intensity of the Balmer emission lines relative to the continuum, the radial velocity of the emission and shell lines, and the presence/absence of emission. An intriguing photometric trend that has been observed in several Be stars (especially μ Cen and π Aqr) is that the V magnitude becomes brighter as the Balmer line emission becomes stronger (Dachs 1982, Nordh and Olofsson 1977). Nordh and Olofsson also showed that there is a tendency for the (B - V) color to be redder, and the (U - B) color bluer, as the Balmer emission increases. The cause for the latter is most likely the increase of continuous emission from hydrogen.

I would like to now discuss some long-term trends that recent coordinated, multiwavelength observations of the Be star 66 Oph have revealed. Hα observations since the early 1970s have shown that the intensity of this emission line has been steadily increasing. The trend from 1982 to 1987 is shown in Fig. 4. Except for one deviant point, which was observed in late June of 1985, a gradual increase in strength from 6.3 to 7.5 I_{cont} was apparent. Throughout the latter interval of time, the FUV continuum and line spectrum of this star have been monitored with the *Voyager* and *IUE* spacecrafts. As one can readily discern from the spacecraft observations shown in Fig. 5, the FUV flux (binned from 950 - 1150 Å) and the strength of the C IV wind line are variable. In fact data acquired to date seem to suggest that these observable quantities are *anticorrelated* and vary quasi-periodically with a period of about 550 days. Now in late June of 1985, when the intensity of the Hα emission line declined, the variable wind in 66 Oph was recovering from a minimum presence that was observed in 1985 April (Peters 1988a). Therefore, in this case there appeared to be an overall decrease of the amount of material contained in the Balmer emitting envelope when the star's wind "turned off".

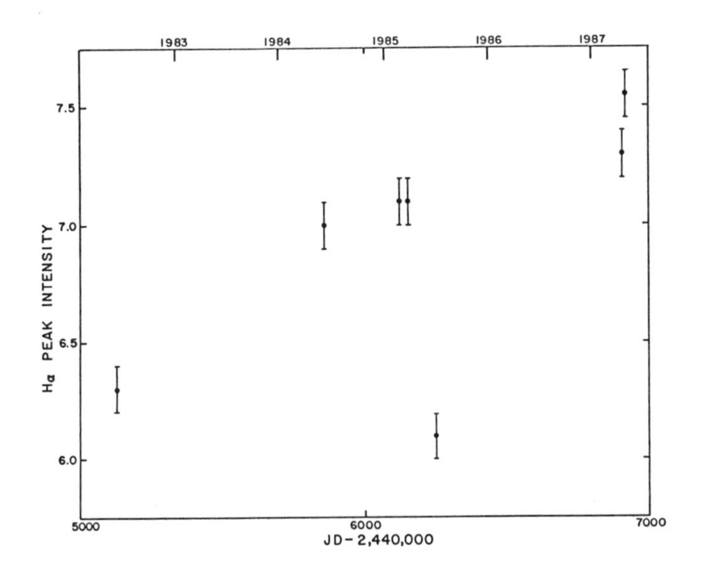

Fig 4. The intensity of the Hα emission feature in 66 Oph observed from mid-1982 to mid-1987 (from Peters 1988a).

Voyager observations of other Be stars suggest that FUV flux variability may be commonplace in this class of objects. But the cause for the flux changes and their importance in the Be phenomenon remain unknown. It will be through coordinated, multiwavelength observations extending over many years that we might begin to understand the stellar physics in these incredibly active stars.

Variations on a time scale of years-decades, especially ones that are quasi-periodic, probably have their origins deep within the star. One possibility is that a change occurs in the low-order $\ell = 2$ mode. The waves associated with this mode, which is frequently observed in Be stars, penetrate well into the star and carry more energy than the higher-order $\ell = 8$, 10 modes, which are surface phenomena. Using an impressive data base acquired at Lick Observatory, Penrod (1987) has found that the amplitudes of the longer-period, low-order nonradial pulsations are largest just before outbursts, they decline during the emission phases, and then slowly recover to their original strength just before the onset of the next emission phase.

If early type stars have magnetic fields, perhaps we are observing something analogous to the solar cycle. It is perhaps possible that there is an interaction between the regions of the star in which magnetic fields are present and those affected by nonradial pulsations (Smith, personal communication).

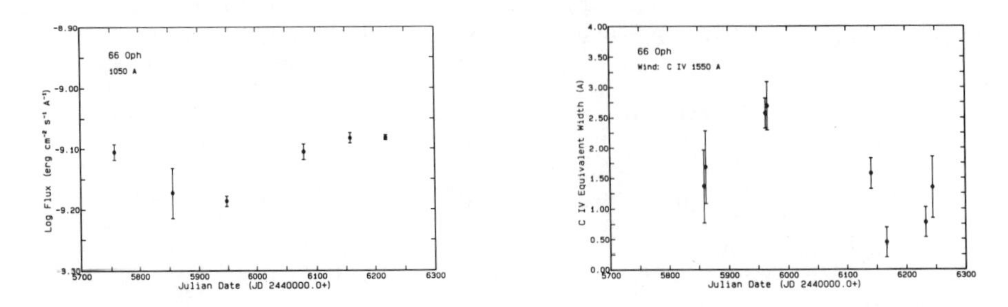

Fig. 5 First panel: *Voyager* UVS observations of the FUV flux in 66 Oph in the interval 950 - 1150 Å during 1984 - 1985. Second panel: Contemporaneous high resolution ($\Delta\lambda \approx 0.2$ Å) *IUE* observations of the equivalent width of the C IV wind line.

Alternatively, some Be stars might be interacting binaries and the causes for the long-term variations might be traced to activity in the mass losing secondary. A long-term light cycle has been discovered in the interacting binary Be star AU Mon (Lorenzi 1985). In this case the V magnitude regularly undergoes a variation of about 0.3 mag with a period of 411 days. *IUE* observations of this star at maximum and minimum light suggest that the mass transfer rate is enhanced when the star is *faint* (Peters 1988c) and, therefore, imply that the source of the activity lies with the secondary.

4. PROSPECTUS

In this paper, selected observations which illustrate some of the newly discovered short-term and long-term phenomena in Be stars have been presented. Those variations which can be studied using low-moderate spectral resolution have been emphasized. Although activity in hot stars, in general, is far more prevalent than one earlier believed, the Be stars stand out as exceptional. Since we have never seen the surface of an early type star, directed, multiwavelength observations are essential to uncover the physical phenomena that are taking place.

Since variability on time scales from days to years has been found, an Automatic Spectrophotometric Telescope (ASPT) would be useful to acquire the ground-based data that comprises an important piece of the multiwavelength puzzle. Daily monitoring of the strengths of selected Balmer lines (e.g. Hα and Hβ), He I lines (e.g. $\lambda\lambda$6678, 7065), and the optical continuum are recommended. Suitably designed interference filters could be used for the lines, while filters with intermediate passbands, such as those for the 13-color system of Johnson and Mitchell (1975), would be best for the continuum. As mentioned in Section 2, it will be necessary to observe acceptable nearby standards regularly in order to correct for varying telluric water vapor absorption around Hα. For an initial program, the number of stars selected for monitoring should be small, about 20 - 50, but eventually this number could be increased to 100 or more as the results of the science demand.

Through daily monitoring, *abrupt* activity could be detected. A network of observers could then be notified to acquire spacecraft data and additional ground-based spectra with higher spectral resolution and signal/noise. Some important information that could come from such a strategy include the time scale for the rise and decay of the emission, the actual behavior displayed by the varying emission lines, whether the onset of emission or a fundamental change in its character is accompanied by a change in the wind, and further information on the apparent correlation between the amplitude of the nonradial waves and the strength/presence of the Balmer emission.

Long-term monitoring would be useful to confirm the correlations which have already become apparent (c.f. Section 3). It would allow us to explore the extent to which the Balmer emission and wind (which is observed in the UV) are linked and determine whether long-term fluctuations in the continuum are a signature of binarity.

The research on the UV spectra of Be stars reported in this paper has been supported in part by NASA grant NSG 5422.

REFERENCES

Baade, D. 1987a in Physics of Be Stars, A. Slettebak and T. P. Snow, eds., Cambridge Univ. Press, Cambridge, p. 361.
Baade, D. 1987b in O, Of, and Wolf-Rayet Stars, P. Conti and A. B. Underhill, eds., NASA/CNRS Monograph Series, in press.
Baliunas, S. L. and Vaughan, A. H. 1985 Ann. Rev. Astron. Astrophys. **23**, 379.
Dachs, J. 1982 in Be Stars, M. Jaschek and H. G. Groth, eds., Reidel, Dordrecht, p. 19.
Doazan, V., Grady, C. A., Snow, T. P., Peters, G. J., Marlborough, J. M., Barker, P. K., Bolton, C. T., Bourdonneau, B., Kuhi, L. V., Lyons, R. W., Polidan, R. S., Stalio, R., Thomas, R. N. 1985 Astron. Astrophys. **152**, 182.
Doazan, V., Marlborough, J. M., Morossi, C., Peters, G. J., Rusconi, L., Sedmak, G., Stalio, R., Thomas. R. N. and Willis, A. 1986 Astron. Astrophys. **158**, 1.
Johnson, H. L., and Mitchell, R. I. 1975 Rev. Mexicana Astron. Astrophys., **1**, 273.
Lorenzi, L. 1985 Info. Bull. Variable Stars, No. 2704.
Nordh, H. L. and Olofsson, S. G. 1977 Astron. Astrophys. **56**, 117.
Penrod, G. D. 1987 in Physics of Be Stars, A. Slettebak and T. P. Snow, eds., Cambridge Univ. Press, Cambridge, p.463.
Peters, G. J. 1986 Astrophys. J. Letters **301**, L 61.
Peters, G. J. 1988a Publ. Astron. Soc. Pacific **100**, 207.
Peters, G. J. 1988b Astrophys. J., in press.

Peters, G. J. 1988c Astrophys. J., submitted.
Peters, G. J. and Polidan, R. S. 1987 in Physics of Be Stars, A. Slettebak and T. P. Snow, eds., Cambridge Univ. Press, Cambridge, p. 278.
Slettebak, A. 1987 in Physics of Be Stars, A. Slettebak and T. P. Snow, eds., Cambridge Univ. Press, Cambridge, p. 24.
Smith, M. A. 1988 in Pulsation and Mass Loss in Stars, R. Stalio and L. A. Willson, eds., Reidel, Dordrecht, in press.
Sonneborn, G., Grady, C. A., Wu, C. -C., Hayes, D. P., Guinan, E. F., Barker, P. K. and Henrichs, H. F. 1988 Astrophys. J. 325, 784.
Stalio, R., Polidan, R. S. and Peters, G. J. 1987 in Physics of Be Stars, A. Slettebak and T. P. Snow, eds., Cambridge Univ. Press, Cambridge, p. 272.
Struve, O. 1931 Astrophys. J. 73, 94.
Underhill, A. B. and Doazan, V. 1982 B Stars with and without Emission Lines, NASA/CNRS Monograph Series.

DISCUSSION

ETZEL: Is there any evidence that 66 Oph is a spectroscopic binary?

PETERS: Currently the only diagnostic to suggest that it may be an interacting binary is the presence of IR Ca II triplet emission, which is often found in the longer period systems. I would like to urge interested members of the audience to attempt to find a radial velocity curve for this star, but it won't be simple because of the emission components seen in most of the stronger spectroscopic features.

JOHNSON: The neutral helium line λ 10830 might be a useful diagnostic. Have you observed any Be stars?

PETERS: No, but I agree it would be interesting to obtain data on this line. Since λ 10830 arises from the metastable 2^3S level, while λ 6678 is a singlet and easily connected to the ground state. Both diagnostics can contribute to our understanding of the physics and geometry of the circumstellar material. Meisel et al. (Astrophys. J. **263**, 759, 1982) report some observations of λ 10830 in Be stars.

ADELMAN: You noted that as Hα increases the star brightens in V. Are there also color changes?

PETERS: Yes, the general trend is that as the Hα emission increases, (B-V) becomes redder while (U-B) becomes bluer.

DETERMINATION OF THE EFFECTIVE TEMPERATURES OF F, G AND K STARS

R. A. Bell

University of Maryland

This talk is based upon two papers, one of which is entitled "Synthetic Strömgren Photometry for F Dwarf Stars" (Bell 1988) and the other of which is "The Effective Temperatures and Colors of G and K Stars" (Bell and Gustafsson 1988).

There are a large number of problems in the field of stellar astronomy which cannot be satisfactorily solved until we have established a reliable temperature scale for stars. These problems include the determination of very accurate stellar abundances and the determination of the ages of globular clusters. I will discuss the use of a particular method for determining stellar temperatures - the infrared flux method (IFRM) of Blackwell and Shallis (1977) - and will compare the results of this method with temperature determinations using stellar colors.

The infrared flux method is based upon measurements of the integrated stellar flux and a monochromatic flux. This latter flux is usually measured at an infrared wavelength, to make the flux ratio relatively independent of surface gravity and abundance and dependent on effective temperature.

To measure the integrated flux we need observational data from a number of sources. The ones I will use are:

IUE observations for $\lambda < 3300$ Å;
narrow band photometry for $3300 < \lambda < 11000$ Å;
broad band infrared measurements for $\lambda > 11000$ Å.

The IUE data are reasonably well suited for our purpose in that it gives continuous wavelength coverage of the stellar fluxes.

The situation in the visible is much less favorable. We do not have an extensive series of measurements of stellar fluxes with continuous wavelength coverage for a large sample of stars. The data of Gunn and Stryker (1983), which are very valuable for some purposes, are for stars which are sufficiently faint as to be reddened to some extent and are fewer in number than needed for this project. The most

A. G. Davis Philip, D. S. Hayes and S. J. Adelman, (eds.)
New Directions in Spectrophotometry 49 - 58

suitable data for our purpose are those of the 13 color photometry of, originally, Johnson, Mitchell and Latham (1967) and, subsequently, Johnson and Mitchell (1975). A plot of the filters used to obtain these data is shown as Fig. 1. This shows that the filters provide direct measurements of the flux in about 50% of the wavelength interval between 3300 and 11000 Å. This raises the question as to how accurately the flux between 3300 Å and 11000 Å can be measured from these data. A check as to how accurate this is has to be made using synthetic spectra, which give a maximum value of the error for G and K stars of about 1.5%. (There is an obvious need here for better data, for a sample of stars which includes some Population II objects).

The situation in the infrared is again not particularly satisfactory. There is the work of Strecker, Erickson and Witteborn (1979, hereafter SEW79) using the Kuiper Airborne Observatory which offers continuous wavelength coverage. However, these data exist only for a few stars. Unfortunately, there is only a little overlap between the IR photometry of Glass (1974a, 1974b), which adds data in the H band, and the 13 color photometry. There are also little data available on the Caltech-CTIO system for the bright stars for which IUE and 13-color data are available. The only large sample of stars for which both 13 color and IR photometry are available is the Johnson system (Johnson et al. 1966) J, K and L photometry. The integrated flux in the IR up to the L band effective wavelength is consequently obtained from the Johnson J, K and L magnitudes using Simpson's rule integration. The result can be compared with the data of SEW79 for a few stars. As a result of this comparison, a correction is applied to the flux obtained from the broad band measurements. While this correction may be large, as much as 34%, it is supported by the model calculations. The integration is extended to infinite wavelength assuming that stars radiate like black bodies beyond the L band.

Now, having obtained the integrated fluxes in this way, we ask what do we do for the reference band? Here we find that there is really no choice for G and K stars although I think there is for F stars. It would be rather tempting to use the relatively narrow 98 or 110 pass bands of the 13 color system as the reference. However, as Table I shows, there is little sensitivity to temperature at the lowest temperature e.g. the flux ratio varies by only 2% between 4000 and 4500 K for the 98 pass band and by only 7% for the 110 pass band. Allowance for a reasonable error of 0.02 mag. in the data suggests that an accuracy of only about 150 K in T_{eff} could be obtained. The J pass band is one which is affected by terrestrial H_2O and fewer stars have L magnitudes than K magnitudes and so we really are forced to use the Johnson K band as the reference.

The point which is quite critical in the use of the IRFM is the accuracy of the conversion of K band magnitudes to flux. In general this is done using either observations of the Sun or a model of Vega, although there have been absolute measurements of the infrared flux of

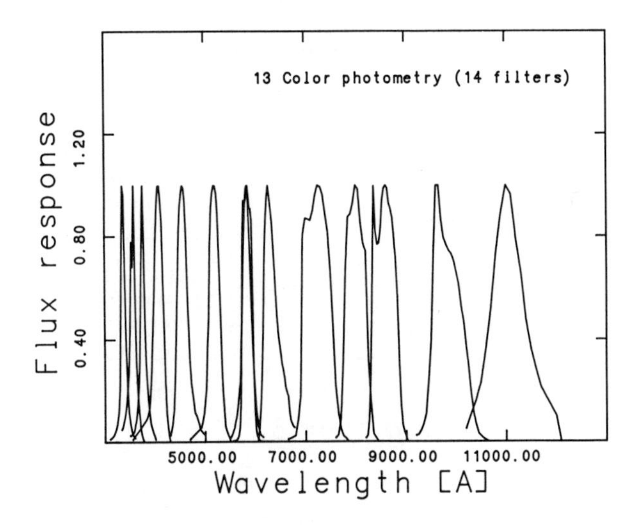

Fig. 1. A plot of the 14 filters used in 13 color photometry.

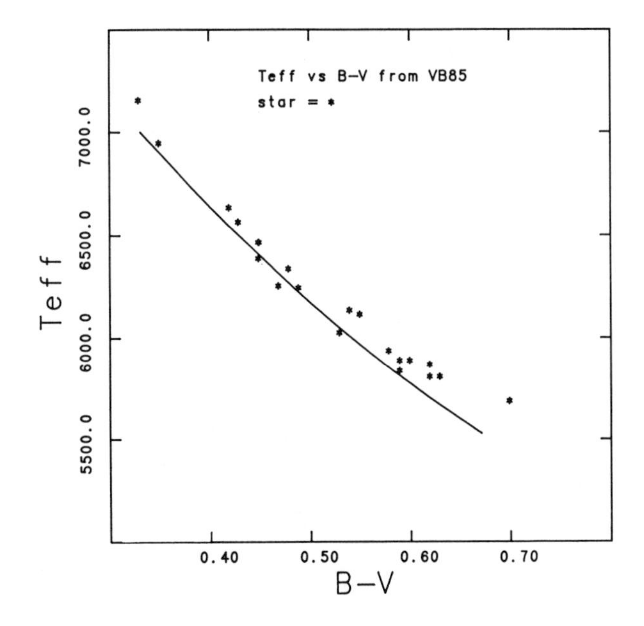

Fig. 2. Effective temperature vs. (B-V) for VB 85 data.

TABLE I

Ratio of total flux to flux in a reference band.

Ratios for Johnson K band as reference

H^- Opacity Source

Giant Stars

Model	Bell et al.	Doughty-Frazer
5500/3.00/0.0	1.384E+05	1.410E+05
5000/3.00/0.0	1.067E+05	1.091E+05
4500/2.25/0.0	0.7994E+05	0.8203E+05
4000/1.50/0.0	0.5830E+05	0.6018E+05

Dwarf Stars

Model	Bell et al.	Doughty-Frazer
6000/4.50/0.0	1.782E+05	1.806E+05
5500/4.50/0.0	1.400E+05	1.421E+05
5000/4.50/0.0	1.077E+05	1.096E+05
4500/4.50/0.0	0.808E+05	0.825E+05

Ratios for Glass K band using Bell et al. opacity

Giant Stars

Model	
5500/3.00/0.0	1.4783E+05
5000/3.00/0.0	1.1390E+05
4500/2.25/0.0	0.8512E+05
4000/1.50/0.0	0.6185E+05

Ratios for 98 and 110 pass bands

Model	98	110
5500/3.00/0.0	1.81E+04	2.30E+04
5000/3.00/0.0	1.70E+04	2.09E+04
4500/2.25/0.0	1.60E+04	1.89E+04
4000/1.50/0.0	1.57E+04	1.77E+04

Vega by Selby et al. (1983). The subject has been discussed in some detail by Saxner and Hammarback (1985). The importance of accuracy in the K band calibration can be seen in Table I - the flux ratio varies by 37% between 4000 and 4500 K. An error of 5% in the flux calibration of the K band, therefore, corresponds to an error of about 70 K in T_{eff}. If the effective wavelength of the Johnson K band sensitivity function is wrong by 0.1 micron, then the K band fluxes are changed by about 18%, the exact value depending on the model being considered. The Glass K band filter has an effective wavelength 0.03 microns longer than that of the Johnson one and so the model fluxes on the Glass system are 7% less than on the Johnson system. This flux difference corresponds to about 80 K in T_{eff}. Another point which must be included is that CO bands occur in the wavelength interval of the K pass band and their effects on the K magnitude are not negligible. The models predict that the K magnitude becomes fainter by 0.07 mag. at 4000 K for standard C and O abundances.

Finally, there is a rather vexing problem. What value do we use for the free-free H^- opacity? The bound-free opacity of H^- is well known - the polynominal fit given by Gray (1976) to the calculations of Geltman (1962) agrees with the work of Doughty, Fraser and McEachran (1966) to about 2% for $4000 < \lambda < 15000$ Å and to within 5% of Wishart (1979) for $3500 < \lambda < 15000$ Å. However, the Bell, Kingston and McIlveen (1975) calculations for the free-free absorption, which are about 4% smaller than those of Stilley and Calloway (1970) at 6300 K, are much different to those of Doughty and Fraser (1966) - differences of up to 25% occur. This uncertainty causes an error in the temperatures of about 50 K for G and K stars. The values used subsequently are obtained from the Doughty and Fraser results.

The two applications of the IRFM that I want to discuss are those of Saxner and Hammerback (1985) and Bell and Gustafsson (1988). The former refers to F dwarfs, the latter to G and K stars.

Saxner and Hammarback (1985) give temperatures for a sample of F dwarf stars. These stars have T_{eff} in the interval 7200-5500 K and a range of abundances and surface gravities. VandenBerg and Bell (1985) published a table of colors as a function of T_{eff}, log g and [A/H]. While these two papers shared some common features - they both use an unpublished grid of models for dwarf stars computed by Eriksson, Gustafsson and Bell - the two papers were written independently and in ignorance of one another until the preprint stage. So here we have a nice confrontation - if one were to derive T_{eff} for the SH85 stars from the VB85 tables of (B-V) and (b-y), how well would they match the SH85 values, which rely on the K band fluxes of these models? To perform this comparison, we use the VB85 data to correct the SH85 stars to log g = 4.25 and only utilize the SH85 results for stars with $|[Fe/H]| < 0.15$. The two plots, i.e. (B-V) versus T_{eff} and (b-y) versus T_{eff}, are shown as Figs. 2 and 3. The fits are really quite good for $7200 > T_{eff}$

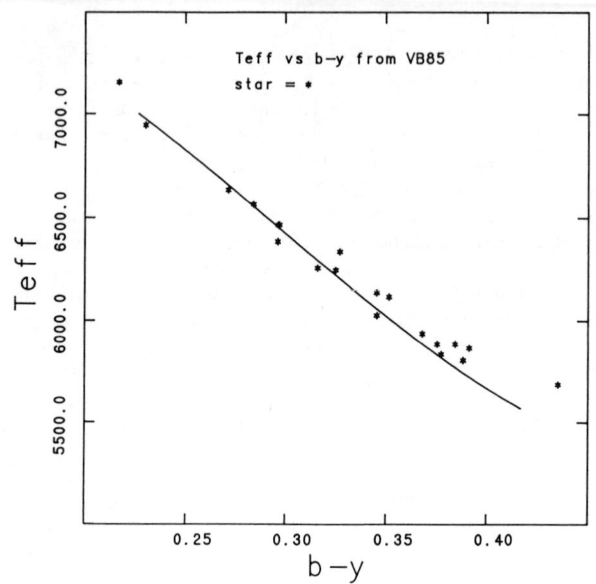

Fig. 3. Effective temperature vs. (b-y) for VB 85 data.

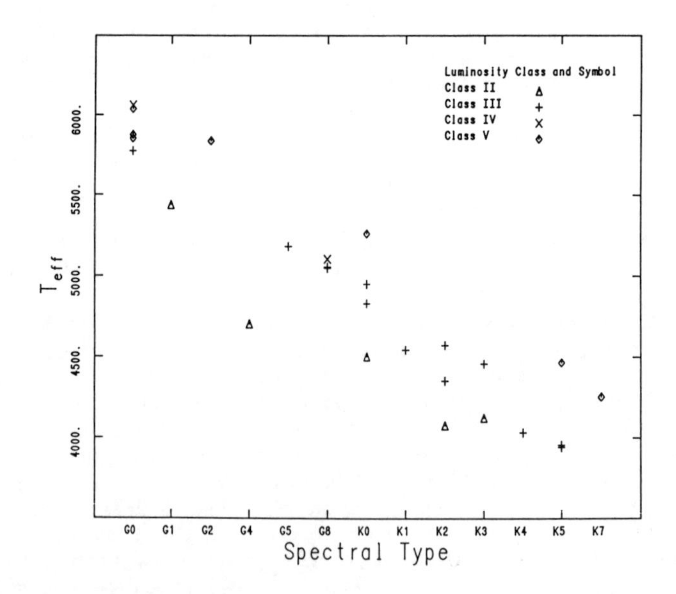

Fig. 4. Temperature-spectral type relation for a sample of G and K standard stars.

> 6000 K, with a possible systematic error of ~0.02 mag. in (B-V) and no obvious systematic error in (b-y).

These results clearly suggest that the T_{eff} - color relations used by VB85 for the conversion of isochrones from the log L, log T_{eff} plane to absolute magnitude, color planes are reasonably correct. As an aside, this has encouraged us to convert isochrones to the filter pass bands of the WF/PC of the Hubble Space Telescope. In addition I have analysed m_1, c_1, and (b-y) observations of subdwarfs to obtain ages for these objects, for comparison with ages deduced for globular clusters.

As a complete aside, we have also used synthetic spectrograms of A - K dwarfs as the basis for an undergraduate laboratory on spectral classification.

A plot of the temperature-spectral type relation for the sample of G and K spectral classification standard stars is shown as Fig. 4. The temperatures used are not those of the IRFM using the Johnson K band but have been reduced by 80 K, to be consistent with results from the Glass K band and from colors. We see that there is a dependence on luminosity class in this figure, with the dwarfs being the hottest at a given spectral type and the bright giants the coolest. One can also see some systematic and understandable patterns. For example, of the two G0 V stars HR 4785 and 4983, the former is more metal poor by 0.5 in [Fe/H] and is cooler by about 150 K. The lower temperature must compensate for the lower abundance when the spectral type is assigned. A similar effect is seen in comparison of Vega and Sirius - the more metal poor Vega is cooler and has an earlier spectral type - and in the K2 III stars.

Fig. 5 shows a plot of (V-K) vs T_{eff} for the standard stars. The relationship is quite a tight one, with the possibility that the late K dwarfs follow a different relation than do the late K giants.

A similar plot, of (B-V) vs T_{eff}, is shown as Fig. 6. This relationship is less tight than the (V-K), T_{eff} one, with the suggestion that the bright giants are redder in (B-V) at a given T_{eff} than are the lower luminosity objects. In addition, the metal poor stars, such as α Boo and ϕ^2 Ori, are bluer than overage for their T_{eff}. Both of these effects are seen in the (B-V) color calculations of Bell and Gustafsson (1978).

A more detailed discussion of these results will be given in Bell and Gustafsson (1988).

ACKNOWLEDGMENTS

The work on G and K stars was carried out in collaboration with Dr. B. Gustafsson. We are grateful for the assistance of Mr. M.

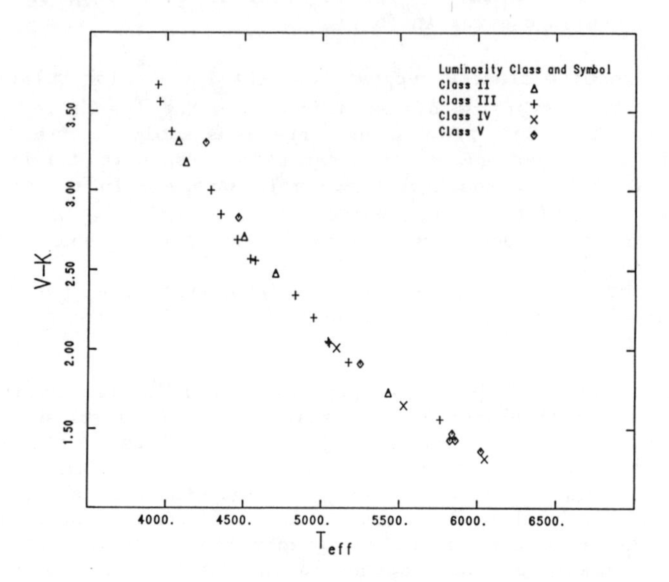

Fig. 5. (V-K) vs. effective temperature for standard stars.

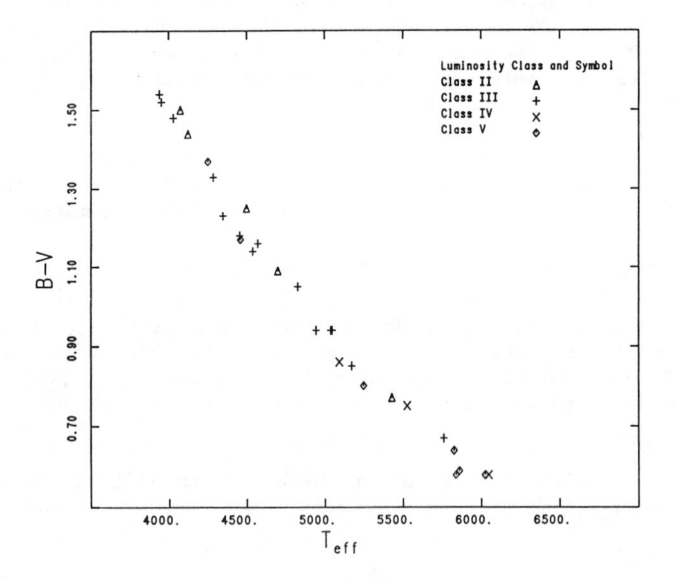

Fig. 6. (B-V) vs. effective temperature for standard stars.

Briley, Dr. K. Eriksson, Dr. R. Gray and Mr. J. T. Ohlmacher. We thank Dr. M. S. Bessell for sending a preprint of his work with Mr. J. M. Brett. The work was supported by the Swedish Natural Science Research Council and by the National Science Foundation, under grants AST80-19570 and AST85-13872. Some of the calculations were made using the Cray X-MP of the San Diego Supercomputer Center, the time being supplied under NSF grant AST85-09915. The interpretation of the results was materially assisted by the use of a Ridge 32C computer and Chromatics CX-1536 image display system, purchased under the above grants.

REFERENCES

Bell, K. L., Kingston, A. E. and McIlveen, W. A., 1975 J. Phys. B. Atom. Molec. Phys. **8**, 358.

Bell, R. A. 1988 Astron. J., in press.

Bell, R. A. and Gustafsson, B. 1988 Monthly Notices Roy. Astron. Soc., submitted.

Blackwell, D. E. and Shallis, M. J. 1977 Monthly Notices Roy. Astron. Soc., **180**, 177.

Doughty, N. A. and Frazer, P. A. 1966 Monthly Notices Roy. Astron. Soc.. **132**, 267.

Doughty, N. A., Frazer, P. A. and McEachran, R. P. 1966 Monthly Notices Roy. Astron. Soc. **132**, 255.

Geltman, S. 1964 Astrophys. J. **141**, 376.

Glass, I. S. 1974a Monthly Notices. Astron. Soc. South Africa. **33**, 54.

Glass, I. S. 1974b Monthly Notices Astron. Soc. South Africa. **33**, 71.

Gray, D. F. 1976 in The Observation and Analysis of Stellar Photospheres, John Wiley and Sons, p 148.

Gunn, J. E. and Stryker, L. L. 1983 Astrophys. J. Suppl. **52**, 121.

Johnson, H. L. and Mitchell, R. I. 1975 Rev. Mex. Astron. Astrofys., **1**, 299.

Johnson, H. L., Mitchell, R. I. and Latham, A. S. 1967 Comm. Lunar Planet. Lab. **6**, 85.

Johnson, H. L., Mitchell, R. I., Iriarte, B. and Wisniewski, W. 1966 Comm. Lunar Planet. Lab. **4**, 99.

Saxner, M. and Hammarback, G. 1985 Astron. Astrophys. **151**, 372.

Selby, M. J., Mountain, C. M., Blackwell, D. E., Petford, A. D. and Leggett, S. K. 1983 Monthly Notices. Roy. Astron. Soc. **203**, 795.

Stilley, J. L. and Calloway, J. 1970 Astrophys. J. **160**, 245.

Strecker, D. W., Erickson, E. F. and Witteborn, F. C. 1979 Astrophys. J. Suppl. **41**, 5.

VandenBerg, D. A. and Bell, R. A. 1985 Astrophys. J. Suppl. **58**, 561.

Wishart, A. W. 1979 Monthly Notices Roy. Astron. Soc. **187**, 59P.

DISCUSSION

JOHNSON: You made an earlier study of the visual surface brightness relation for several colors and found the best overall relation for (V-R). Have you done a similar study for these new models?

BELL: The only recent work that I have been associated with that has used the visual surface brightness, (V-R) relation is a study of the absolute magnitudes of Cepheids by my former student Robert Hindsley.

WING: The higher effective temperatures you get from (V-K) for the G and K stars relative to the occultation scale of Ridgway, et al. (Ridgway, S., Joyce, R., White, N. and Wing, R., Ap. J. **235**, 126, 1980) seems to confirm very nicely the results of Wing, Gustafsson and Eriksson (IAU Symp. 111: Calibration of Fundamental Stellar Quantities, Hayes, Pasinetti and Philip, eds, Reidel, Dordrecht, 1985, p. 571), which were based on an analysis of eight-color photometry by spectral synthesis. The occultation data were very meager for G and K stars but are much stronger for the M stars.

BELL: I quoted the result of Ridgway et al. because it has been used by a number of people.

HORNE: Based on your experience, can you say whether the difficulty in fitting synthetic and observed photometric data is due to uncertainties in the model atmospheres, in the passband functions, or in the observations?

BELL: This is a very hard question to answer. Sometimes it is the synthetic spectra e.g. the ultraviolet region of the Sun. Sometimes it is the pass band function - only now is the U profile of the UBV system reasonably well known. Finally, it can be the observations e.g. models indicated that the (V-K) colors given by Johnson for Sirius and Vega could not be self-consistent. This was confirmed by later data.

VARIABILITY IN LATE-TYPE GIANTS: SPECTROPHOTOMETRY AS A POWERFUL TOOL BESIDES PHOTOMETRY

Francois R. Querci and Monique Querci

Observatoire Midi-Pyrénées

1. INTRODUCTION

Usually, photometry gives stellar parameters such as spectral type, gravity, absolute magnitudes, metal deficiencies, etc. (e.g., see the review by Wing 1985). Photometry also turns out to be a very powerful tool for monitoring the variability of stars.

We have to recall that, starting with the UBV system originally devised by Johnson and Morgan, an extension to the infrared was attempted and many intermediate band photometric systems, such as the DDO, uvbyHβ, Geneva and Vilnius, were developed for the investigation of various astrophysical problems of cooler stars.

However, the filter widths of these systems are too large when applied to the crowded M, S and C spectra. So, the observed flux variations obtained with these filters are due to the contribution of many stellar features which could have different time behavior from each other. Consequently, the Wing (1971) near-IR system (8 filters) and the Querci and Querci (1985a) system (22 filters for the whole spectral range) were proposed for the M and C type stars, respectively. Piccirillo (1976, 1977) modified the 8-color Wing system to include filters for ZrO and LaO features characteristic of S stars. These systems are deduced from a spectroscopic analysis of stellar spectra and from model atmosphere synthetic spectra.

In the visible and near-infrared spectral ranges, the red giant spectra are so crowded with spectral lines that it is practically impossible to find regions free of lines except around 1.04 μm where Wing (1967) created the I 104 filter in which only a few faint molecular lines eat away the stellar continuum. So, whereas many studies endeavor to determine stellar parameters, only a few analyses of the brightness changes and of the spectrum variations have been made with the various systems quoted above. Eggen (1972a,b) and Vetesnik (1984a,b) with the UBVRI filters, Lockwood and Wing (1971) with I 104 and the Wing 8-color system (see also Lockwood and McMillan 1970), Baumert (1972) with the 8-color Wing system, Gillet (1981) with the

A. G. Davis Philip, D. S. Hayes and S. J. Adelman, (eds.)
New Directions in Spectrophotometry 59 - 72

visible filters of the Querci system and Alksne et al. (1983) with the Dzervitis 6-color system, have investigated the stellar flux variations.

The many narrow filters of the above systems can be used to follow the variation of the characteristic spectral features of M, S, and C stars. Their features are formed in different layers of the atmosphere as is shown by model calculations. For example, the He I 10830 Å filter follows the evolution of high excitation waves, the TiO, CN, C_2 filters follow the deep photospheric behavior and the Na-D filter follows the evolution of the circumstellar layers.

Finally, the narrowness of the filters makes it necessary to use telescopes of 80 cm to 1 meter in diameter and the high number of filters significantly increases the observing time so that spectroscopy seems to be a more rapid solution producing a calibrated spectrum with all the required features which at the same time contains other potentially interesting features.

In the past, the main goal of spectroscopy was the determination of the physical stellar parameters by the comparison of the observed fluxes with those deduced from static model atmosphere calculations (Querci 1974, Gustafsson et al. 1975, Bell et al. 1976, Querci and Querci 1976, Strecker et al. 1979) and the main effort was concentrated on the atomic and molecular opacity codes. Temporal observations of variable stars were rarely done with this technique to study the variability itself. However, in some early studies, scans of the same star were made two or three times inadvertently. For example Gow (1975) showed noticeable differences in two scans of R CMi with the Indiana rapid spectrum scanner (Fig. 1).

Nowadays, the problem of running static model atmospheres with good opacities is solved (Johnson 1986). Progress still remains to be made with various theoretical problems such as, introducing realistic convection and grain nucleation in the models. However, this does not prevent theoretical and observational efforts devoted to the interpretation of red giant variations through radial pulsations and chaotic and stochastic motions, and through non-radial pulsations. A large field of research is open in this area.

2. STATUS OF KNOWLEDGE ON THE THEORETICAL ASPECTS OF THE LATE-TYPE STAR VARIABILITY

2.1 The Radial Global Stellar Pulsation

Turbulent convection at the top of the hydrogen convection zone produces the radial global stellar pulsation by driving acoustic waves, which turn into shock waves. In fact, the gas dynamics is characterized by quantities such as velocity, pressure, and density; strong discontinuities of these quantities represent shock waves. The

visible filters of the Querci system and Alksne et al. (1983) with the Dzervitis 6-color system, have investigated the stellar flux variations.

The many narrow filters of the above systems can be used to follow the variation of the characteristic spectral features of M, S, and C stars. Their features are formed in different layers of the atmosphere as is shown by model calculations. For example, the He I 10830 Å filter follows the evolution of high excitation waves, the TiO, CN, C_2 filters follow the deep photospheric behavior and the Na-D filter follows the evolution of the circumstellar layers.

Finally, the narrowness of the filters makes it necessary to use telescopes of 80 cm to 1 meter in diameter and the high number of filters significantly increases the observing time so that spectroscopy seems to be a more rapid solution producing a calibrated spectrum with all the required features which at the same time contains other potentially interesting features.

In the past, the main goal of spectroscopy was the determination of the physical stellar parameters by the comparison of the observed fluxes with those deduced from static model atmosphere calculations (Querci 1974, Gustafsson et al. 1975, Bell et al. 1976, Querci and Querci 1976, Strecker et al. 1979) and the main effort was concentrated on the atomic and molecular opacity codes. Temporal observations of variable stars were rarely done with this technique to study the variability itself. However, in some early studies, scans of the same star were made two or three times inadvertently. For example Gow (1975) showed noticeable differences in two scans of R CMi with the Indiana rapid spectrum scanner (Fig. 1).

Nowadays, the problem of running static model atmospheres with good opacities is solved (Johnson 1986). Progress still remains to be made with various theoretical problems such as, introducing realistic convection and grain nucleation in the models. However, this does not prevent theoretical and observational efforts devoted to the intèrpretation of red giant variations through radial pulsations and chaotic and stochastic motions, and through non-radial pulsations. A large field of research is open in this area.

2. STATUS OF KNOWLEDGE ON THE THEORETICAL ASPECTS OF THE LATE-TYPE STAR VARIABILITY

2.1 The Radial Global Stellar Pulsation

Turbulent convection at the top of the hydrogen convection zone produces the radial global stellar pulsation by driving acoustic waves, which turn into shock waves. In fact, the gas dynamics is characterized by quantities such as velocity, pressure, and density; strong discontinuities of these quantities represent shock waves. The

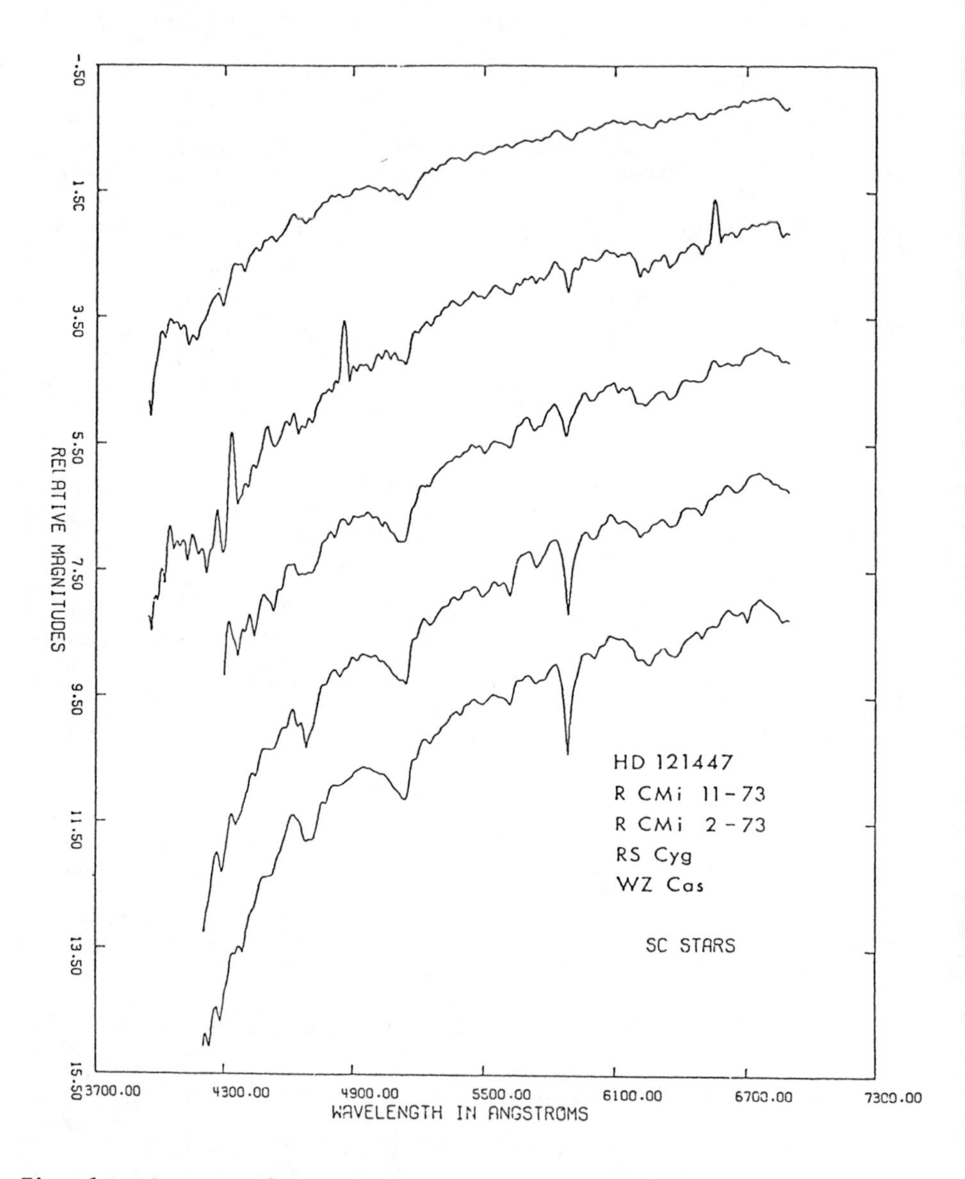

Fig. 1. Scanner plots of SC-type stars. The line variations over three months for R CMi are striking (from Gow 1975).

laws of mass, momentum and energy conservation are the basis of studies of these discontinuities (e.g., Zeldovitch and Raiser 1966). From the shock-wave model the following parameters can be determined: 1. the energy balance of the atmosphere, and 2. the modes of pulsation (which induces shocks). Shocks are considered as a mechanism which produces mass loss.

Three types of red giant and supergiant variables are recognized: Miras, the semi-regulars and the irregulars (e.g., F. Querci 1986, p. 28). Miras are regular long-period variables with large amplitude variations (over 2 mag), with a characteristic emission spectrum around their maximum (mainly the hydrogen series). The semi-regulars have amplitudes smaller than those of Miras and shorter periods; alternative irregularities and some constancy of brightness are observed in the light curves. In the third group, the true irregular variables, some stars have large brightness variability (over some mag range) while others present variations of 0.5 mag or even less.

Mira type variability is presently the most studied. It is explained in the context of the shock-wave interpretation: Wood (1979) by the adiabatic shock approach and Willson and Pearce (1982) by the isothermal shock approach which fails at very low densities. A detailed structure of shock waves is presented by Gillet and Lafon (1983, 1984) and Fox et al. (1984) and studies discussed in de la Reza (1986) (Fig. 3).

Willson (1976) illuminates the interpretation of the complex spectra of the long-period variables. She mentions the existence of two types of lines: the primary lines of Mg II, Ca II, Fe II and Ti II which are formed in the region behind the shock, and the secondary lines (or fluorescent or pumped lines) of Sc I, Mn I and Fe I formed in front of the shock. The latter are formed by radiative excitation emitted by the primary lines which are formed in the velocity discontinuity region (in front of the shock).

Willson and Hill (1979), Hinkle (1978), Hinkle et al. (1982 and references therein) and Wood (1979) demonstrate that the global, radial, stellar pulsation drives compression acoustic waves, generated at the top of the hydrogen convection zone, which develop into shocks before emerging into the photosphere, dissipating energy and heating the stellar layers. The discontinuous S-shaped radial velocity curve is a characteristic of the shock-wave model first shown by Wallerstein (1959) in W Vir type stars. The shock progression in the photosphere is also followed as are the appearance and disappearance of various emission lines and their radial velocity variations with phase (e.g., see the review by M. Querci 1986, p. 140 and her Fig. 2-1).

In the outer layers, Tielens (1983) investigates the Mira variable flows driven by radiation pressure on dust grains. He considers the direct coupling of the cooling of the gas with the velocity gradients in a two-component fluid composed by dust and gas interacting by

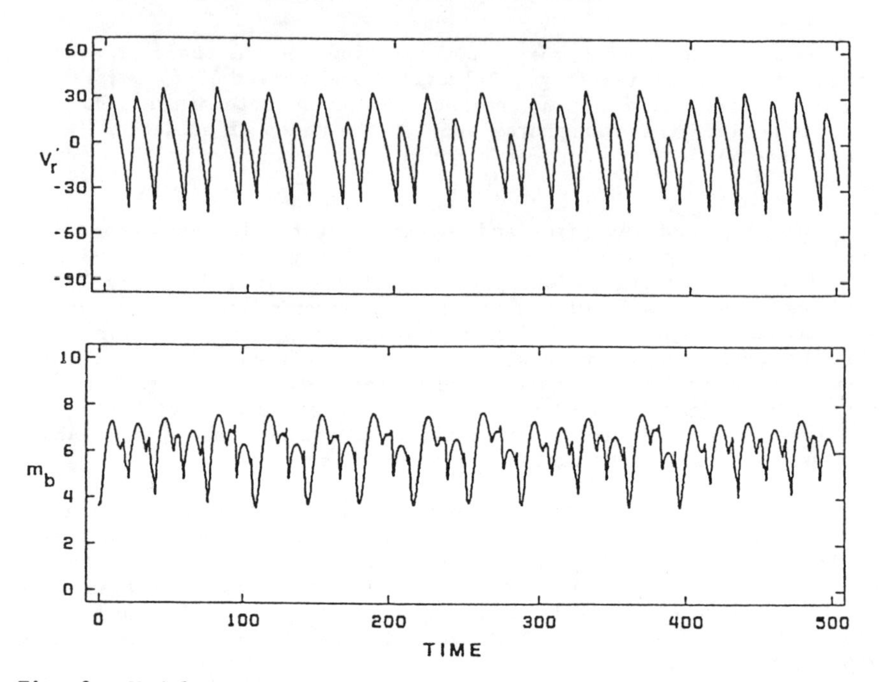

Fig. 2. Modeled temporal behavior of the radial surface velocity (top) and of the relative bolometric magnitude (bottom) for a model undergoing chaotic oscillations (time in days) (from Buchler and Kovacs 1987).

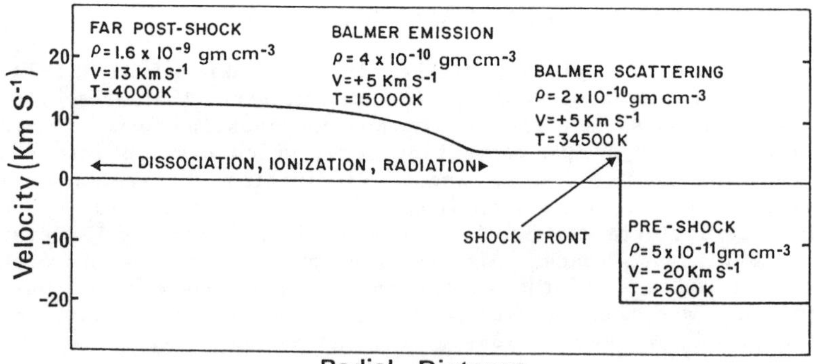

Fig. 3. Schematic diagram of shock-wave sturucture in a typical Mira variable near maximum light (velocities in the rest frame of the star) (from Fox et al. 1984).

collisions in a stationary layer where the maser SiO appears, located at 10 R_* as proposed by Hinkle et al. (1982).

There is a paucity of observational data for semiregulars and irregulars. However, Mira-like shock waves progressing in the photosphere of semiregulars are deduced: 1. from S-shaped radial velocity curves for absorption lines observed by Hinkle et al. (1984) in some M-type semiregulars and discovered by M. Querci (1986, see her Fig. 2-3) in a few C-type ones, RR Her, T Cnc and V Hya, and 2) from Hα line behavior, in emission at maximum light and in absorption at minimum light, in a few examples (e.g., McLaughin (1946) for μ Cep, and Sanford (1950) for RR Her and V Hya). However, the shocks have lower amplitudes than in Miras. They can be drastically damped in the photospheric layers or have originally a low strength and dissipate their energy high in the atmosphere, creating a chromosphere (e.g., the C star, TW Hor (Querci and Querci 1985), the M supergiant, α Ori (Dupree et al. 1987). The theory of short-period acoustic waves (Ulmschneider and co-workers: e.g., Schmidt and Ulmschneider 1981) might be applied in this case.

The observed line variability might be due to spatial and temporal fluctuations of the chromospheric thermodynamical state, or to a change in the dynamics of the convective zone, giving a variable transit time for the wave pulse to reach the stellar chromosphere.

Cuntz (1987) gives a detailed study of the development and the interaction of acoustic wave packets assuming that the wave period changes stochastically in the short period range. As the atmospheric density gradient and the radiation damping work differently on shock waves of different periods (Ulmschneider 1977), the shocks attain different shock speeds depending on the wave period. So, an overtaking shock appears with combined strength and increasing speed, overtaking more and more shocks by merging. This mechanism of overtaking and merging shocks creates episodic mass loss.

In Miras as well as in semiregulars, the efficiency in the line excitation, i.e., in the heating of the various layers by the dissipating shock acoustic waves, changes with time in a given star and from one star to another and appears to be bound to the light curve behavior and amplitude.

Recent hydrodynamical models show that the irregular behavior of the pulsation motion of semiregular red giants and supergiants has its origin in an underlying chaotic dynamic (e.g., Perdang and Blacher 1982, Whitney 1985, Buchler and Kovacs 1987, Buchler et al. 1987). The effective temperature acts as a control parameter (Fig. 2).

2.2 The Nonradial Oscillations

A mechanism of energy transport other than convection (giving a global radial stellar pulsation) is to be considered. This mechanism is bound to the nonradial oscillations.

Exploratory models for late-type giants and supergiants were made by Ando (1976). The structure of red giants and supergiants allows nonradial modes which propagate as p-modes (pressure) near the surface, and as g-modes (gravity) below the convection zone with very high radial wave number (Dziembowski 1971, 1977, Shibahashi and Osaki 1976). Keeley (1980) demonstrates that these two kinds of modes can propagate independently in extreme giant branch models, whereas the p- and g-mode regions may interact in hotter stars,

The first example in which nonradial oscillations are advocated, is the hot-helium star, BD -9°4395. This star appears as a variable both in light output and in radial velocity (there is no discernible color variation) by Jeffery et al. (1985), through photometry and Fourier analysis of data in which gaps occur, and spectroscopy (at 8, 10 and 16 Å/mm). The variability is attributed to nonradial g-mode oscillations (see Cox 1980 for the characteristics of these modes). However a generalization for the hot-helium stars is to be ruled out as BD +13°3224 appears to be a radial pulsator (Hill et al. 1981) with $P = 0.108$ days and $\Delta m < 0.1$ mag.

Using a spectropolarimeter, Stanford et al. (1988) observed the 1986 minimum of R CrB. They suggest that the ejection of dust clouds occurs in a preferred plane, in contrast to the standard model of RCB stars (ejection from all parts of the stellar surface by tubes of matter). The most likely explanation of this preferred plane is an ejection mechanism connected with nonradial oscillations. The mode described by the spherical harmonic parameters $l = 1$ and $m = 0$ would produce a bulge around the equator of the star (e.g., Cox 1980), once every pulsation period. The surrounding layers above this bulge would be cooler than the other parts of the photosphere and grains may form. The ejection is driven by radiative pressure on the grains. From this example, let us look at the importance of spectropolarimetry in the stellar environment analysis.

3. A BRIEF REVIEW OF THE CHARACTERISTIC LINES OF LATE-TYPE GIANTS

What lines are to be investigated to help the theories developed here? Selected lines have to come from all the different stellar layers; namely from the layers in which the shock waves dissipate energy. Generally speaking, the observations show that these layers are not the same for the different variable type stars; for the Miras they are the photospheric layers; for the semiregulars, they are the chromospheric layers.

So we need to study lines formed in the following layers:

1. the deep photosphere - lines around 1.6 μm
2. the mid-photosphere - atomic lines as well as molecular bands
3. the outer photosphere or high photosphere - low excitation atomic energy lines such as those from Ti I, V I, Al I; low energy molecular electronic lines, and some polyatomic bands such as HCN, C_2H_2
4. the chromosphere - Ca II H and K lines, Hα, Hβ, Mg II h and k lines
5. the envelope - Li I, K I, Na-D lines and the fundamental vibration-rotation molecular bands.

The large scale energy variations involved in molecules, coming from the jump of electrons, the vibration and rotation of their atomic compounds, and the various molecular dissociation energies, make them a powerful tool for probing the whole huge atmosphere. Querci and Querci (1977) have shown good examples. Nevertheless, the analysis of the visible and IR ranges can only give a part of the solution of the present problems.

To be efficient, we start by choosing to study the bright atomic and molecular lines selected in the narrow band photometry systems. Be aware that some of them are formed simultaneously in a large number of layers: TiO, ZrO, CN, etc. In consequence, a careful analysis must be made to discriminate between low and high energy bands (cool and hot bands) coming from the top or from the bottom of the photosphere, respectively. We retain also: Hα, Hβ, He I, Na-D, CH lines, as well as the I 104 region free of lines. Then, faint lines may be added, though a minimum spectral resolution is needed (see below).

A few observational examples of line variability explained and not yet explained in terms of hydrodynamics, follow.

3.1 Explained Line Variations

Pioneer studies on the Balmer series by Merrill (1945) and Joy (1947), and more recent studies by Fox et al. (1984) and Gillet et al. (1983) in Miras allow one to depict the behavior of shock waves in their atmospheres. Other lines used for this purpose are the UV Fe II and Mg II h and k lines, the intensity variations of which are explained, for example in the Mira S star, χ Cyg, by the progression of shock waves through the whole atmosphere (Cassatella et al. 1980). The Fe I λ4202 line and the Mg I λ4571 fluorescent lines pumped by Mg II λ2795 are also used by Fox et al. (1984) for their detailed analysis of the shocks.

In the M semiregular supergiant, α Ori, Sonneborn et al. (1986) find a 1.1-year periodicity in the ratio of the Mg II h and k lines. Malki (1986) finds a period of variation of the Hβ profile of 1.1

years, whereas Dupree et al. (1987) shows a 1.15-year periodic modulation in the Mg II h and k lines and in the UV (λ3000) and the blue (λ4530) continuum regions. This implies periodic pulsations that become shocks traveling through the chromosphere of the star, heating and extending it. In this star, using Fe II UV lines, Ca II lines and molecular (CO) lines, Querci and Querci (1986) and M. Querci (1986) show the decoupled dynamical behavior of the various layers, from the photosphere to the outer circumstellar layers.

Periodic radial velocity variations have also been found in Arcturus by Smith et al. (1987) using a high spectral resolution system (interferometrically calibrated spectrometer). The observed period of the variations is $P = 1.842 \pm 0.005$ days, whereas the theoretical value is 1.82 days. The fundamental radial mode period is ~7 days (Cox et al. 1972).

3.2 Not Yet Explained Observations

Some examples of surprising features not clearly interpreted, follow:

1. the Al H emission spectrum which appears near the minimum light in some M-type Miras,
2. [Fe II], [Mn I], [S II] and [O I] forbidden emission lines which appear near the light minimum in U Ori, R Leo, R Hya but never simultaneously with Al H,
3. the Al O emission spectrum which is observed in abnormally low maxima of o Ceti (Herbig 1956),
4. the CaCl absorption bands which are found in the red region of C stars, but only during the light minima (Sanford 1950),
5. the atomic absorption line radial velocity variations from one cycle to the next in o Ceti (Joy 1954),
6. the Ba II variations in UU Aur, RV Mon and ST Cam (Oganesyan et al. 1985),
7. the UV depression around 4000 Å which is so different among the C stars. For example, it increases rapidly over a few years in UU Aur blocking the photospheric flux completely so that very faint and narrow low energy emission lines of Ti I, V I etc., emerge.

These features should be an indicator of the available energy or of its damping at each particular cycle. The question to be answered is: What hydrodynamic phenomena sustain their behavior?

The line changes on short time-scales are also puzzling:

1. In the SR carbon star, TW Hor, the Al II U1 lines at λ2670 are not visible on an IUE image on March 1982, while suspected one hour later, and becoming much stronger one day later (Querci, and Querci 1985).
2. Again in the C Star, TW Hor, Fe II U1 lines appear and disappear

with a characteristic time of a day (Bouchet et al. 1983).

3. Variations with a time scale of about 1 hour have been detected on the M8 II star, R Crt, around the Ca I λ4227 line and over TiO bands (Livi and Bergmann 1982).
4. Such rapid variations are also mentioned by de la Reza (1986) in L_2 Pup in the Ca I λ4227 region.

De la Reza (1986) suggests that such rapid variations are linked to nonradial oscillations. Late-type star seismology is a promising but difficult task!

4. CONCLUSION: THE SPECTROPHOTOMETRIC CONTRIBUTIONS

What kinds of contributions may the spectrophotometry and/or the spectropolarimetry give to the red giant variability analysis? First, the spectrophotometry has to be run on faint lines not taken into account in the photoelectric approach. Consequently, the chosen resolution is evidently a prime parameter. With a resolution of 3 to 5 Å, a line variation analysis is possible, but a line profile analysis requires a higher resolution. Second, observations at various phases of the light curve have to be monitored. Continuous observations as done with the Automatic Photoelectric Telescopes, are planned.

REFERENCES

Alkne, Z., Alknis, A. and Dzervitis, V. 1983 Properties of Carbon Stars of the Galaxy, Riga Acad. of Science.

Ando, H. 1976 Publ. Astron. Soc. Japan **28**, 517.

Baumert, J. H. 1972 Ph. D. Thesis, Ohio State University

Bell, R. A., Eriksson, K., Gustafsson, B. and Nordlund, A. 1976 Astron. Astrophys. Suppl. **23**, 37.

Buchler, J. R. and Kovacs, G. 1987 Astrophys. J. **320**, L 57.

Bouchet, P., Querci, M. and Querci, F. 1983 The Messenger, ed. P. Veron, ESO **31**, 7.

Buchler, J. R., Goupil, M. J. and Kovacs, G. 1987 Physics Letters A **126**, 177.

Cassatella, A., Heck, A., Querci, F., Querci, M. and Stickland, D. J. 1980 in Proc. Second European IUE Conference, ESA SP-157, p. 243.

Cox, J.. P. 1980 Theory of Stellar Pulsation, Princeton Univ. Press, Princeton.

Cox, J. P., King, D. S. and Stellingwerf, R. F. 1972 Astrophys. J. **171**, 93.

Cuntz, M. 1987 Astron. Astrophys. **188**, L 5.

De La Reza, R. 1986 in The M-Type Stars, H. R. Johnson and F. Querci, eds. NASA/CNRS Monograph Series on Nonthermal Phenomena in Stellar Atmospheres, p. 373.

Dupree, A.K., Baliunas, S. L., Guinan, E. F., Hartmann, L., Nassiopoulos, G. E. and Sonneborn, G. 1987 Astrophys. J. Letters **317**, L 85.

Dziembowski, W. 1971 Acta Astron. **21**, 289.

Dziembowski, W. 1977 Acta Astron. **27**, 95.
Eggen, O. J. 1972a Astrophys. J. **174**, 45.
Eggen, O. J. 1972b Astrophys. J. **177**, 489.
Fox, M. W., Wood, P. R. and Dopita, M. A. 1984 Astrophys. J. **286**, 337.
Gillet, D. 1981 Etoiles Froide, Raies D'Emission et Ondes de Choc, These de 3ieme cycle, Universite de Paris VII.
Gillet, D. and Lafon, J. P. J. 1983 Astron. Astrophys. **128**, 53.
Gillet, D. and Lafon, J. P. J. 1984 Astron. Astrophys. **139**, 401.
Gillet, D., Maurice, E. and Baade, D. 1983 Astron. Astrophys. **128**, 384.
Gow, C. E. 1975 Ph.D. Thesis, Indiana University.
Gustafsson, B., Beel, R. A., Eriksson, K. and Nordlund, A. 1975 Astron. Astrophys. **42**, 407.
Herbig, G. M. 1956 Publ. Astron. Soc. Pacific **68**, 204.
Hill, P. W., Kilkenney, D., Schönberner, D. and Walker, H. J. 1981 Monthly Notices Roy. Astron. Soc. **197**, 81.
Hinkle, K. H. 1978 Astrophys. J. **220**, 210.
Hinkle, K. H., Hall, D. N. B. and Ridgway, S. T. 1982 Astrophys. J. **252**, 697.
Hinkle, K. H., Scharlach W. W. G. and Hall, D. N. B. 1984 Astrophys. J. Suppl. **56**, 1.
Jeffery, C. S., Skillen, I., Hill, P. W., Kilkenny, D., Malaney, R. A. and Morrison, K. 1985 Monthly Notices Roy. Astron. Soc. **217**, 701.
Johnson, H. R. 1986 in The M-Type Stars, H. R. Johnson and Querci, F., eds., NASA/CNRS Monograph Series on Nonthermal Phenomena in Stellar Atmospheres, p. 323.
Joy, A. H. 1947 Astrophys. J. **106**, 288.
Joy, A. H. 1954 Astrophys. J. Suppl. **1**, 39.
Keeley, D. 1980 Highlights of Astronomy **5**, P. A. Wayman, ed., p. 497.
Livi, S. H. B. and Bergmann, T. S. 1982 Astron. J. **87**, 1783.
Lockwood, G. W. and McMillan, R. S. 1970 in Proc. Tucson Conf. on Late-Type Stars, G. W. Lockwood and H. M. Dyck, KPNO Contribution **554**, 171.
Lockwook, G. W. and Wing, R. F. 1971 Astrophys. J. **169**, 63.
Malki, M. 1986 Memoire de Rapport de Stage de DEA, Univ. Paul Sabatier, Toulouse.
McLaughin, D. B. 1946 Astrophys. J. **103**, 35.
Merrill, P. W. 1945 Astrophys. J. **102**, 347.
Oganesyan, R. Kh., Nersisyan, S. E. and Karapetyan, M. Sh. 1985 Astrofizika **23**, 99.
Perdang, J. and Blacher, S. 1982 Astron. Astrophys. **112**, 35.
Piccirillo, J. 1976 Publ. Astron. Soc. Pacific **88**, 680.
Piccirillo, J. 1977 Ph.D. Thesis, Indiana University.
Querci, F. 1986 in The M-Type Stars, H. R. Johnson and F. Querci, eds., NASA/CNRS Monograph Series on Nonthermal Phenomena in Stellar Atmospheres, p. 1.
Querci, F. and Querci, M. 1977 in Proc. 21th Liege International Astrophysical Colloq. Les Spectres des Molecules Simples au

Laboratoire et en Astrophysique, Univ. de Liege, p. 206.
Querci,F. and Querci, M. 1985a in Proc. Conf on Cool Stars with Excesses of Heavy Elements, M. Jaschek and P. C. Keenan, eds., Reidel, Dordrecht, p. 99.
Querci, M. 1974 These de Doctorat D'Etat, Univ. de Paris VII.
Querci, M. 1986 in The M-Type Stars, H. R. Johnson and F. Querci, eds., NASA/CNRS Monograph Series on Nonthermal Phenomena in Stellar Atmospheres, p. 113.
Querci, M. and Querci, F. 1976 Astron. Astrophys. **49**, 443.
Querci, M. and Querci, F. 1985b Astron. Astrophys. **147**, 121.
Querci, M. and Querci, F. 1986 in Proc. Fourth Cambridge Workshop on Cool Stars, Stellar Systems and the Sun, M. Zeilik and D. M. Gibson, eds., p 492.
Sanford, R. F. 1950 Astrophys. J. **111**, 270.
Schmitz, F. and Ulmschneider, P. 1981 Astron. Astrophys. **93**, 178.
Shibahashi, H. and Osaki, Y. 1976 Publ. Astron. Soc. Japan **28**, 199.
Smith, P. H., McMillan, R. S. and Merline, W. J. 1987 Astrophys. J. Letters **317**, L 79.
Sonneborn, G., Baliunas, S. L., Dupree, A. K., Guinan, E. F. and Hartmann, L. 1986 in New Insights in Astrophysics, ESA SP-263, p.221.
Stanford. S. A., Clayton, G. C., Meade, M. R., Nordsieck, K. H. Whitney, B. A., Murison, M. A., Nook, M. A. and Anderson C. M. 1988 Astrophys. J. Letters **325**, L 9.
Strecker, D. W., Eriksson, E. F. and Wittenborn, F. C. 1979 Astrophys. J. Suppl. **41**, 501.
Tielens, A. G. G. M. 1983 Astrophys. J. **271**, 702.
Ulmschneider, P. 1977 Memoirs of Astr. Soc. It. **48**, 439.
Vetesnik, M. 1984a Bull. Astron. Inst. Czech. **35**, 65.
Vetesnik, M. 1984b Bull. Astron. Inst. Czech. **35**, 74.
Wallerstein, G. 1959 Astrophys. J. **130**, 560.
Whitney, Ch. A. 1985 in Proc. 25th Liege International Astrophysical Colloq., Theoretical Problems in Stellar Stability and Oscillations, Univ. de Liege, p. 454.
Willson, L. A. 1976 Astrophys. J. **205**, 172.
Willson, L. A. and Hill, S. J. 1979 Astrophys. J. **228**, 854.
Willson, L. A. and Pearce, J. N. 1982 in Second Cambridge Workshop on Cool Stars, Stellar Systems, and the Sun, M. S. Giampapa and L. Golub, eds., Smithsonian Astrophys. Obs. Special Report **392**, 147.
Wing, R. F. 1967 in Proc. Trieste Colloquium on Astrophysics, Late-Type Stars, M. Hack, ed., Osservatorio Astronomico di Trieste, p. 205.
Wing, R. F. 1971 in Proc. Conf. on Late-Type Stars, G. W. Lockwood and H. M. Dyck, eds., Kitt Peak Obs. Cont., n. 554.
Wing, R. F. 1980 in Proc. Conf. on Current Problems in Stellar Pulsation Instabilities, D. Fischer, J. Rountree Lesh and M. W. Sparks, eds., NASA TM 80625, p. 533.
Wing, R. F. 1985 in Proc. Conf. on Cool Stars with Excesses of Heavy Elements, M. Jaschek and P. C. Keenan, eds., Reidel, Dordrecht, p.61.

Wood, P. R. 1979 Astrophys. J. **227**, 220.
Zeldovich, Ya. B. and Raizer, Yu. P. 1966 Physics of Shock Waves and High-Temperature Hydrodynamic Phenomena, Academic Press.

DISCUSSION

JOHNSON: Pulsation-initiated shocks are clearly observed in the photospheric spectra of Mira-type variables. Did you mean to imply that shocks were not seen in semi-regular and irregular variables?

QUERCI: Yes, shocks are also seen in semi-regular and irregular stars, but they are much more highly damped or have less energy originally. In the Miras, the acoustic waves give shocks from the top of the convective zone; in the semiregulars the shocks are preferably developed in the chromospheres.

ETZEL: If non-radial pulsations are responsible for the formation of a disk-like structure of dust around the star, could this promote the future development of a bi-polar planetary nebula?

QUERCI: Perhaps, but I have not seen any calculations of it. We do know that the formation of a bi-polar nebula appears in a binary system with matter ejection and disk formation.

THE DETECTION OF STELLAR COMPANIONS FROM RADIAL VELOCITY MEASURES

A. R. Upgren and K. A. Gloria

Van Vleck Observatory

A program to determine radial velocities for dwarf stars with high precision transverse velocities from parallaxes and proper motions of the Van Vleck Observatory has been started and a first list of radial velocities of 225 stars has been completed and published. The procedures of observation and reduction are given by Latham (1985) and by Wyatt (1985). The details of the observation, reductions and analysis are given in a paper by Upgren and Caruso (1988) and will not be repeated here.

In the course of the program it was noticed that additional information was obtained from the use of a speedometer and two standard masks used in a number of programs of the Center for Astrophysics. The masks are centered at about 5200 Å and cover a range of about 50 Å. They are most closely matched to solar-type and M-type dwarf stars. The output includes a correlation coefficient which corresponds closely with the goodness of fit of the mask to the tracing of the star, and hence, to the error in the velocity.

The ratio between the logarithms of the correlation coefficients in the stellar spectral analysis is defined as log[R(SX1)/R(MX1)] where R is the correlation coefficient and the symbols in parentheses, SX1 and MX1, represent the masks matching solar type and M type stellar spectra, respectively. This ratio is useful since it was found by Upgren and Caruso to be very closely correlated to the (R-I) colors of the stars. The (R-I) index is known to be a nearly linear measure of stellar surface temperature, and thus the ratio is a measure of temperature as well. In this paper, the ratio is also shown to be a good predictor of duplicity among these stars. Fig. 1 shows the relation between the ratio of the coefficients produced by the two masks, and the (R-I) color index. The known close binaries and the stars which show variations in radial velocity, are individually indicated. In this paper, all values of (R-I) are on the Kron system.

A total of 142 stars is included in the plot; this includes all with more than one radial velocity observation. Of this total, 21 are known close doubles for which both the ratio and the color apply jointly to both components and 15 others were found to vary

A. G. Davis Philip, D. S. Hayes and S. J. Adelman, (eds.)
New Directions in Spectrophotometry 73 - 77

substantially in radial velocity. The doubles are divided into those with the difference in magnitude between the component stars, Δm, greater than and less than 1.0 magnitude and are designated by open and filled circles, respectively. The 15 suspected doubles are indicated by crosses. The remaining 106 stars, indicated by smaller dots, appear to be either single or members of wide, fully-resolved binaries or multiple stars. They do not appear to show variations in radial velocity; however, the limited number of observations per star does not preclude such variation, and a few of them can be expected to be close binaries with undetected companions on the basis of the expected probability.

It can be seen that the stars presumed to be single define a very narrow and nearly linear sequence. The close binaries, both known and detected from velocity variability, form a conspicuously wider sequence with the exception of the binaries with Δm less than 1.0 magnitude which fall along the sequence as expected since these pairs are composed of nearly identical stars. The implication is that the ratio between the correlation coefficients derived from the use of the two masks, is a useful indicator of binarity among low-mass stars, at least for those with a difference in mass between the components. It suggests that the occasional single star falling well outside the sequence is a good candidate for variation and merits further observation.

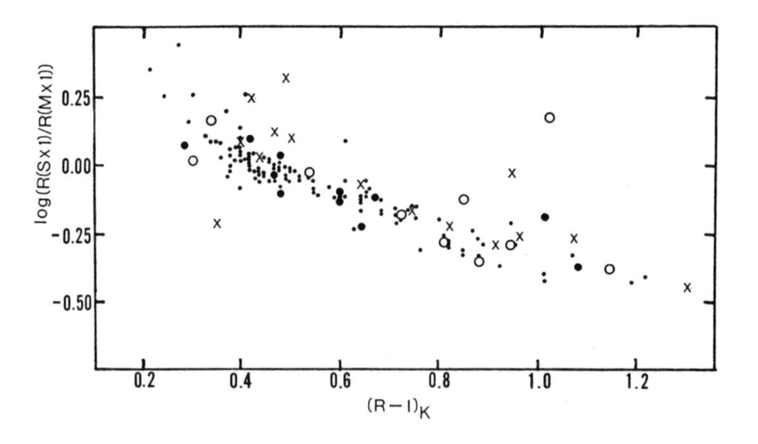

Fig. 1. The relation between the correlation coefficient ratio and (R-I) color. The definitions of the symbols are explained in the text.

The unresolved binaries and suspected binaries based on variable radial velocity would be expected to fall along the sequence formed by the single stars if they are formed of two almost identical stars; that is, if Δm is close to zero. Those which lie at a considerable distance from the sequence are more likely to be composed of two component stars that are unequal in color or luminosity. This premise can be tested since the stars shown in Fig. 1 have accurate parallaxes and proper motions made in the course of the astrometric program of the Van Vleck Observatory. Fig. 2 shows the M_v, (R-I) diagram with the main sequence indicated by the curved line. It was derived by Gliese and Jahreiss (1986) from 404 stars whose trigonometric parallaxes produce absolute magnitudes with standard errors not exceeding ±0.20 magnitude. The individual stars have the same symbols as they do in Fig. 1. It is

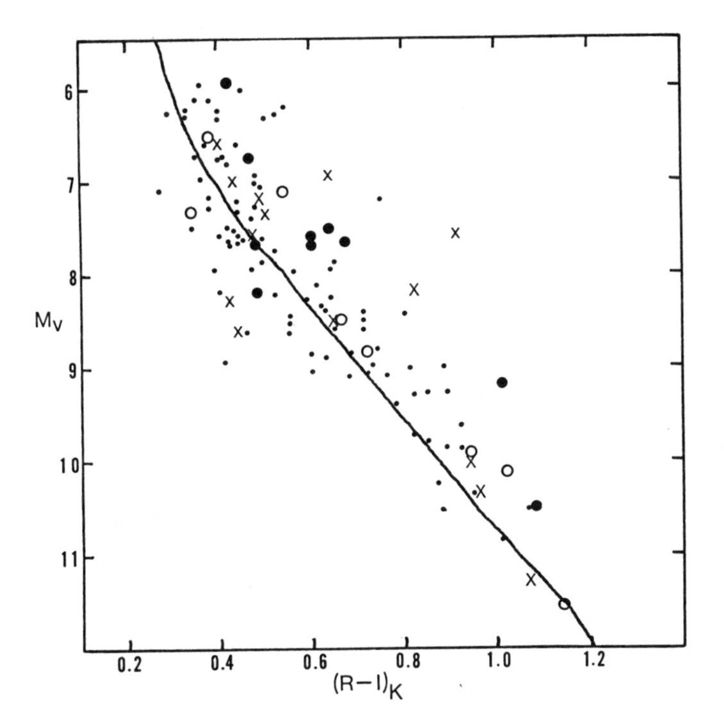

Fig. 2. The M_v, (R-I) diagram showing the stars included in Fig. 1. Each star is represented by the same symbol in both figures.

apparent that most of the close binaries and the stars showing variations in radial velocity lie above the main sequence formed by the single stars and members of well resolved binaries. It is suggestive of the presence of an unseen or unresolved companion for those stars which fall well off the sequence shown in Fig. 1, that are not already known to be members of binaries. The close binaries composed of nearly identical component stars lie generally farthest above the main sequence as expected.

This study is supported by Research Grant AST-8610424 from the National Science Foundation.

REFERENCES

Gliese, W. and Jahreiss, H. 1986, Private Communication.

Latham, D. W. 1985 in IAU Colloquium No. 88, Stellar Radial Velocities, A. G. D. Philip and D. W. Latham, eds., L. Davis Press, Schenectady, p. 21.

Upgren, A. R. and Caruso, J. R. 1988 Astron. J. **96**, in press.

Wyatt, W. F. 1985 in IAU Colloquium No. 88, Stellar Radial Velocities, A. G. D. Philip and D. W. Latham, eds., L. Davis Press, Schenectady, p. 123.

DISCUSSION

AKE: Are you finding any double-lined spectroscopic binaries? For two stars of the same magnitude, the combined light will be 0.75 mag. brighter. How much of the cosmic dispersions in your M_v vs. (R-I) plot could be due to unrecognized binaries? Do your variable velocity stars tend to fall above the main sequence line?

UPGREN: We have not been looking for double-lined spectroscopic binaries. Most stars falling well above the main sequence in the M_v, (R-I) plane are known close binaries or are variable radial velocity stars implying duplicity. But not all velocity variables fall above the sequence if only because pairs with Δm as low as even 2 mag. show a negligible increase in M_v, although the fractional mass is still considerable.

BELL: Have you thought of: (a) asking for speckle observations of stars in the M_v, (R-I) diagram; and (b) fitting isochrones to the M_v, (R-I) diagram?

UPGREN: I have not yet sought to obtain speckle observations of this group of stars because the results are still very new and the total stellar sample is not yet completely observed. The same applies to the fitting of isochrones.

Certainly this is work to be done since the stars are so crucial to the first step in determining the distance scale.

SPECTROPHOTOMETRY OF ECLIPSING BINARY STARS

Paul B. Etzel

San Diego State University

ABSTRACT: Spectrophotometry of eclipsing-binary stars has the potential of yielding the spectral-energy distributions of both components. This paper deals with some of the practical aspects of determining the energy distributions with an emphasis on totally eclipsing systems of the Algol type. The discussion is extended to the partially eclipsing case, which is also the usual situation for detached binary systems. Some suggestions are offered concerning observational and analysis procedures in anticipation of more precise observations.

1. INTRODUCTION

An important goal of stellar spectrophotometry is to derive surface temperatures and gravities. (In some regimes other data are needed to specify the two uniquely). Anomalies are frequently observed in various objects. The actual interpretations of these anomalies differ from object to object; chemical peculiarities circumstellar material, and binarity are examples of possible causes.

The observations of binary stars may be decomposed to provide the spectral energy distributions of the individual components. For the non-eclipsing case (composite binaries), this decomposition requires the assumption of some stellar parameters to infer an optimal description of the individual components. However, in the case of an eclipsing system, the changes in the observed spectral energy distribution may be exploited to describe better the individual components.

There have been relatively few systematic investigations of eclipsing-binary stars with grating scanners. A notable effort was that made by a group at the University of Iowa. Davidson (1973, 1974) compared spectrophotometry of the detached system WW Aur to predictions from stellar atmospheres. Rhombs and Fix (1976, 1977) searched for excess ultraviolet light in the Algol-type binaries U Cep, U Sge, and SX Cas. Clements and Neff (1979) extended this study to partially eclipsing systems of the detached (U Oph, AG Per) and Algol types (λ Tau, V 356 Sgr).

A. G. Davis Philip, D. S. Hayes and S. J. Adelman, (eds.)
New Directions in Spectrophotometry 79 - 89

More recently, a group at UCLA has employed image tube scanner (ITS) spectra (scans) in conjunction with fluxes determined with the IUE satellite. Plavec, Weiland, and Koch (1982) confirmed for SX Cas the excess ultraviolet light observed at totality, and also determined that the primary star is more like a B7 V rather than A5 III as quoted for many years in the literature. Similar studies have since been made on other Algol- and W Serpentis-type binaries including: U Cep (Plavec 1983); U Sge, RW Per, and RX Gem (Dobias and Plavec 1985, 1987a,b); and, RS Cep and RY Gem (Plavec and Dobias 1987a,b). Direct subtractions of the totality scans from those outside eclipse were frequently employed to determine the spectrum for the hotter component in the optical region. The large light-ratios of the stellar components and large baseline of wavelength coverage minimized systematic scan-subtraction errors for the process of fitting model atmosphere flux distributions. The paper on RY Gem by Plavec and Dobias is notable in that grating scanner observations (by the author of this paper) were used to extend the ultraviolet and red ends of their ITS scan obtained during totality.

In spite of the relative inefficiency of grating scanners compared to more modern wavelength-multiplexing spectrometers, the former are still difficult to surpass in terms of wavelength coverage and precision. It is hoped that one product of this conference will be the specification of a design for a new generation of spectrometers, which will combine the best attributes of both. If these new spectrometers are to be exploited to their maximum extent, then more refined treatments will be required in the reduction and analysis of their data. In this regard, I would like to discuss some of the procedures employed by Etzel and Olson (1985) and Etzel (1988) in their analyses of the Algol-type binary stars S Cnc and TT Hya, respectively.

2. INSTRUMENTATION AND OBSERVATIONS

I obtained photoelectric spectrum scans, in the wavelength range of 3200 to 8090 Å, as part of an observational program on Algol-type eclipsing binaries. Observations were made at Lick Observatory with the Wampler (1966) spectrophotometer (scanner) on the 0.9 m Crossley telescope. Other observations were obtained with Harvard College Observatory scanners (similar to the Wampler design) at Kitt Peak National Observatory (KPNO) on the 0.9 m telescope and at Cerro Tololo Inter-American Observatory (CTIO) on the 1.5 m telescope. Additional information on the instrumentation and selection of standard stars, are given by Etzel and Olson (1985) and Etzel (1988). The observational procedures, however, deserve special attention.

Great pains were taken to retain absolute fluxes on the system of Hayes and Latham (1975) by choosing realistic values of $m_\nu(\lambda\ 5556)$ for all standards relative to a value of $m_\nu\ (\lambda\ 5556) = 0.00$ for Vega. This was done so that, for a given binary, inside-eclipse observations could be compared directly to those outside eclipse to yield the energy distributions for both components. The commonly-adopted procedure of using mean atmospheric-extinction coefficients (employed by most observers for the reduction of ITS or similar data) was avoided. Multiple observations of two standard stars were used to measure extinction nightly, with the same two standards in dominant use during any observing run. (The Iowa group also determined atmospheric extinction nightly). This procedure also provides a direct mechanism for evaluating relative values of m_{ν} (λ 5556) for the standards. The subsequent transformation coefficients are also very well determined.

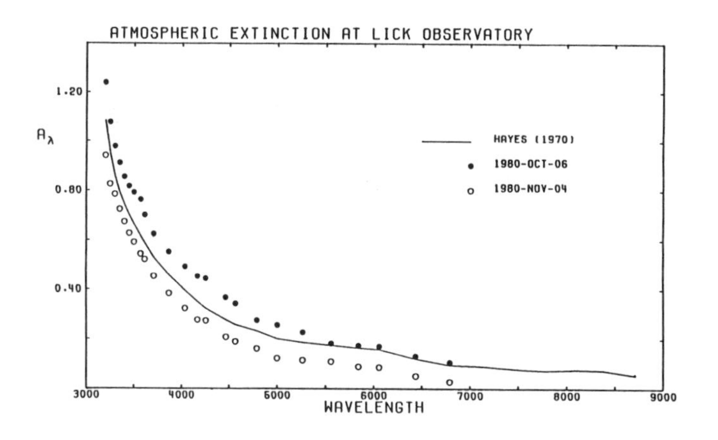

Fig. 1. Atmospheric extinction variations at Lick Observatory.

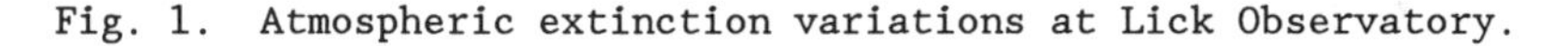

The need for nightly determination of extinction was clearly stated by Hayes (1970) and is demonstrated by Fig. 1. Shown here is the atmospheric extinction at Lick Observatory determined on two representative nights from two observing runs in the autumn of 1980. Also plotted for comparison is the oft-quoted mean extinction curve of Hayes. On October 6 UT, the atmosphere contained smoke from regional brush fires hundreds of miles from the observatory. On November 4 UT the air was clean and crisp. During a stable run, the nightly variations are small. The large variations in extinction presented here illustrate the necessity of nightly extinction determinations versus the procedure of adopting an observatory mean. Such variations, especially in the ultraviolet, should serve as a caution to observers who desire accurate fluxes (e.g., searching for moderate ultraviolet excesses in late-type stars, or measuring Balmer jumps in early-type stars). The problem would be exacerbated when comparing or subtracting observations obtained on different nights or observing runs.

3. THE TOTALLY ECLIPSING CASE

In the case of a totally eclipsing system, it should be possible to subtract the energy distribution observed at totality from an observation outside of eclipse to obtain the energy distribution of the eclipsed star. A comparison of the two stellar energy distributions could then be made to those of standard stars and/or to model predictions from stellar atmospheres theory. The temperatures for the stellar components derived in this manner are more astrophysically meaningful than those solved as parameters by light-curve synthesis. Additionally, anomalies in the resultant energy distributions might provide evidence of circumstellar material, chemical peculiarites, interstellar reddening, or other phenomena.

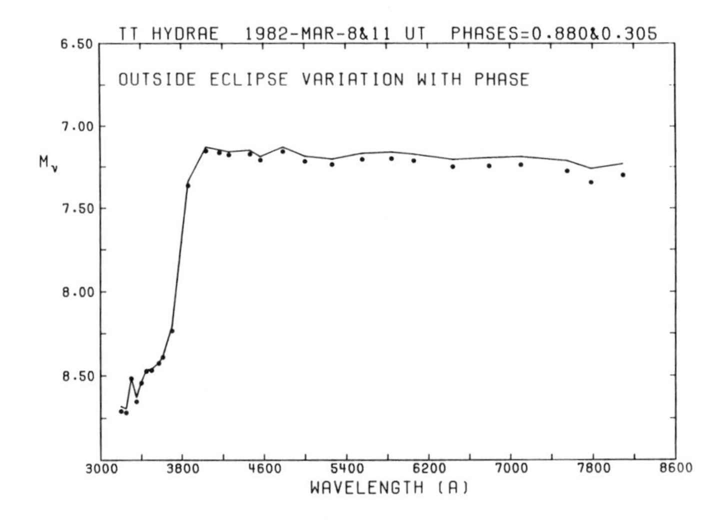

Fig. 2: Outside-eclipse variation in spectral energy distribution.

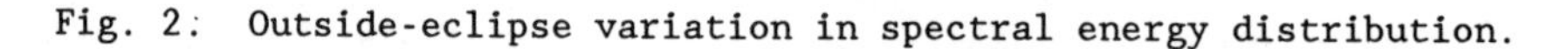

The light curves of Algol-type systems show variations with phase outside of eclipse that increase in amplitude with longer wavelengths. These luminosity variations are dominated by the secondary component (usually the least massive, cooler, and larger star). They are caused by the effects of stellar "ellipticity" (a combination of projected surface area, limb darkening, and gravity darkening) and "reflection" (actually atmospheric heating and reradiation). These variations can be modeled with a suitable light-curve synthesis code, such as WINK by Wood (1971 - 1978), as updated to Status Report Number 10 (Etzel and Wood 1982), or the Wilson-Devinney code (Wilson and Devinney 1971, Wilson 1979).

Fig. 2 illustrates the analogous variation in spectral energy distribution with phase for TT Hya outside eclipse. Shown are observations at phase 0.880 (dark circles) and at phase 0.305 (line segments). These variations are greatest in the red, and almost unobservable in the ultraviolet. Their sense and direction are consistent with the changes expected in the apparent luminosity of the cool component as a consequence of ellipticity and reflection effects. That the variations are small in the ultraviolet again demonstrates the necessity for nightly extinction determinations. Since night-to-night extinction deviations can increase dramatically from red to ultraviolet (Fig. 1), systematic errors in the resultant stellar energy distributions could be similarly increased by the use of a mean extinction curve.

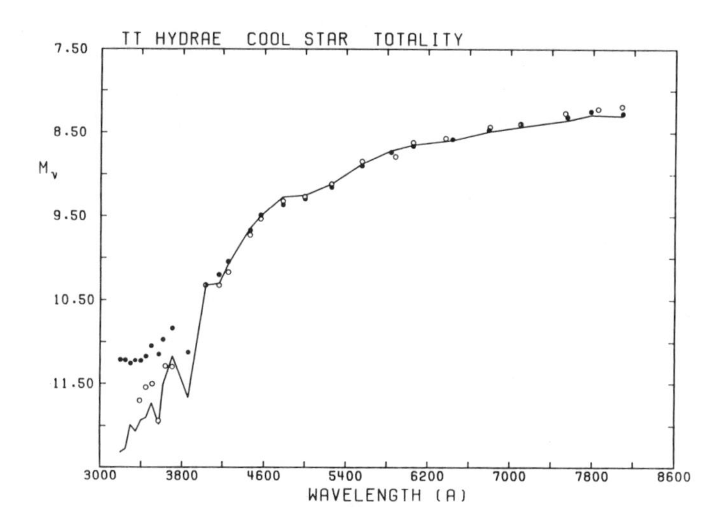

Fig. 3. Spectral energy distribution during totality.

Plotted in Fig. 3 are the spectral energy distributions of TT Hya as seen during totality (solid circles), the K0.5 III star 110 Vir (line segments) shifted by +4.605 mag., and the Bell and Gustafsson (1978) theoretical fluxes for T_{eff} = 4500 K and log g = 2.63 (open circles). The TT Hya data were obtained just after mid-totality to avoid the complications observed at second contact in the U-filter light curve (Etzel 1988). This energy distribution for TT Hya represents the outer hemisphere of the cool component (slightly cooler than 110 Vir) with additional flux emitted by unocculted circumstellar material surrounding the hot primary component. Notice the marked ultraviolet excess for λ < 3862 Å which averages about 0.7 mag.

Preliminary efforts to determine the spectral energy distribution for the hot star alone, by direct subtraction of the totality scan from outside eclipse scans, did not produce consistent results. The phase-variable luminosity of the cool star is at its minimum during totality. At this phase, there is no flux contribution by the reflection effect, and the ellipticity factor is minimal since the cool star projects its smallest area. Thus, residual flux from the cool star will remain in the direct scan differences, flattening the slope of the Paschen continuum. The Kurucz (1979) model-atmosphere flux distribution for T_{eff} = 9000 K and log g = 4.5 (similar to an A1 V star) approximately describes the average scan difference found by <u>direct</u> subtraction.

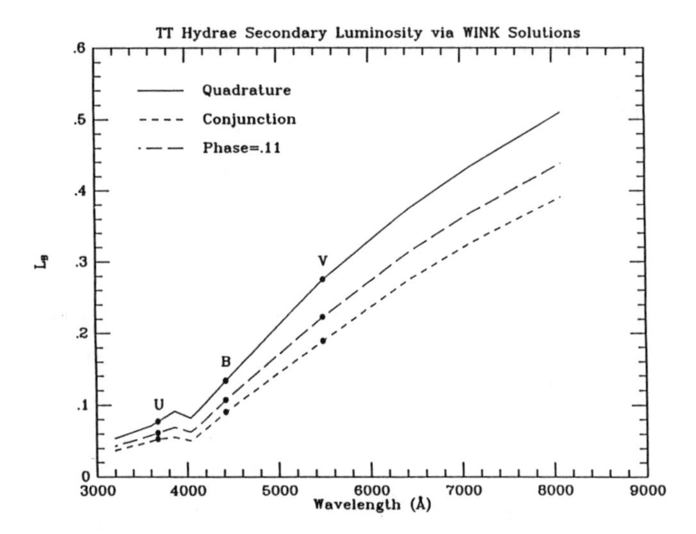

Fig. 4. Light-curve synthesis model of cool-star luminosity.

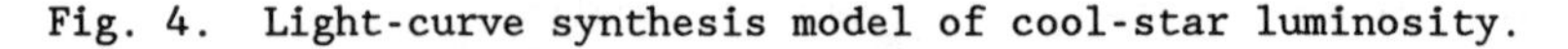

To improve the inference of the energy distribution for the hot star, we should subtract the energy distribution for the cool star as it would be seen (alone) at any phase, θ, outside eclipse. A set of wavelength-dependent corrections was applied to the observed totality scan (at primary conjunction), to predict the energy distribution of the cool star for each θ. The corrections are simply the ratio $L_B(\theta)/L_B$ (conj.) as computed from the adopted light-curve solution (Etzel 1988) for the wavelengths of the scanner observations. Limb-darkening and gravity-darkening coefficients were assumed from theory. Plotted in Fig. 4 is the variation with wavelength of the luminosity contribution of the cooler star, L_B, in TT Hya as calculated at quadrature, primary conjunction, and phase 0.11 (just past primary eclipse). The marked points came directly from the light-curve solutions.

These corrections have only a second-order effect on the resultant energy distributions for the hot star. They are effectively scaled by the light of the cooler secondary and hence are more important in the red than in the ultraviolet. Fig. 5 illustrates the importance of the corrections to the scan subtraction process, and how they affect the inferred energy distribution of the hot star. Shown are the observed fluxes for the combined light of both stars (line segments), the same scan with a simple subtraction of the totality scan (solid circles), and the result of subtracting the predicted energy distribution for the cool star at the phase of observation (open circles). The last of these energy distributions is assumed to represent the hot star best.

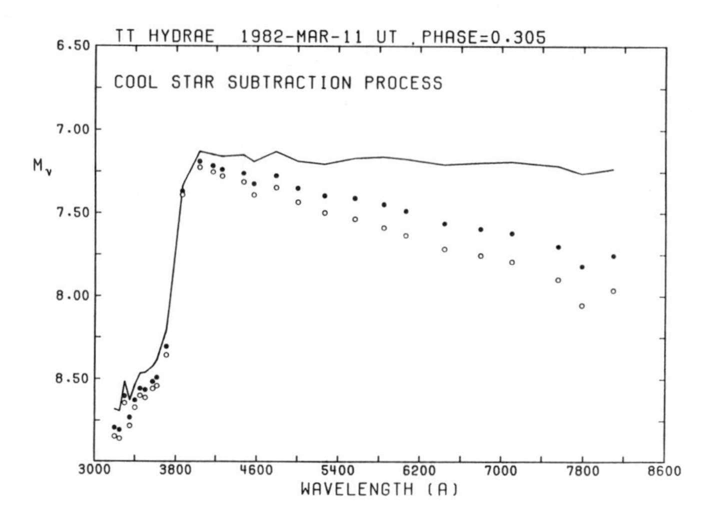

Fig. 5. Scan subtraction process with and without corrections.

This refined process of scan subtraction was applied to all outside-eclipse scans to establish a well-determined spectral energy distribution for the hotter star in TT Hya. The standard deviation of one flux point is 0.008 mag. averaged over the entire wavelength range. The Kurucz model atmosphere fluxes interpolated for T_{eff} = 9800 K and log g = 4.25, which is similar to a B 9.5 V star, match the observations fairly well (Etzel 1988). This fit was made in conjunction with IUE satellite ultraviolet fluxes. The result is consistent with the star's mass and radius, and the interpretation of the optical spectrum being produced in an extended, optically thin shell. Other studies produced ambiguous results because of assumed photospheric spectral types from the literature as late as A 5 III for the hotter star (Kondo et al. 1981, McCluskey 1982). The stellar components of TT Hya are actually quite similar to those in S Cnc (Etzel and Olson 1985), a system which is relatively uncomplicated by circumstellar material.

4. THE PARTIALLY-ECLIPSING CASE

The analysis of spectrophotometry for a totally eclipsing system requires only a second-order correction from a light-curve synthesis code to extract the spectral energy distributions for the individual components. However, such binary systems are frequently of the Algol-type and subject to complications (e.g., mass-transfer effects, variations by the cool star, etc.). To provide for fundamental tests of stellar atmosphere models, it would be better to extract similar information from suitable unevolved, detached binary systems with well-determined masses and radii (c.f., Popper 1980). However, the vast majority of such systems exhibit partial eclipses. Therefore, the extraction of spectral energy distributions for the individual components will have a first-order dependence upon the light-curve synthesis model. Fortunately, the intrinsic complications of such systems are generally nil. Unfortunately, the photometric solutions for such systems frequently exhibit indeterminacy to various degrees even with hundreds of precise observations and ample phase coverage (Popper and Etzel 1981, Popper 1984).

Wilson (1970) pointed out the usefulness of spectrophotometry obtained outside eclipse to improve the determination of the elements of partially-eclipsing binaries. He suggested matching the observed, composite energy distribution with standard star fluxes to provide limits on the light ratio of the binary system. Thus, if limb darkening were also assumed, then one could solve for the ratio of the radii, k, and the geometrical depth at central eclipse, p_o, with a higher degree of accuracy than from the light curves alone for indeterminant systems.

Clements and Neff (1979) employed an earlier blackbody-model version of WINK (Wood 1971, 1972) to calculate theoretical fluxes at various phases to match the observed spectral energy distributions of U Oph and AG Per. They found it impossible to separate the effects of temperature and limb darkening for these two detached systems, which have approximately equal components in each case. Limb-darkening coefficients were, therefore, adopted from stellar atmospheres theory (a tactic frequently required in similar light-curve synthesis solutions). Popper and Etzel found general agreement (especially in color-dependent differences) between the derived limb-darkening coefficients and those predicted by theory in their photometric analyses of several detached systems.

What about working the problem in reverse and solving for the spectral energy distributions of the individual components if the light-curve solution is well determined? Such a direct approach was suggested by Popper (1956) in his discussion of multicolor photometric observations of Z Her. He reminds us that the color of light lost at any phase during eclipse is simply the color of the eclipsed star. Small corrections must be made for complicating effects such as limb darkening, reflection, and ellipticity. In an optimal situation, we should be able to impose the geometry from suitable light-curve solutions (nominally fix the limb-darkening coefficients from theory) and solve for the luminosity ratio at each wavelength for each observed phase by allowing the surface brightness ratio (temperature difference for some models) to be a free parameter. The derived luminosity ratio could then be used to estimate the spectral energy distributions for the individual components directly from the observations rather than solving for them as solution parameters.

5. THE FUTURE

We anticipate more efficient and precise spectrophotometry in the future that may approach the precision of filter photometry. As suggested by Clements and Neff (1979), differential spectrophotometry should be carried out to minimize observational errors just as is done for light-curve studies. Eventually, differential spectrophotometry could eliminate the need for separate light-curve observations, particularly if the observations are binned over large wavelength intervals for light-curve synthesis. However, a large commitment of resources would be required.

The bold attempt by Davidson (1973, 1974) to model the observed variations in spectral energy distribution through eclipse by employing limb-darkening curves predicted from stellar atmospheres theory should be duplicated anew with more precise spectrophotometry, better light-curve solutions, and the results of more modern stellar atmospheres theory. In this regard, particular attention must be given to the selection of candidate systems according to the results and suggestion given by Popper (1984).

REFERENCES

Bell, R. A. and Gustafsson, B. 1978 Astron. Astrophys. Suppl. **34**, 229.
Clements, G. L. and Neff, J. S. 1979 Astrophs. J. Suppl. **41**, 1.
Davidson, J. K. 1973 M.S. thesis, University of Iowa (unpublished).
Davidson, J. K. 1974 Bull. Amer. Astron. Soc. **&6**, 222.
Dobias, J. J. and Plavec, M. J. 1985 Publ. Astron. Soc. Pacific **97**, 138.
Dobias, J. J. and Plavec, M. J. 1987a Publ. Astron. Soc. Pacific **99**, 159.
Dobias, J. J. and Plavec, M. J. 1987b Publ. Astron. Soc. Pacific **99**, 274.
Etzel, P. B. 1988 Astron. J. **95**, 1204.
Etzel, P. B. and Olson, E. C. 1985 Astron. J. **90**, 504.
Etzel, P. B. and Wood, D. B. 1982 WINK Status Report No. 10. (private circulation).

Hayes, D. S. 1970 Astrophys. J. **159**, 165.
Hayes, D. S. and Latham, D. W. 1975 Astrophys. J. **197**, 593.
Kondo, Y., McCluskey, G. E. and Wu, C. -C. 1981 Astrophys. J. Suppl. **47**, 333.
Kurucz, R. L. 1979 Astrophys. J. Suppl. **40**, 1.
McClusky, G. E. 1982 In Advances in Ultraviolet Astronomy: Four Years of IUE Research, NASA Conference Publ. 2238, Y. Kondo, J. M. Mead and R. D. Chapman, eds., NASA, Washington DC, p. 102.
Plavec, M. J. 1983 Astrophys. J. **275**, 251.
Plavec, M. J. and Dobias, J. J. 1987a Astron. J. **93**, 171.
Plavec, M. J. and Dobias, J. J. 1987b Astron. J. **93**, 440.
Plavec, M. J., Weiland, J.L. and Koch, R. H. 1982 Astrophys. J. **256**, 206.
Popper, D. M. 1956 Astrophys. J. **124**, 196.
Popper, D. M. 1980 Ann. Rev. Astron. Astrophys. **18**, 115.
Popper, D. M. 1984 Astron. J. **89**, 132.
Popper, D. M. and Etzel, P. B. 1981 Astron. J. **86**, 102.
Rhombs, C. G. and Fix, J. D. 1976 Astrophys. J. **209**, 821.
Rhombs, C. G. and Fix, J. D. 1977 Astrophys. J. **212**, 446.
Wampler, E. J. 1966 Astrophys. J. **144**, 921.
Wilson, R. E. 1970 Publ. Astron. Soc. Pacific **82**, 146.
Wilson, R. E. 1979 Astrophys. J. **234**, 1054.
Wilson, R. E. and Devinney, E. J. 1971 Astrophys. J. **166**, 605.
Wood, D. B. 1971 Astron. J. **76**, 701.
Wood, D. B. 1972 Goddard Space Flight Center Report X-110-72-473.
Wood, D. B. 1973 to 1978 WINK Status Reports Nos. 1 - 9. (private circulation).

DISCUSSION

PETERS: Between phases 0.8 - 0.9, you should see a depression of the flux and TT Hya due to absorption in the gas stream. Did you?

ETZEL: Please refer to Etzel (1988, Table VI). A scan obtained at phase 0.88 appeared as expected allowing for variations of the cool star. There were no anomalous depressions. A scan obtained the previous night at phase 0.75 was about 0.05 mag lower in the UV than expected. However, those observations are untrustworthy since the night was broken by clouds, which prevented an adequate extinction solution.

PETERS: In the UV, the gas stream is definitely seen at phase 0.9.

ETZEL: I see evidence for gas stream effects in the high dispersion, absorption-line spectroscopy (Etzel, 1984 Bull. Am. Astron. Soc. **16**, 504). A detailed account is in preparation.

HORNE: Is the physics of the reflection effect really well enough known to permit you to interpret the UV measurements with light-curve synthesis programs? For example, it is not clear that the effect of irradiating the cool star's photosphere is simply to raise its effective temperature, as the light curve synthesis program assumes.

ETZEL: Historically, the physics of the "reflection effect" has been less well established than its geometry. WINK actually has a relatively sophisticated bolometric model. Significant advances have recently been made by Vaz and Nordlund (1985 Astron. Astrophys. **147**, 281). I use multicolor light-curve solutions to predict small corrections to the observed totality spectrum. A glance at Fig. 4 illustrates that these corrections are relatively small in the UV. Fortunately, almost all of the variation outside of eclipse seen in Algols is due to the cool star, which is "described" by the light-curve synthesis.

WARREN: Is the network of spectrophotometric standard stars dense enough to allow the determination of adequate extinction easily?

ETZEL: No, not the network itself. You must select two standards at the beginning of the night (one on meridian, one in east) and observe them at various air masses. Extinction should be determined independently of transformation. The two are separable.

APPLICATIONS OF SPECTROPHOTOMETRY IN THE INFRARED

Robert F. Wing

The Ohio State University

ABSTRACT: The infrared spectral region is rich with applications for spectrophotometry. Many classes of astronomical objects have strong spectral features in the infrared which can be measured advantageously by spectral scanning or by narrow-band photometry at a few selected wavelengths; this review emphasizes the late-type stars. Molecular bands strong enough to measure at low spectral resolution are present over a wide range in spectral type and can be used both for classification and for abundance determinations of the light elements. There are also stretches of relatively clean continuum in most stars, and calibrated photometry in these regions is useful for determinations of effective temperature and studies of continuous opacity sources. Realistic model atmospheres can now be computed for all but the very coolest stars (and good progress is being made on those) so that infrared spectrophotometry can serve as the basis for quantitative analyses of the chemical and physical parameters of stellar atmospheres.

1. GENERAL REMARKS

Spectrophotometry is a technique that combines the opportunities of spectroscopy and photometry by allowing the measurement of spectral features in the same data set that is used to determine the brightness and energy distribution of the source. The data set may provide an uninterrupted record of the spectrum over a narrow or wide range of wavelength, or it may consist of brightness measurements at a limited number of points chosen to include or avoid specific spectral features. In the latter case, the data set may be referred to as "narrow-band photometry" or "spectrophotometry" depending on whether the bandpasses are defined by narrow-band filters or by some kind of spectrometer, but this distinction is not important. I will use the term "spectrophotometry" to mean any data set that has been (or can be) calibrated onto an absolute flux scale, and that has sufficient resolution to be sensitive to the presence of spectral features. The importance of the absolute calibration is that it allows comparisons with theoretical energy distributions, such as blackbody curves or flux distributions computed from model atmospheres.

A. G. Davis Philip, D. S. Hayes and S. J. Adelman, (eds.)
New Directions in Spectrophotometry 91 - 108

Spectrophotometry blossomed as a technique with widespread applications in the 1960s, at a time when the distinction between spectroscopy and photometry was much clearer than it is today. Photometry was then mostly of the wideband variety, done to provide accurate magnitudes and colors but with little concern for spectral features; and spectroscopy was mainly the photographic kind, carried out with emphasis on spectral resolution but without a way of knowing the true shape of the continuum. Photometrists and spectroscopists tended to be suspicious of each other's methods, and most observational astronomers were one or the other. Those of us who steered the middle course into spectrophotometry tended not to be fully accepted by either group, and despite the great promise of the technique, its spectroscopic applications often suffered from the relatively low resolution and its photometric applications tended to become bogged down in calibration problems.

Today the situation is considerably different, and the old distinctions between spectroscopy and photometry have largely disappeared. Spectra are often recorded digitally with linear detectors and are routinely calibrated onto an absolute flux scale. And photometry - even the broadband variety - is now often interpreted through calculations of synthetic spectra, so that the observer is likely to be very much aware of the spectral content of each bandpass. In other words, a large part of modern spectroscopy <u>and</u> photometry can be called "spectrophotometry", and it becomes difficult to know what should be included (or rather, what shouldn't) in a review of spectrophotometric observations.

When a spectrophotometric data set is compared to a theoretical energy distribution, two types of spectrophotometrists can be distinguished: those who are interested in seeing which curve fits, and those who wonder why some of the points don't. The former use the data to derive effective temperatures, to measure the interstellar reddening, and to investigate non-grey continuous opacity sources. The latter measure depressions relative to the continuum and use them for spectral classification, abundance analyses, and the recognition of spectral peculiarities. But as more observers become accustomed to using spectrophotometric data, there is a greater awareness of the advantages of using <u>all</u> of it. One of the new directions in spectrophotometry, then, is that it is now commonly being done by observers who previously would have been content to do spectroscopy <u>or</u> photometry.

In the infrared, nearly all our knowledge of the spectra of astronomical objects has come from spectrophotometry of one kind or another. Some of the main techniques used are narrow-band filter photometry, scans with circular variable filters, scans with movable-grating spectrometers, and Fourier-transform spectroscopy in which a linear detector is used to scan the interferometric fringe pattern as a mirror is moved to alter the path length of one of the two beams. These techniques give widely varying spectral resolutions, and they have been used in a wide range of wavelength regimes and applied to just about every known kind of astronomical object. Since there is no possibility of reviewing these applications comprehensively in a short paper, I will limit myself to describing a number of projects that illustrate several types of infrared spectrophotometry. For practical reasons I will emphasize the applications that I know best, namely studies of cool stars.

The infrared sky, from the far red out to about 5 μm, is dominated by late-type stars. Lists of the brightest stars (Wing 1971a; Neugebauer and Becklin 1973) have the same character throughout this interval, being dominated by K and M giants, M supergiants, and Mira variables. But beyond about 5 μm, both the instrumentation and the scientific objectives change as the lists become populated by extended, dusty objects.

The reasons for observing late-type stars in the infrared go beyond the fact that they are bright there. Most important is the extent to which it is possible to separate out the various spectral features. The optical spectrum is so cluttered with overlapping lines and bands that it is simply impossible, even with high spectral resolution, to make good quantitative measurements of any of them. In the infrared, atomic lines are almost negligible: with a few exceptions they are limited to weak, high-excitation lines well separated from one another. Molecular bands are present but they, too, are relatively weak, since in general the rotation-vibration bands that occur in the infrared are intrinsically much weaker than the electronic band systems at shorter wavelengths. Although each molecular band has a large number of lines, the infrared bands are sufficiently spread out that it is usually possible to measure them one at a time, and to measure clean stretches of continuum between them. This makes the infrared bands useful for abundance determinations, and indeed they provide our best information on the abundances of the light elements H, C, N and O in cool stars.

For the purposes of the discussion that follows, the infrared is divided into three regions: the one-micron region, the 1-4 μm region, and the ten-micron region. This division is suggested not only by the way the observations are made but also by the different problems that can be studied in the different parts of the infrared spectrum.

2. THE ONE-MICRON REGION

With regard to instrumentation, the one-micron region - which we will generously define to mean the entire interval from 7000 to 11,000 Å, since many observing programs make use of this whole range - is more similar to the visible region than to the farther infrared. The detectors used include photographic plates, image tubes, CCDs, and photomultipliers; all of these are used in the visible region as well, but in using them in the one-micron region one is frequently pushing them to the long-wavelength limit of their response. The wavelength

11000 Å is a natural and nearly absolute boundary to the region, since the onset of strong atmospheric H_2O absorption coincides there with rapidly decreasing detector sensitivity.

Spectroscopic work in the 9000 - 11000 Å region, whether done with type M or Z photographic emulsions or with S-1 image tubes, has seldom been absolutely calibrated and will not be considered here; the early work has been reviewed by Spinrad and Wing (1969). In the 7000 - 9000 Å region, the type N photographic emulsion has been used extensively for uncalibrated spectroscopy but has been replaced at many observatories by CCD detectors, which are both more efficient and easier to calibrate. Some examples of recent work showing calibrated CCD spectra in the 7000 - 9000 Å region are the study of the dwarf nova Z Cha and a set of M dwarf comparison stars by Wade and Horne (1988) and the classification of faint M supergiants by MacConnell, Wing, and Costa (1987). Calibrated reticon spectra for symbiotic stars and a large set of MK classification standards have been presented by Schulte-Ladbeck (1988). The N emulsion continues to be used for astronomical spectroscopy, but primarily for objective-prism survey work for which their availability in large sizes is important.

Most applications of spectrophotometry in the one-micron region - including both filter photometry and scanner measurements - have employed photomultipliers and have dealt with the M-type stars and their peculiar-composition cousins, the S and C stars. Bandpasses of 50 Å or even more are effective in studying the molecular bands that occur in these stars because many of them have such dense rotational structure that they depress the spectrum continuously over substantial spectral intervals. Molecules that can be studied in this way include TiO, VO and CN in normal stars and a number of additional molecules such as ZrO and LaO which sometimes appear strongly in S stars. The C_2 molecule is present in carbon stars but its bands in the one-micron region are difficult to find and measure among the much stronger CN bands.

The spectra of three normal M giants from 3500 to 11000 Å, covering nearly the whole range accessible to photomultipliers on the ground, are shown in Fig. 1. The observations were made with the HCO scanner at Kitt Peak and consisted of integrations every 20 Å through a 40 Å bandpass. The data were transformed to an absolute flux system through direct or indirect comparisons to Vega, for which a model-atmosphere flux distribution was assumed. An important advantage of the absolute calibration is that it permits comparisons with blackbody curves, shown as dashed lines. These spectra were used by Smak and Wing (1979) to evaluate the effect of TiO absorption on the B, V, R and I bandpasses.

Virtually all the absorption that can be seen longward of 4500 Å in Fig. 1 is due to TiO, which is strongest in the visual region. As a result of the spectral distribution of the TiO absorption, a

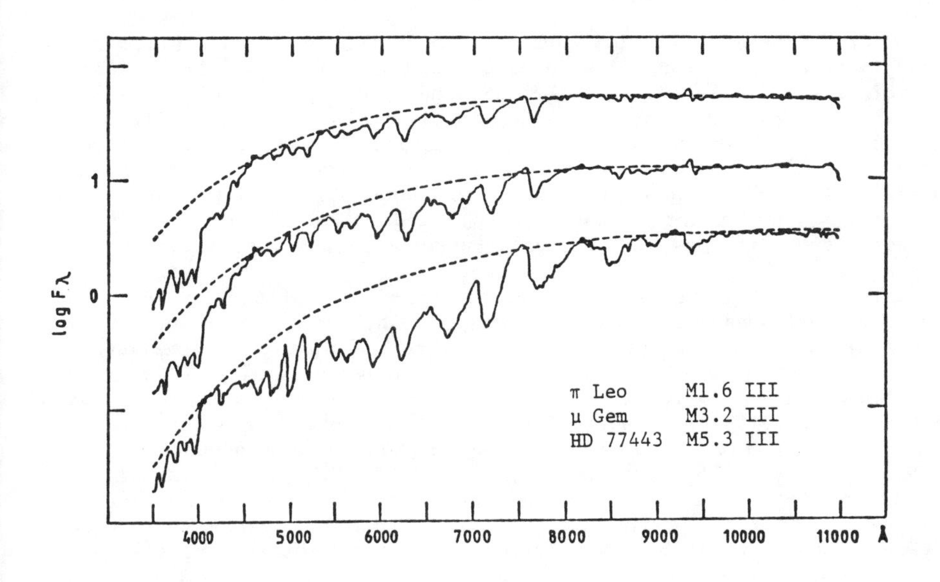

Fig. 1. Calibrated spectra of three giant stars of approximate types M 1, M 3 and M 5 (top to bottom) obtained with the HCO scanner at Kitt Peak (resolution 40 Å). Nearly all the absorption longward of 4500 Å is due to TiO. Fluxes per unit wavelength are plotted on a logarithmic scale against wavelength. Blackbody curves are shown for comparison as dashed lines. From Smak and Wing (1979).

fundamental change in the character of an M- type spectrum occurs at about 9000 Å, and the 1000-Å stretch from 10000 to 11000 Å is free of significant absorption in types as late as M 5. It is this circumstance that makes the one-micron region particularly useful for measuring color temperatures and indices of molecular band strength.

The CN molecule is present in nearly all late-type giants and supergiants but is strong only in carbon and SC stars. In the M giants shown in Fig. 1, only the strongest CN band - the (0,0) band of the Red System - can be seen, depressing the last few data points near 11000 Å. Note, however, that the spectra are plotted on a very small vertical scale, with tick marks at intervals of 2.5 magnitudes. Several of the infrared CN bands produce depressions of about 0.1 mag in giants (and about twice that amount in supergiants), and useful CN indices can be extracted from data having the usual 1 percent photometric accuracy.

It is not efficient to record spectra continuously with a single-channel scanner, and any program intending to reach large numbers of faint stars must limit the number of wavelengths measured. It is, however, possible to choose the wavelengths in such a way that the observations contain nearly all the useful information of continuous scans. From 1965 to 1967 I used the movable-grating scanner at Lick Observatory to obtain exploratory spectrophotometry of late-type stars at 26 (sometimes 27) wavelengths between 7500 and 11000 Å; the bandpass was 30 Å. The data were reduced as multicolor photometry; that is, extinction and transformation coefficients were measured each night at each wavelength through observations of standard stars which in turn had been compared to Vega, the absolute flux standard. More than 1300 observations of 314 different stars were recorded in this way and are given in my thesis (Wing 1967).

Some examples of the Lick 26-color photometry are shown in Fig. 2. The first panel shows six giants of types G 8 to K 1, in which virtually all the absorption seen is due to CN. Note that two of the stars - η Dra and β Gem - have weaker CN than do the other stars. The CN bands are greatly strengthened in the stars of panel (b), and it is noteworthy that they are as strong in the SC star as in the two carbon stars. The (0,0) band of the CN Red System, which depresses the point of longest wavelength, is clearly visible in giants as late as M5 (panel (c)), although the other CN bands are lost whenever TiO is present. The TiO bands grow impressively from M 0 to M 5, and the VO band at 1.05 μm takes over as the most useful temperature criterion in the coolest stars.

When spectrophotometric data are recorded with a large entrance aperture rather than a slit, the stellar magnitude at any program wavelength can be reduced by standard photometric procedures. This is a great advantage in studies of variable stars: one can, for example, determine the spectral type at each point on the light curve, or one can construct a light curve for each spectral point. Shore and Adelman (1984), for example, have studied seven RS CVn systems with spectrophotometric measurements from 4032 to 10,800 Å and good phase coverage. Variables of the Mira type received special attention in my program of 26-color scanner photometry, and several were followed through two or more cycles. Unfortunately, no straightforward interpretation of the Mira data proved possible since these stars do not show a well- defined relation between band strength and color temperature. In fact, in plots of band strength vs. color, the Miras execute enormous loops which differ greatly from star to star and from cycle to cycle (Spinrad and Wing 1969). These loops have never been modeled, but they have been discussed qualitatively in terms of a greatly extended atmosphere in which the molecular bands and the continuum are formed in well-separated layers whose temperature variations are out of phase (Wing 1980).

Most of the information provided by the 26-color photometry can be obtained with simpler instrumentation and far fewer wavelength points, and narrow-band photometric systems using interference filters have been designed for this purpose by Lockwood (1972) and the writer (Wing 1971b). My eight-color system is designed to measure the strongest bands of TiO, VO, and CN in order to provide two-dimensional (temperature/luminosity) classifications for M stars; it includes the best available continuum points in stars showing strong bands of both TiO and CN (e.g. the M supergiants) and is especially successful in providing indices of TiO and CN that are independent of each other. Fig. 3 shows a sequence of normal M giants observed on the eight-color system, and Fig. 4 contains observations of several IRC objects, illustrating how reddened M supergiants can be distinguished from M giants on this system. Note that the eight-color system extends to a shorter wavelength than does the 26-color scanner photometry in order to include the strong (0,0) band of the γ System around 7100 Å. The principles of two-dimensional classification on the basis of molecular band measurements in the near infrared have been discussed by Wing and White (1978).

The ability of the eight-color system to distinguish supergiants from giants and to determine the interstellar reddening from comparison of TiO strength and color has led to its adoption as a follow-up technique for an objective-prism survey of the southern Milky Way being conducted by MacConnell, and we have already more than doubled the number of M supergiants known in the southern hemisphere (Wing, MacConnell, and Costa 1987). Whether these stars will improve our knowledge of spiral-arm structure depends upon the accuracy of the distances that can be assigned to them, and our observing program is currently emphasizing luminous M stars in clusters and associations which can serve to calibrate the relation between CN strength and absolute magnitude.

The quantitative interpretation of the photometry through calculation of synthetic colors opens up a wide range of additional applications for the eight-color system, including tests of model atmospheres, determinations of chemical abundances, and determinations

Fig. 2. Examples of 26-color scanner photometry from Wing (1967). Absolute fluxes per unit wavelength, on a magnitude scale, are plotted against wavelength in microns.

(a) G and K giants. Top to bottom: κ Gem, G 8 III; ϵ Vir, G 8 IIIab; η Dra, G 8 III; ϵ Tau, K 0 III; β Gem, K 0 IIIb; ψ UMa, K 1 III. The main depressions are caused by the (3,1), (1,0) and (0,0) bands of the CN Red System.

(b) Carbon and SC stars. Top to bottom: T Lyr, C 5,3 (1965 Oct 18); R Ori, SC (1965 Oct 28); WZ Cas, C 9,1 (1965 Oct 17). Again nearly all the absorption is due to CN. Note that the scale is 2x smaller than in panel (a).

(c) Early M giants. Top to bottom: β And, M 0.5 III; β Peg, M 2.5 III; ρ Per, M 4 III; R Lyr, M 5 III. The main absorptions here are due to TiO but the (0,0) band of CN is still visible.

(d) Late M giants. Top to bottom: RX Boo, M 8.0 (1966 May 17); o Cet, M 9.0 (1965 Sep 9). Note the growth of the VO band near 1.05 μm.

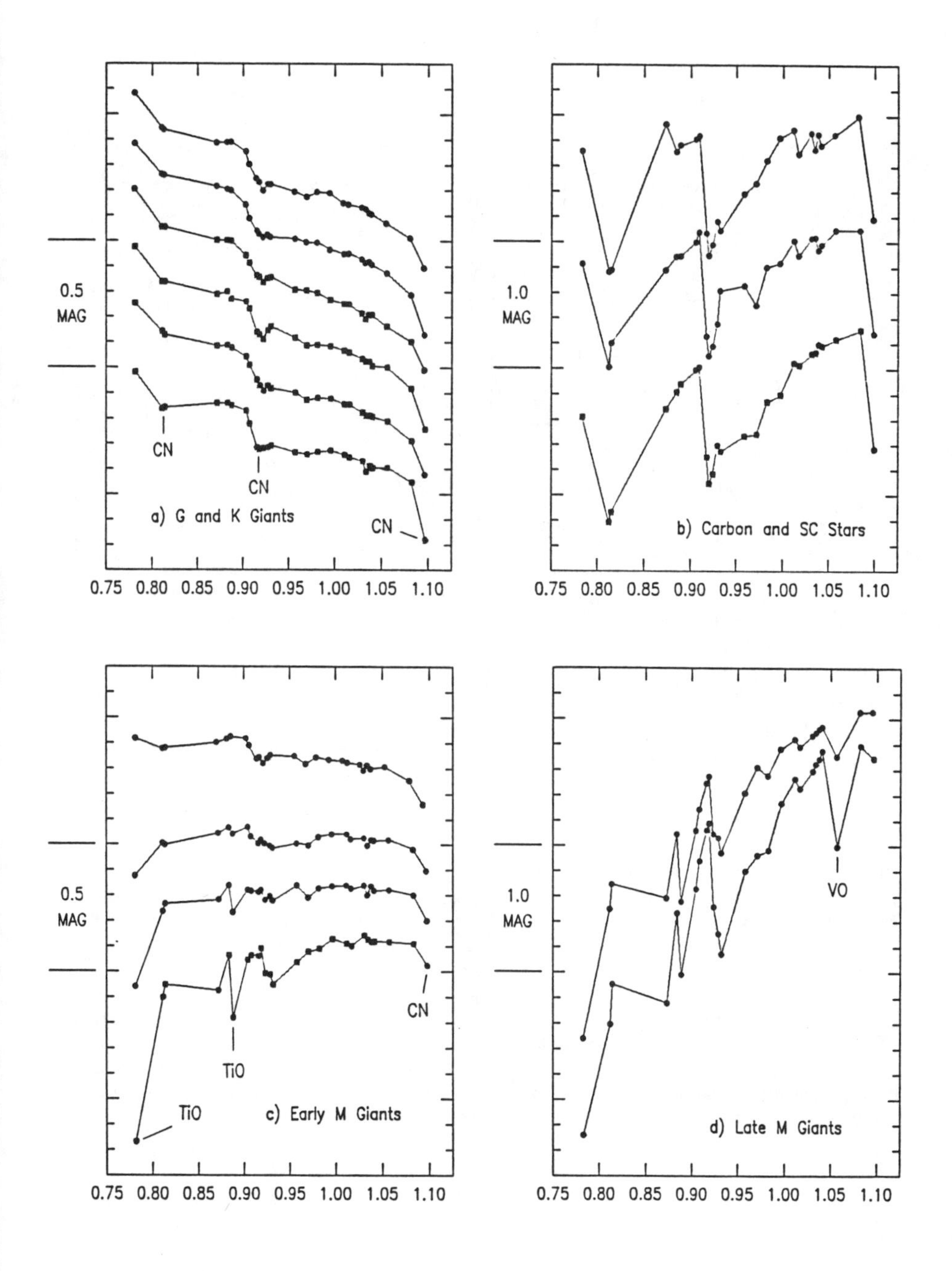
0.5
MAG
CN
CN
CN
a) G and K Giants
0.75 0.80 0.85 0.90 0.95 1.00 1.05 1.10
1.0
MAG
b) Carbon and SC Stars
0.75 0.80 0.85 0.90 0.95 1.00 1.05 1.10
0.5
MAG
TiO
TiO
CN
c) Early M Giants
0.75 0.80 0.85 0.90 0.95 1.00 1.05 1.10
1.0
MAG
VO
d) Late M Giants
0.75 0.80 0.85 0.90 0.95 1.00 1.05 1.10

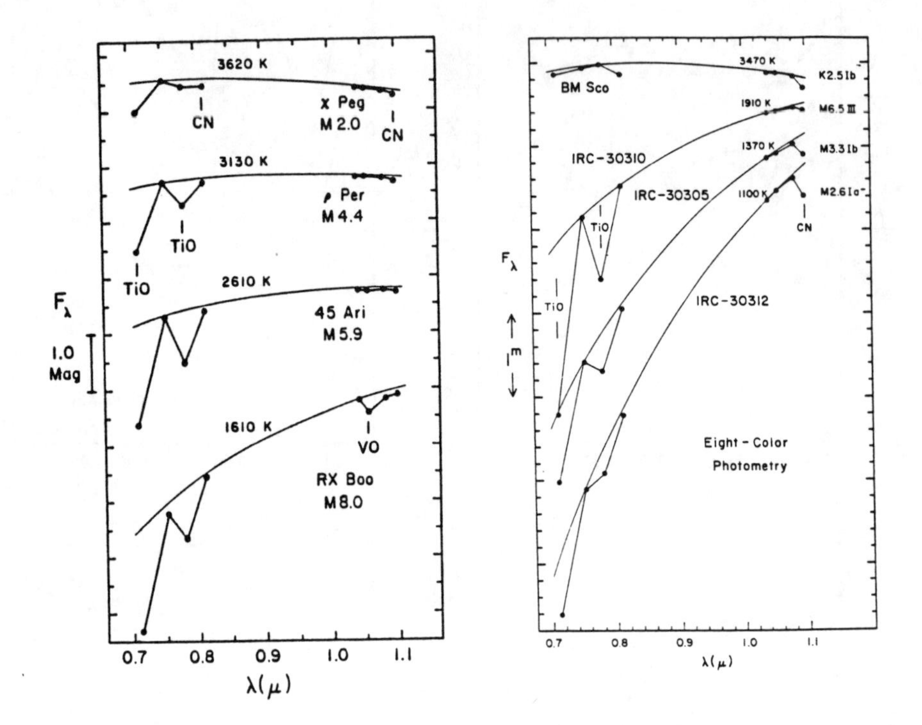

Fig. 3. (left) Photometry on the writer's eight-color narrow-band system for typical giants of approximate types M 2, M 4, M 6 and M 8. Absolute flux per unit wavelength is plotted against wavelength in microns. Molecules responsible for the depressions are labeled, and blackbody curves are shown for comparison.

Fig. 4. (right) Eight-color photometry for three IRC objects in the vicinity of the open cluster M 6 and for BM Sco, a K supergiant member of the cluster. Two of the IRC objects are seen to be heavily reddened M supergiants, much more distant than the cluster, while IRC-30310 is a moderately-reddened giant. From Warner and Wing (1977).

of effective temperatures. An improved effective temperature scale for G and K giants has been determined in this way (Wing, Gustafsson, and Eriksson 1985), while for the M giants, calculations by Piccirillo, Bernat, and Johnson (1981) including the effects of TiO have shown that the eight-color photometry and its calibration is consistent with the lunar-occultation effective temperature scale of Ridgway et al. (1980).

Red giants of peculiar composition are difficult to classify in a useful manner but nevertheless have been favorite targets of spectrophotometrists in the near infrared. Here we should mention Baumert's (1972) eight-color photometry of carbon stars and the study of S stars by Piccirillo (1977), for which he modified the eight-color system by replacing two of the filters with ones designed to measure bands of ZrO and LaO. An absolutely calibrated spectrum of the S-type variable R Cyg near minimum light has been published by Wing (1972); several of the unidentified bands appearing in that spectrum, as well as a set of strong bands discovered by Wing and Hinkle (1987) in the region of the J filter, have recently been attributed to the ZrS molecule (Hinkle, Lambert, and Wing 1988).

Although the determination of isotope ratios is usually considered a problem for high-resolution spectroscopy, spectrophotometry can play a useful role here, too. The isotope shift between ^{13}CN and ^{12}CN is large enough at the (2,0) band that scans of 20 Å resolution allow the determination of the $^{12}C/^{13}C$ ratio from the absorption profile. A six-color photometric system defined with the HCO scanner at Kitt Peak has been used successfully for this purpose and is described elsewhere in this volume by Little (1988).

3. THE 1-4 μm REGION

To observe longward of 11,000 A one must use a different type of detector, usually a PbS or InSb photovoltaic device which has useful sensitivity to about 4 μm. These detectors are marginally useful in the one-micron region, providing some overlap with work done with photomultipliers; thus the 1-4 μm region is often treated as a unit. In this region telescopes with chopping secondary mirrors are favored but not essential, and circumstellar dust shells begin to be detected but rarely dominate over photospheric radiation.

Absorption by the Earth's atmosphere divides the 1-4 μm region into four regions or windows corresponding to the J, H, K and L bands of Johnson's photometric system. Most of the absorption defining these windows is caused by water vapor, but it is a strong band of CO_2 that imposes the long-wavelength limit to the L band.

The molecules that can be studied in the 1-4 μm region include some of the most abundant molecules in stellar atmospheres: CO, CN, NH, OH, H_2O, H_2 and SiO. Of these, NH, OH and H_2 are best studied at high resolution, but the others are prime targets for spectrophotometric investigations.

In the K and H bands, the most conspicuous spectral features are usually the first and second overtone rotation-vibration bands of CO. The first-overtone bands are clearly seen in the calibrated spectra of two stars shown in Fig. 5, and the presence of the isotopic ^{13}CO bands provides an important opportunity to measure carbon isotope ratios in oxygen-rich stars. The differences between the M 3III and M 8e spectra shown are almost entirely due to the presence of H_2O absorption in the latter. The water bands do not appear in giants and supergiants until about type M 6, but in dwarfs they are seen at much warmer effective temperatures (Mould 1978).

Merrill and Stein (1976a,b) have presented spectrophotometry of an assortment of late-type stars in the 2 to 4 μm region and also in the 8-13 μm window. Both carbon-rich and oxygen-rich stars are included. At the low resolution of these observations, which were made with a circular variable filter, the spectra are nearly devoid of features apart from the first-overtone band sequence of CO and the strong polyatomic 3.1 μm absorption in carbon stars. The spectra are, however, absolutely calibrated, and the effect of stellar water vapor on the shapes of the later-type spectra is well shown.

The CN molecule is also a significant absorber in the J, H and K windows, but its nearly featureless structure (its main bandheads occurring in regions of telluric absorption) makes it hard to identify, even in carbon stars. Wing and Spinrad (1970) have provided a guide to the identification of CN in this region.

In the L band relatively little spectroscopy has been done, but a magnificent atlas of Fourier-transform spectra of cool stars, with detailed molecular identifications, has been published by Ridgway et al. (1984). Spectrophotometric observations in this region include scans of the first-overtone SiO bands by Rinsland and Wing (1982); these bands are important as the best indicator of the temperature class of M stars that can be measured at low spectral resolution in the 1-4 μm region.

Airborne observations provide nearly uninterrupted spectral coverage throughout the infrared. The water bands in cool Mira variables are nicely illustrated by Strecker, Erikson, and Witteborn (1978), and absorptions by the polyatomic molecules HCN and C_2H_2 in a cool carbon star are shown in Goebel et al. (1981). The infrared spectrophotometry of standard stars used to reduce these Kuiper Airborne Observatory scans is given in Strecker et al. (1979).

The cooled grating scanner known as "Audrey", which operates at the KPNO 1.3 m telescope, can be programmed to make integrations at precisely-defined wavelengths throughout the 1-4 μm region. In collaboration with C. P. Rinsland, the writer has used this instrument

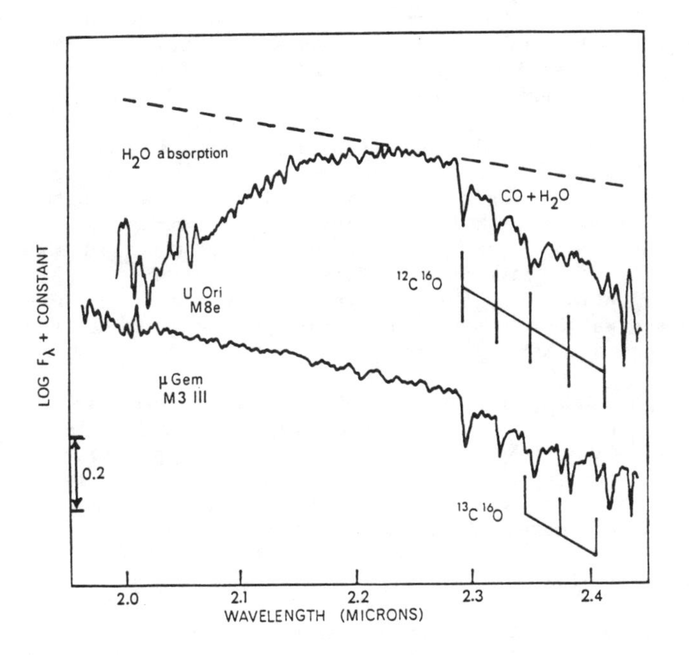

Fig. 5. Calibrated spectra in the K band for U Ori, a cool Mira variable showing strong water vapor absorption, and μ Gem, a normal M 3 giant. Both stars show the first-overtone CO bands, and the isotopic species ^{13}CO is clearly visible in the M 3 star. The ordinate is flux per unit wavelength, and the interval between tick marks is 0.2 dex = 0.5 mag. From Hyland (1974).

to define a 13-color photometric system consisting entirely of continuum points - avoiding both stellar and telluric absorption features - and extending from 1.04 to 4.00 μm. By using an adequately large entrance aperture and following normal photometric procedures, we have found that an accuracy of 1 percent or better can be achieved as easily with an InSb cell as with photomultipliers in the visible region. One objective of the observations has been to test the continuum fluxes calculated from model atmospheres and to determine the relations between the various color temperatures (which are strongly affected by H^- opacity) and the effective temperature (Wing and Rinsland 1981). Also, by including measurements of a blackbody source with the help of Ridgway, Joyce and Hayes at Kitt Peak, we attempted to determine a new absolute flux calibration at these 13 wavelengths. My current observing program with this 13-color system emphasizes the red giants of peculiar composition. Comparison of the observed shape of the continuum with flux distributions from model atmospheres permits a quantitative evaluation of the effects of H^-, and it will be interesting to see whether cases of hydrogen deficiency among late-type stars can be detected in this manner.

Spectrographs using arrays of InSb detectors are starting to become available and no doubt will soon replace the single-channel scanners used to date. Nevertheless, it can be expected that the accurate observations of standard stars made with single-channel instruments such as Audrey will play an important role in the absolute calibration of spectra obtained with array detectors.

4. THE TEN-MICRON REGION

Beyond 4 μm, the next atmospheric window is the M band centered at about 5 μm. Unfortunately, it is so cluttered with both terrestrial water lines and stellar lines from the fundamental band of CO that there is little to recommend it for spectrophotometric studies. The main contribution to astrophysics from this window has been the determination of oxygen isotope ratios through high-resolution spectroscopy of CO lines (Geballe et al. 1977, Harris and Lambert 1984, Harris, Lambert and Smith 1988).

The 10 μm window, which follows an interval of very heavy water absorption, is both broader and cleaner than the M band, extending from 8 to 13 μm and interrupted only by a rather innocuous band of ozone. Here we are near the flux peak for thermal radiation from bodies at a temperature of a few hundred degrees, and the large excess fluxes shown by many stars in this region (over what could be expected from the stellar photosphere) have been attributed to radiation from circumstellar dust grains filling a volume much larger than the star and kept warm by absorption of optical radiation from the star. It appears that nearly all stars that are sufficiently cool and/or sufficiently luminous are surrounded by detectable quantities of circumstellar dust.

Spectrophotometry of late-type stars in the 10 μm region, which has mostly been done either with sets of narrow-band filters or with variable-bandpass filters, has been carried out almost exclusively for the purpose of defining the shape of the circumstellar dust emission spectrum. By good fortune, this atmospheric window is wide enough to include nearly the entire "hump", although additional measurements in the 18-25 μm region can be helpful. There are significant differences in the shapes of the humps appearing in the spectra of M stars, which are relatively broad and peaked near 9.8 μm, and in carbon stars, which are sharper and peaked near 11.2 μm. Examples of ground-based spectrophotometry of the dust bumps of M and C stars are given by Forrest, Gillett and Stein (1975).

The low-resolution spectra acquired by the IRAS satellite are well suited to studying the circumstellar dust bumps of late-type stars and indeed represent by far the largest set of 8-22 μm spectrophotometry currently available. In general, IRAS sources that were observed spectroscopically can be classified as M stars or C stars on the basis of such spectra, but a number of cases have been noted in which a carbon star (as classified optically) shows a dust bump of the shape normally associated with M stars (Little-Marenin and Wilton 1986). It remains to be seen whether these objects should be interpreted as M+C binaries (Little-Marenin 1986) or as single stars whose photospheres have undergone a change in the O/C ratio since the ejection of the material now found in the grains (Chan and Kwok 1988).

Many other kinds of objects - e.g. emission-line stars, planetary nebulae, and active galaxies - have been observed in the 10 μm region but cannot be considered here. I would, however, like to mention that the 10 μm spectrophotometry of SN 1987A reported recently by Aitkin et al. (1988) shows several emission lines, but no evidence of dust.

I will end this brief discussion of 10 μm spectrophotometry with one thought: despite the obvious interest in circumstellar grains, we should not fail to observe stars which are dust-free. Spectra of stellar photospheres in the 10 μm region, when not filled in by grain emission, may provide a great deal of new spectroscopic information. Many metallic oxide molecules, for example, have their fundamental rotation-vibration bands in this region. A concerted effort to identify late-type stars without dust bumps, and to record their 10 μm spectra in as much detail as possible, might prove extremely rewarding.

REFERENCES

Aitken, D. K., Smith, C. H., James, S. D., Roche, P. F., Hyland, A. R. and McGregor, P. J. 1988 Monthly Notices Roy. Astron. Soc. **231**, 7p.

Baumert, J. H. 1972, Ph.D. dissertation, The Ohio State University.

Chan, S. J. and Kwok, S. 1988 Astrophys. J., in press.

Forrest, W. J., Gillett, F. C. and Stein, W. A. 1975 Astrophys. J. **195**, 423.

Geballe, T. R., Wollman, E. R., Lacy, J. H. and Rank, D. M. 1977 Publ. Astron. Soc. Pacific **89**, 840.
Goebel, J. H., Bregman, J. D., Witteborn, F. C., Taylor, B. J. and Wilmer, S. P. 1981 Astrophys. J. **246**, 455.
Harris, M. J. and Lambert, D. L. 1984 Astrophys. J. **285**, 674.
Harris, M. J., Lambert, D. L. and Smith, V. V. 1988 Astrophys. J. **325**, 768.
Hinkle, K. H., Lambert, D. L. and Wing, R. F. 1988, in preparation.
Hyland, A. R. 1974 in Highlights of Astronomy, Vol. 3, G. Contopoulos, ed., Reidel, Dordrecht, p. 307.
Little, S. J. 1988 in New Directions in Spectrophotometry, A. G. D. Philip, D. S. Hayes and S. J. Adelman, eds, L. Davis Press, p. 109.
Little-Marenin, I. R. 1986 Astrophys. J. Letters **307**, L 15.
Little-Marenin, I. R. and Wilton, C. 1986 in Proc. Fourth Cambridge Workshop on Cool Stars, Stellar Systems, and the Sun, M. Zeilik and D. M. Gibson, eds., Springer-Verlag, p. 420.
Lockwood, G. W. 1972 Astrophys. J. Suppl. **24**, 375.
MacConnell, D. J., Wing, R. F. and Costa, E. 1987 Revista Mexicana Astron. Astrofis. **14**, 367.
Merrill, K. M. and Stein, W. A. 1976a Publ. Astron. Soc. Pacific **88**, 285.
Merrill, K. M. and Stein, W. A. 1976b Publ. Astron. Soc. Pacific **88**, 294.
Mould, J. R. 1978 Astrophys. J. **226**, 923.
Neugebauer, G., and Becklin, E. E. 1973 Scientific American **228**, No. 4, p. 28.
Piccirillo, J. 1977, Ph.D. dissertation, Indiana University.
Piccirillo, J., Bernat, A. P. and Johnson, H. R. 1981 Astrophys. J. **246**, 246.
Ridgway, S. T., Carbon, D. F., Hall, D. N. B. and Jewell, J. 1984 Astrophys. J. Suppl. **54**, 177.
Ridgway, S. T., Joyce, R. R., White, N. M. and Wing, R. F. 1980 Astrophys. J. **235**, 126.
Rinsland, C. P. and Wing, R. F. 1982 Astrophys. J. **262**, 201.
Schulte-Ladbeck, R. E. 1988 Astron. Astrophys. **189**, 97.
Shore, S. N. and Adelman, S. J. 1984 Astrophys. J. Suppl. **54**, 151.
Smak, J., and Wing, R. F. 1979 Acta Astron. **29**, 187.
Spinrad, H. and Wing, R. F. 1969 Ann. Rev. Astron. Astrophys. **7**, 249.
Strecker, D. W., Erikson, E. F. and Witteborn, F. C. 1978 Astron. J. **83**, 26.
Strecker, D. W., Erikson, E. F. and Witteborn, F. C. 1979 Astrophys. J. Suppl. **41**, 501.
Wade, R. A. and Horne, K. 1988 Astrophys. J. **324**, 411.
Warner, J. W. and Wing, R. F. 1977 Astrophys. J. **218**, 105.
Wing, R. F. 1967, Ph.D. dissertation, University of California, Berkeley.
Wing, R. F. 1971a Publ. Astron. Soc. Pacific **83**, 301.

Wing, R. F. 1971b in Proc. Conf. on Late-Type Stars, G. W. Lockwood and H. M. Dyck, eds., Kitt Peak National Observatory Contrib. No. 554, p. 145.

Wing, R. F. 1972 Mem. Soc. Roy. Sci. Liege, 6th ser., **3**, 123.

Wing, R. F. 1980 in Current Problems in Stellar Pulsation Instabilities, D. Fischel, J. R. Lesh and W. M. Sparks, eds., NASA TM-80625, p. 533.

Wing, R. F., Gustafsson, B. and Eriksson, K. 1985 in IAU Symposium No. 111, Calibration of Fundamental Stellar Quantities, D. S. Hayes, L. E. Pasinetti and A. G. D. Philip, eds., Reidel, Dordrecht, p. 571.

Wing, R. F. and Hinkle, K. H. 1987 Bull. Amer. Astron. Soc. **19**, 645.

Wing, R. F., MacConnell, D. J. and Costa, E. 1987 Revista Mexicana Astron. Astrofis. **14**, 362.

Wing, R. F. and Rinsland, C. P. 1981 Revista Mexicana Astron. Astrofis. **6**, 145.

Wing, R. F. and Spinrad, H. 1970 Astrophys. J. **159**, 973.

Wing, R. F. and White, N. M. 1978 in IAU Symposium No. 80, The HR Diagram, A. G. D. Philip and D. S. Hayes, eds., Reidel, Dordrecht, p. 451.

DISCUSSION

BOHLIN: What is the fundamental reference for absolute flux in the infrared?

WING: I've generally assumed a model-atmosphere energy distribution for Vega and have tied everything to that. Since the publication of the paper by Schild, Peterson, and Oke (Astrophys. J. **166**, 95, 1971), I've used the model they recommended as best representing the absolute photometry of Vega. This works very well, except that some artwork may be necessary in regions affected by hydrogen lines. There have been more recent studies of Vega but they haven't changed the infrared. In the case of the 13-color system in the 1 - 4 μm region, Don Hayes and I, in collaboration with others at Kitt Peak, attempted an absolute calibration by observing a blackbody along with the stars, but we ran into instrumental difficulties and were not able to come up with a calibration that we trusted more than the Vega model.

TAYLOR: I might add that Bob's thesis standards agree very well with the "consensus extension" of the Breger standards which I published in 1984 (Taylor, Astrophys. J. Suppl. **54**, 259).

BELL: If using regions which contain TiO lines to get T_{eff}, I think you will have to be very careful to get the stellar composition since TiO band strength is clearly composition dependent.

WING: That's true, although the relation between TiO strength and color changes only slightly between field and globular cluster stars (Mould and McElroy, Astrophys. J. **221**, 580, 1978). Anyway, with synthetic spectra we can force-fit the TiO bands where they are explicitly measured, and then use that amount of TiO to correct the continuum points for TiO contamination. It's the continuum points that we have to get right to determine T_{eff}.

A DISCUSSION OF THE SIX-COLOR SYSTEM TO MEASURE CARBON ISOTOPE RATIOS

Stephen J. Little

Bentley College

ABSTRACT: I will discuss the six-color narrow-band infrared (7820-9048 Å) system that was developed by R. F. Wing and I. R. Marenin at Ohio State University in 1971. The system was originally intended to measure the $^{12}C/^{13}C$ ratio in G and K giant and supergiant stars in which the 8000 Å ($\Delta v=2$) CN band is strong and relatively uncontaminated. We later included carbon stars in our observations. The system was designed with 20 Å bandwidths and the HCO scanner, equipped with an S-1 phototube at Kitt Peak National Observatory, was used for our observations.

We observed several hundred stars using this system. I will discuss some of our results and some of the observing problems we encountered. The discontinuation of the HCO scanner by KPNO has made the original six-color system somewhat moot, but the proposed spectrophotometric telescope could be an excellent instrument for use with a similar system.

1. THE PHOTOMETRIC SYSTEM

The six-color system was designed to measure relative $^{12}C/^{13}C$ ratios in the $\Delta v=2$ CN band located at about 8000 Å in the near infrared. The 8000 Å CN band is ideal for this purpose since it is relatively uncontaminated with atomic lines and because the isotope shift for ^{13}CN is about 30 Å to the red. The $^{12}C/^{13}C$ ratio in giant and supergiant stars is an indicator of the internal production of ^{12}C and subsequent CNO-cycle processing as the ^{12}C is mixed to the surface layers of the stars. The objective was to be able to obtain rough $^{12}C/^{13}C$ ratios for large numbers of stars with a more modest telescope aperture and a shorter observing time per star than was possible for a complete high-dispersion spectral analysis.

The system is composed of six narrow (20 Å) bandpasses (see Fig. 1) chosen to include two continuum bands (nos. 1 and 6), two bands in which the contamination of the ^{12}C by ^{13}C is minimal (D2 and D5), and two bands in which the effect of ^{13}CN contamination is large (D3 and D4). These bands were also chosen to avoid, as much as possible, atomic line contamination. We estimate that atomic lines contribute

A. G. Davis Philip, D. S. Hayes and S. J. Adelman, (eds.)
New Directions in Spectrophotometry 109 - 112

less than 1% to any of the six bands. The spectral line contamination was estimated from the Sun and from α Bootis. A set of photometric standard stars, chosen from Wing eight - color standards, was observed frequently on all our observing runs.

We used the No. 1 and 6 continuum bands to define a local black body continuum for each set of observations. We then defined the depression below that continuum in the D2-5 bands in magnitudes. We planned to use the ratio of (D2+D5)/(D3+D4) as the indicator of relative $^{12}C/^{13}C$ ratios. The continuum temperatures defined by the continuum bands for our stars were quite close to the temperatures defined by the Wing eight - color system.

2. OBSERVATIONS AND RESULTS

We made most of our observations with the HCO scanner at KPNO with the No. 1 0.9 m telescope (although some observations were made at CTIO with a similar instrument). The relatively slow S-1 phototube was necessitated by the 9046 Å continuum point, but we were able to observe carbon stars to magnitude 9 with several 20-second integrations in each band. The HCO scanner was ideal for our survey because of its photometric accuracy; we needed precise relative photometry (±0.01 mag.) to make the system work. We calibrated the scanner for wavelength by use of a narrow exit slot and by making scans of the Hγ line in A stars. We felt these scans calibrated the wavelength within about ±1Å.

We observed several hundred stars with the system, and were able to obtain a good set of standard-star observations. We observed nearly every G and K supergiant in the Bright Star Catalog and observed about 50 carbon stars. We encountered some problems in observing on the system because the HCO scanner was unable to hold a fixed wavelength calibration for even part of a night on many occasions. Sporadically-varying wavelength errors were fatal to our system, and many nights of data were lost because of this problem. Seeing effects also tend to dilute the data by smearing the edges of the bandpass, and nights of large, variable seeing excursions were not useable. The D2 bandpass was also found to be of marginal use because it was bordered on the blue side by a steeply varying spectral gradient. We ended up using the ratio of (D3+D4)/D5 to define our $^{12}C/^{13}C$ ratios.

Fig. 2 shows data for carbon stars calibrated against the recent results of Lambert et al. (1976). The ratios we measured allow us to define two groups of carbon stars; the low $^{12}C/^{13}C$ group and a high group with ratios above 40. There are no carbon stars with ratios between about 10 and 30.

REFERENCE

Lambert, D. L., Gustafsson, B., Eriksson, K. and Hinkle, K. H. 1986 Astrophys. J. Suppl., 62, 373.

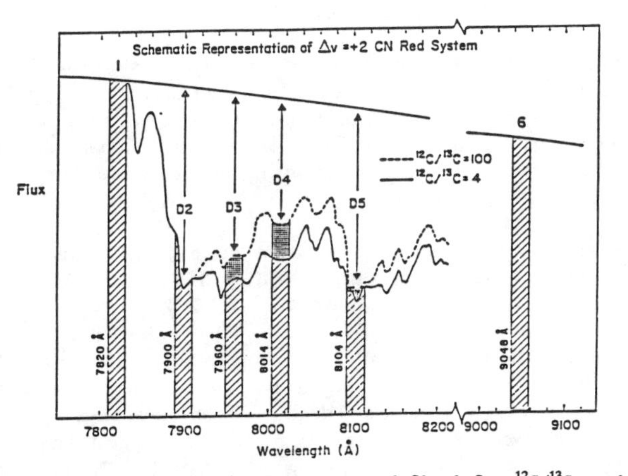

Fig. 1. The CN six-color bandpasses are defined for $^{12}C/^{13}C$ ratios of 100 and 4. The CN absorption was calculated with a Milne-Eddington model.

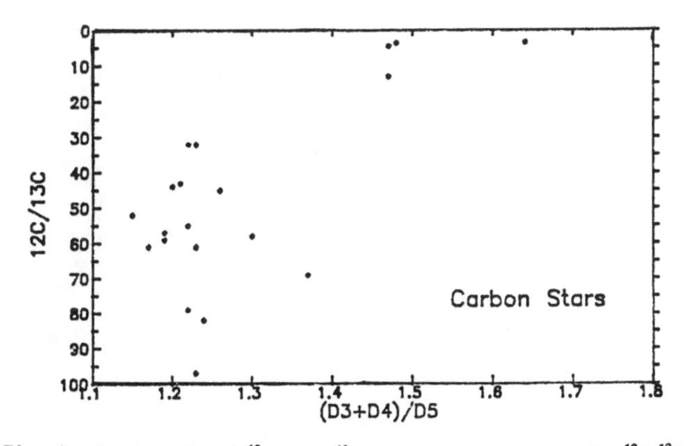

Fig. 2. Ratios of the ^{12}CN and ^{13}CN depressions versus the $^{12}C/^{13}C$ ratios of Lambert et al. (1986).

DISCUSSION

WEAVER: What is the effect of small wavelength shifts on the accuracy of your intended 1% photometry? D2 and D5 seem like they might be especially sensitive.

LITTLE: There is a definite effect. D2 proved to be too sensitive to wavelength shifts in practice and was not used. We did run model atmosphere simulations of wavelength shifts in 1974 (R. Pitts and D. Carbon, unpublished), and D5 seemed, both from these results and in practice, to be less seriously affected.

WING: We did compute synthetic colors on this system for wavelength shifts up to ± 4 Å, as well as for the nominal wavelengths. Band 2 definitely has problems, but the fluxes in other bands were as steady as one could hope for in a line-filled spectrum. This work was carried out by Ron Pitts when he was a summer student assistant for Duane Carbon at Kitt Peak.

ETZEL: I have an observation more than a question. The recent work by Hawkins and Jura (1987 Astrophys. J. **317**, 926.) indicates an interstellar abundance ratio of about 43 for C^{12}/C^{13} from CH^+.

LITTLE: I agree, however the importance of this $^{12}C/^{13}C$ ratio in stellar evolution lies in its changes by interior processing in stars during giant branch evolution and later. Both decreases (due to mixing of ^{12}C into the CNO processing zone due to deep dredging on the giant branch) and increases (due to mixing of "new" ^{12}C on the AGB) are expected.

OPTICAL SPECTROPHOTOMETRY OF THE CHEMICALLY PECULIAR STARS OF THE UPPER MAIN SEQUENCE: AN OVERVIEW AND A PROSPECTUS

Diane M. Pyper

University of Nevada, Las Vegas

Saul J. Adelman

The Citadel

ABSTRACT: We present the results of our spectrophotometric survey of the hot non-magnetic and magnetic Ap stars of the upper main sequence.

1. INTRODUCTION

With R. E. White, we conducted a major program of photoelectric scanner spectrophotometry of magnetic Ap, Mercury-Manganese, and normal B, A, and early F stars in the late 1970's and early 1980's (see, e.g., Adelman and Pyper 1983a,b, Pyper and Adelman 1985, and references therein) at Mt. Wilson, Palomar, and Kitt Peak National Observatories. In this presentation we summarize our findings (given in more detail in Adelman and Pyper 1988) and indicate some directions this type of research could take with the next generation of spectrophotometric instruments such as the one proposed by Hayes, Adelman and Genet (1988). We also consider observations performed by other investigators and discuss briefly the λ Bootis, hot Am, and Am (metallic-line) stars.

The chemically peculiar stars of the upper main sequence are B, A, and early F type dwarfs which exhibit substantial abundance anomalies compared to the Sun. The two major types are the magnetic Ap stars, the spectra of which show observable magnetic fields when examined with Zeeman analyzers, and the non-magnetic Ap stars. The subtypes of magnetic Ap stars, corresponding to early B to early F dwarfs, respectively, are the helium-strong, the helium-weak, the silicon, and the rare-earth or SrCrEu stars. These stars exhibit magnetic, spectral, and photometric variability. The various subtypes are not sharply defined, showing much overlap. The non-magnetic Ap stars lack observable magnetic fields and spectrum and/or photometric variations. Their spectral anomalies were discovered usually on moderate-or high-dispersion spectrograms. The mercury-manganese (HgMn) stars correspond to mid to late B stars while the metallic-line (Am) stars correspond to late A-to early F-stars according to optical colors

A. G. Davis Philip, D. S. Hayes and S. J. Adelman, (eds.)
New Directions in Spectrophotometry 113 - 119

and optical spectrophotometry. Between these two groups are the hot Am stars of early A spectral types. Also known are a few very hot non-magnetic Ap stars like 3 Cen A, the metal-weak λ Boo stars, and the metal-weak A stars whose abundance anomalies are reminiscent of the HgMn stars. How or if these groups are related is a subject of current investigation (see, e.g., Adelman 1986).

In the early 1970's, the typical Ap star investigator told his colleagues that the energy distributions of these stars were quite similar to those of normal stars of similar broad-band colors except for some incidental effects of heavier line blanketing. Indeed Ap stars have broad-band UBV colors like those of normal B and A dwarfs while those which are spectrum variables show complicated (U-B) and (B-V) color variations with rotational period. However, in 1969 Kodaira reported that the energy distribution of one extreme magnetic Ap star, HD 221568, showed three possibly variable broad continuum features in its optical region energy distribution. Subsequent investigations by us and others (especially Leckrone 1973), both in the optical and ultraviolet regions, have shown the earlier point of view to be greatly oversimplified. In fact, many Ap stars have large ultraviolet depressions resulting in backwarming which causes their optical continua to resemble less metal-rich stars with higher effective temperatures.

The initial goals of our spectrophotometric programs were to derive effective temperatures (and surface gravities) particularly for the Ap stars and to explain the complicated variations of the broad-band colors in spectrum variables. A number of normal stars were included for comparisons. We realized that the ultraviolet opacity variations would lead to variable backwarming and that the line and continuum opacities would depend on the local values of the photospheric abundances. But we did not fully appreciate the role that the broad, continuum features might play.

We felt it prudent to compare our spectrophotometry with the best available photometric colors. We followed Breger (1976) in synthesizing the colors of Strömgren uvby photometry from our scans. Since we sampled the energy distributions only for selected wavelength intervals, we had to interpolate the fluxes for the missing wavelength values. Usually we avoided measuring regions affected by Balmer lines. Photometric colors for which one or both of the magnitudes included Balmer lines usually could not be synthesized. Our "colors" thus were constructed from a different sampling of the energy distributions than was the photometry. This was unfortunate, but to have obtained continuous energy distributions of many stars at a number of different photometric phases with reasonable resolution would not have been practical with the scanners then at our disposal. Nevertheless we consider complete optical region observations a necessity for the next generation of spectrophotometric instruments both to study the energy distributions and to synthesize broad, intermediate, and narrow band colors properly. Our progress compared with that of previous investigators was due in large measure to our spectrophotometry including relatively more magnitudes per scan.

2. RESULTS OF OUR STUDIES

We used the normal star data for comparison with those of the Ap stars. If necessary the data were dereddened. For the Ap stars, this was done only if there was a high probability that reddening was present since dereddening Ap star fluxes using the same techniques as for normal stars leads to considerably greater uncertainties. Application of the Q method of Johnson UBV photometry to hot Ap stars, especially the magnetic types, probably does not yield quite the correct results.

We created several indices to investigate the presence of the three optical region broad, continuum features centered at 4200, 5200, and 6300 Å, respectively. As we were interested in the differences relative to the normal stars as a means of detection and as a measure of peculiarity, we also found the values of the same indices for the normal stars and then determined fits through these values as a function of colors such as (b-y) which are correlated with temperature. Ideally one should use an index unaffected by the 5200 Å feature. Thus, for example, (B2-G) is a better index than (B2-V1) in the Geneva system. We then established criteria of presence using the deviations of our normal stars from the mean normal star fits. Some normal stars had larger deviations than the others. This indicates that within our group of normal stars there may be some with slight hitherto unrecognized abundance peculiarities. Elemental abundance analyses of such stars should be performed to see what is happening. The ability to discriminate between stars with normal and slightly peculiar energy distributions should be enhanced in spectrophotometric instruments which obtain the complete spectrum. As we could not justify discarding the most "peculiar" normal stars from our sample, our sensitivity in detecting spectrophotometric anomalies in slightly peculiar stars, such as HgMn stars, was probably somewhat impaired.

Our spectrum scans revealed that the broad continuum features, centered at 4200 and 5200 Å were found in almost all magnetic Ap stars, while the one near 6300 Å was found in only some magnetic Ap stars. Weaker analogs of the first two features were seen in the HgMn stars, but not in the normal B and A stars. These broad features are situated so that they affect the broad-band colors. If they are variable they could cause variations in these colors in addition to variations caused by changes in the continuum flux. With many data points, explaining the variability became a far more complicated problem than we had imagined from broad-band colors. The complexity of the phenomena had been hidden by the simplicity of the broad-band data.

The lack of continuous coverage and the presence of Balmer line wings hampered determining the full wavelength coverage affected by the broad, continuum features. This was particularly acute for the 4200 Å feature, the shortward wing of the 5200 Å feature, and the longward wing of the 6300 Å feature. The 5200 Å feature appears to be composed of several components (see also Maitzen and Muthsam 1980 and Maitzen and Seggewiss 1980) and the centroid of such a feature in the hottest helium variables may be offset from those of the silicon and cool Ap stars. We attempted a classification of 5200 Å feature profiles which resulted in a most interesting grouping of the magnetic Ap stars. We found evidence for profile differences between these features in the magnetic Ap stars and their weaker analogs in the HgMn stars. This subject needs to be reexamined with continuous wavelength coverage at higher resolution.

Even outside the spectral regions most affected by the broad continuum features, the Ap star energy distributions differ from those of the normal stars. This was evident when we tried fitting the spectrophotometry with the predictions of Kurucz's (1979) model atmospheres. For the normal stars, this fitting procedure and comparison of observed and predicted Balmer line profiles leads to the selection of the best model atmospheres. With only the continuum measurements, one finds a locus of effective temperature and surface gravity values which can be further constrained by knowledge of the stellar luminosity.

For the HgMn stars, when we tried to fit their energy distributions with the predictions of solar composition model atmospheres, we often could not find a model whose value of log g was between 3.2 and 4.0, which subsequent analysis has shown to be appropriate (see Adelman 1987). For some HgMn stars such differences might be due to companions, as many HgMn stars are single-lined spectroscopic binaries. But even those stars whose radial velocities have been well observed and found to be constant show this problem. When the predictions of metal-rich model atmospheres are used instead, the fitting problems are only slightly reduced. However, the model atmospheres are probably somewhat underblanketed for the assumed abundances. These considerations suggest that we are dealing with intrinsic differences produced by the anomalous photospheric abundances. The presence of weak analogs of the 4200 and 5200 Å features of the Ap stars supports this conclusion. We must emphasize that the continua of the HgMn stars outside the regions slightly affected by the weak broad continuum features resemble those of the normal stars and they show only slight ultraviolet depression.

For the Ap stars, this lack of ability to fit the entire optical region with the predictions of a single model atmosphere with an appropriate surface gravity becomes more severe. Further, for some of the cooler stars, the continua show substantial differences compared with the normal stars. This is what one expects with severe line blanketing. The flux redistribution from the ultraviolet is enhanced due to the overall enhanced line and continuum opacities, and increased line opacity is also present in the optical region for some stars. This tends to change the shape of the optical flux distribution.

The origins of the broad, continuum features are still a subject of some controversy. Line blanketing, bound-free discontinuities, and autoionization features are still candidates. A similar feature near 1400 Å is due to silicon autoionization (Artru 1986). The optical region features, especially the one near 5200 Å, may have more than one origin.

When one tries to connect the ultraviolet fluxes with the optical observations, there are problems in joining the optical and ultraviolet fluxes. IUE observations, for example, become noisy especially longward of 3000 Å. It has been very difficult to do good spectrophotometry shortward of 3400 Å. To join the observations one can consider using model fluxes as a guide. Observations from both regions need to be extended so that better matches can be achieved.

3. ADDITIONAL COMMENTS

In the course of our program we obtained measurements of a few hot Am stars. Their fluxes in the optical were similar to those of normal stars. This is not very surprising as their major abundances are on the whole only slightly different from solar.

Oke (1967) measured the optical region absolute spectral energy distributions of several λ Boo stars. These are a group of "broad-lined" metal-weak A stars. He found that two of the seven were double stars in which the secondary component contributes a substantial fraction of the light. The other stars resembled normal A-type luminosity class III-V stars. So far the peculiarities of these stars have only been seen in their spectra, except for a broad absorption feature near 1600 Å (Dworetsky 1986). Continuous wavelength coverage in the optical region may reveal other subtle peculiarities.

Lane and Lester (1985, 1987) derived lower effective temperatures from spectrophotometric scans of Am stars than did previous determinations. They argued that the older values were too high due to the combined errors produced by using unblanketed model atmospheres, by the way the energy distributions were analysed and by the use of an older calibration of Vega. But Dworetsky and Moon (1986), on the basis of empirically calibrated ubvy and Hβ photometry, found higher temperatures which are comparable to the older determinations, and larger values of the surface metallicity than are found by Lane and Lester. In response, Lester (1987) suggested that the effective temperatures, surface gravities and metallicities derived for the Am stars from Strömgren and Hβ photometry are all affected by a region near 4760 Å where there is an enhanced brightness in the Am stars compared with normal stars or with model atmospheres. He claims that the atmospheric parameters derived from the full energy distributions are not so distorted, as much more spectral information is used and as all parts of the spectrum are treated equally. Unfortunately, it is impossible to say at present who is correct, as Lane and Lester's spectrophotometry is not sufficiently continuous and as Dworetsky and Moon's method can be affected by the Hβ profile. Dworetsky has also expressed concern that the present line broadening theory is not adequate to calculate the line profile adequately (see Adelman and Lanz 1988). In solving this problem it would also help if we had the fluxes of the models computed with the same bandpasses as the observations.

ACKNOWLEDGEMENTS

We thank our colleague Dr. Richard E. White for his help with this project. SJA's work was supported in part by grants from The Citadel Development Foundation.

REFERENCES

Adelman, S. J. 1986 Astron. Astrophys. Suppl. **67**, 353.
Adelman, S. J. 1987 Monthly Notices Roy. Astron. Soc. **228**, 573.
Adelman, S. J. and Lanz, T., eds. 1988 Elemental Abundance Analyses, Institute for Astronomy of the University of Lausanne, in press.
Adelman, S. J. and Pyper, D. M. 1983a Astrophys. J., **266**, 732.
Adelman, S. J. and Pyper, D. M. 1983b Astron. Astrophys. **118**, 313.
Adelman, S. J. and Pyper, D. M. 1988, in preparation.
Artru, M. -C. 1986 Astron. Astrophys. **168**, L 5.
Breger, M. 1976 Astrophys. J. Suppl. **32**, 1.
Dworetsky, M. M. 1986 in Upper Main Sequence Stars With Anomalous Abundances, C. R. Cowley, M. M. Dworetsky and C. Megessier, eds., Reidel, Dordrecht, p. 397.
Dworetsky, N. M. and Moon, T. T. 1986 Monthly Notices Roy. Astron. Soc. **220**, 787.
Hayes, D. S., Adelman, S. J. and Genet, R. M. 1988 in New Directions in Spectrophotometry, A. G. D. Philip, D. S. Hayes and S. J. Adelman, eds., L. Davis Press, Schenectady, p. 311.
Kodaira, K. 1969 Astrophys. J. Letters **157**, L 59.
Kurucz, R. L. 1979 Astrophys. J. Suppl. **40**, 1.
Lane, M. C. and Lester, J. B. 1980 Astrophys. J. **238**, 210.
Lane, M. C. and Lester, J. B. 1987 Astrophys. J. Suppl., **65**, 137.
Leckrone, D. S. 1973 Astrophys. J. **185**, 577.
Lester, J. B. 1987 Monthly Notices Roy. Astron. Soc. **227**, 135.
Maitzen, H. M. and Muthsam, H. 1980 Astron. Astrophys. **83**, 334.
Maitzen, H. M. and Seggewiss, W. 1980 Astron. Astrophys. **83**, 328.
Oke, J. B. 1967 Astrophys. J. **150**, 513.
Pyper, D. M. and Adelman, S. J. 1985 Astron. Astrophys. Suppl., **59**, 369.

DISCUSSION

LODEN: Do you think this will be a more accurate and sensitive method for the classification of CP stars than the "classical" ones? In the introduction to the Michigan Catalogue, for instance, there is an explicit warning that a majority of the HgMn stars are lost. Do you expect a higher percentage of revelation here? There are corresponding problems with the other CP stars for which there is frequently discordance between classifications presented by different observers.

PYPER: Many HgMn stars do not have the 5200 Å feature, so it would be difficult to identify them using Maitzen's Δa photometry. They are slightly flux deficient in the UV, and that might be used to detect them.

AKE: The marginally peculiar stars are somewhat easier to identify in the UV due to the 1400 Å depression. Brown and Shore have been finding the Bp stars in the Orion OB associations. Unfortunately the 1400 Å dip is not well correlated with the 4200 Å and 5200 Å features.

PYPER: Many of these stars do not have UV observations at the present time.

ADELMAN: The HgMn stars are slightly flux deficient in the UV. The 1400 Å features are seen only in the silicon stars, not in the non-magnetic Ap stars.

SPECTROPHOTOMETRIC STANDARDS FOR HST AND THE UV CALIBRATION OF IUE

Ralph C. Bohlin

Space Telescope Science Institute

ABSTRACT: The Hubble Space Telescope (HST) will require many spectrophotometric and photometric standards for the calibration of the six scientific instruments, which can detect photons from 1050 - 11000 Å. The standard stars and standard fields for this wavelength range are specified in three of the documents of the six part series entitled "STANDARD ASTRONOMICAL SOURCES FOR HST." In particular, the UV flux standards in the 1150 - 3200 Å range are established by a joint NASA/ESA program of observation and recalibration of the low dispersion spectrophotometry from the International Ultraviolet Explorer (IUE). The recalibration includes corrections for the change in IUE sensitivity with time, as well as a new absolute flux calibration based on a comprehensive set of recent observations of the OAO-2 standard stars. The IUE sensitivity loss was as much as 20% in the 2200 - 2300 Å range, while the absolute flux scale changes by more than 10% in the 1400 Å region. The goal of the UV calibration program is a 10% external accuracy and a 2% internal consistency over a brightness range of more than five orders of magnitude.

1. INTRODUCTION

The Hubble Space Telescope (HST) will require many types of standard sources for a diverse range of calibrations to be performed after launch. The scientific instruments are sensitive to a wide range of wavelengths from 1050 to 11,000 Å and encompass a broad range of measurement capabilities including astrometry, photometry, imaging, polarimetry, and spectroscopy. To verify proper operations of each instrument and to provide quantitative calibrations, a diverse range of standard sources and fields are required.

The background material for the choice of calibration targets comes from the individual instrument teams, which have the responsibility for calibrating their scientific instrument during the first six months after launch. At the end of the six-month period, calibration becomes the responsibility of the STScI. Therefore, cooperation and joint planning of calibration efforts is imperative for all concerned parties. Through an intercomparison of all of the team

A. G. Davis Philip, D. S. Hayes and S. J. Adelman, (eds.)
New Directions in Spectrophotometry 121 - 137

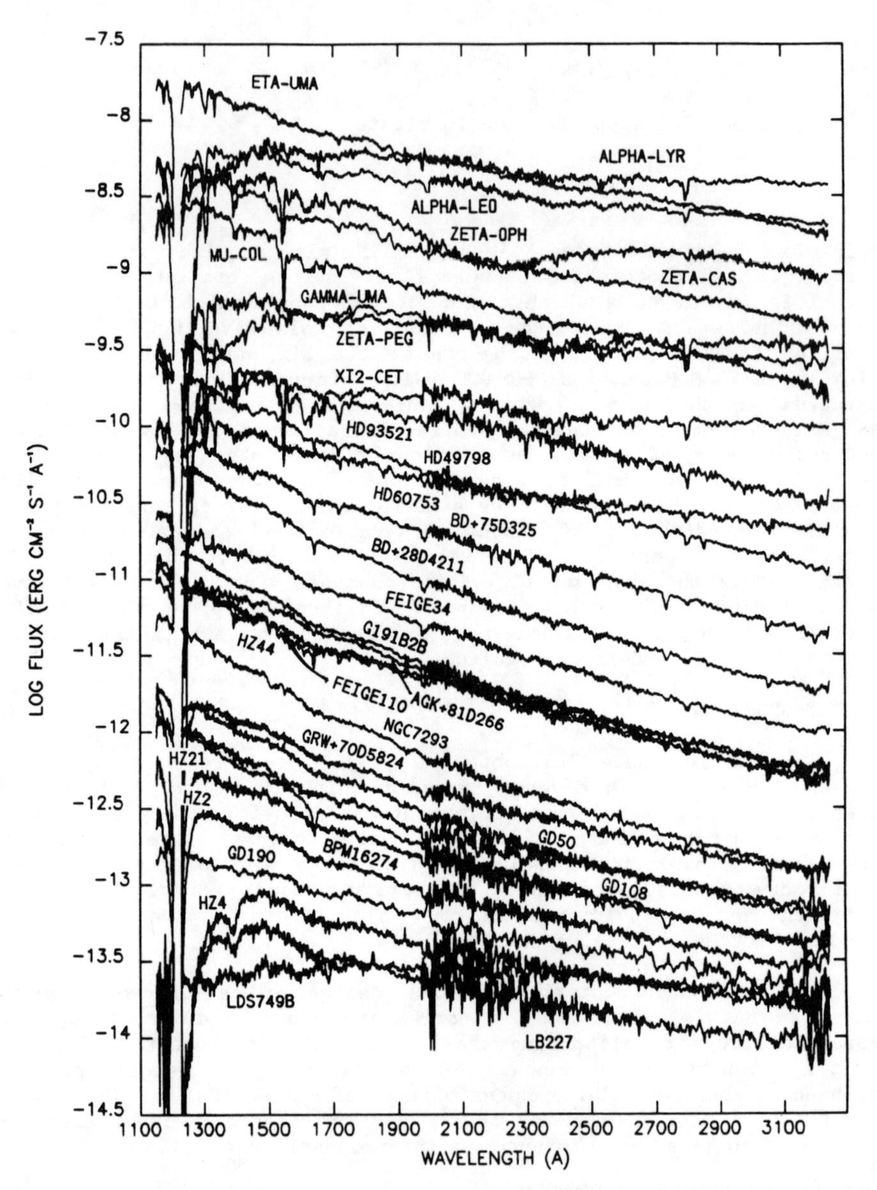

Fig. 1 Preliminary UV spectrophotometry of ST standard stars from the IUE spacecraft. The fainter stars were obtained at the Vilspa site using ESA time, while the brighter targets were observed at GSFC on NASA time.

plans, the centralization of the calibration planning at the STScI leads to the identification of omissions and the establishment of a common basis for calibrations among the different instruments.

Table I contains the list of calibration target categories and the scientists who are contributing. The first name listed is the scientist at the STScI who is leading the effort to obtain a consensus on the requirements for each instrument and to identify a minimal set of targets that satisfy the complete matrix of requirements.

For three of the target categories in Table I, UV, optical and polarimetric, there are requirements for accurate spectrophotometry. J. B. Oke will present some of the work being done on the HST optical spectrophotometric standards, while K. Horne will explain how to use the standard star spectrophotometry to calibrate the HST instruments on a common photometric basis. I am the leader of the effort to obtain a set of UV standards using the IUE spacecraft.

2. REQUIREMENTS FOR UV STANDARDS

Five of the six HST scientific instruments are sensitive in the ultraviolet and will observe UV spectrophotometric standard stars for absolute photometric calibration and other reasons. An important goal is to identify a minimum set of stars that will meet all HST calibration requirements. The aim is to make the most efficient use of HST observing time while minimizing the effort required to collect standard star data in advance of launch. In the magnitude range of 12 to 16 that is accessible to all five instruments, a special effort was made to include high-ecliptic-latitude stars that would be available to HST at any time. New optical spectrophotometry of the proposed standards is being obtained (see Oke 1988). We have avoided reliance on any individual star and have also avoided using strong-lined stars, where practical. The proposed standards cover a large magnitude range to allow linearity checks.

Because of the requirements of HRS and HSP for brighter targets, the complete IUE program is to obtain spectrophotometry of hot stars in low dispersion from the bright limit of second mag to the faint limit of 16 mag with a spacing of about one mag. Fig. 1 shows the current set of IUE spectra for 30 of the hot standard stars. Multiple spectra have been co-added in all cases, but the fainter stars still have relatively poor signal-to-noise ratios, especially in the 2000 - 2500 Å region where the long wavelength IUE cameras have poor sensitivity. A more complete rationale for the UV standard star program is in Bohlin et al. (1987) along with a complete target list and finding charts.

TABLE I

STANDARD ASTRONOMICAL SOURCES FOR HST

CATEGORY	AFFECTED INSTRUMENT	STUDY TEAM MEMBERS
1. UV Spectrophotometric	FOC, FOS, HRS, HSP, WFPC	R. Bohlin, J. Blades, A. Holm, B. Savage, D. Turnshek
2. Optical Spectrophotometric and Photometric	FGS, FOC, FOS, HSP, WFPC	D. Turnshek, W. Baum, R. Bohlin, J. Dolan, J. Koornneef, K. Horne, J. Oke, R. Williamson
3. Wavelength	FOC, FOS, HRS, WFPC	H. Ford, L. Hobbs, D. York, D. Duncan
4. Astrometric	ALL 6	A. Fresneau, R. Bohlin, P. Hemenway, B. Marsden, K. Seidelman, W. van Altena
5. Polarimetric	FOC, FOS, HSP, WFPC	O. Lupie, H. Stockman, R. Allen, A. Code, J. Dolan, D. Turnshek, R. White
6. Spatially Flat Fields	FOC, FOS, HRS, HSP, WFPC	C. Cox, R. Bohlin, R. Griffiths, T. Kelsall

Notes to Table I.

FGS - Fine Guidance Sensor. This instrument was built by the Perkin Elmer Corporation for the primary purpose of fine pointing the Space Telescope. The FGS can also be used for astrometry, and these scientific uses are directed by an astrometry team led by W. Jeffreys at the University of Texas.

FOC - The Faint Object Camera has been developed by the European Space Agency. P. Jakobsen is the Project Scientist, replacing F. Macchetto, who is now the science team leader.

FOS - Faint Object Spectrograph. The principal investigator is R. Härms from the University of California at San Diego and the Applied Research Corporation.

HRS - High Resolution Spectrograph. The principal investigator is J. Brandt of the University of Colorado.

HSP - High Speed Photometer. The principal investigator is R. Bless of the University of Wisconsin.

WFPC - Wide Field and Planetary Camera. The principal investigator is J. Westphal of the California Institute of Technology.

3. INTERNAL CONSISTENCY OF IUE SPECTROPHOTOMETRY

IUE routinely obtains spectra of point sources in a small three arcsec aperture (S), a large 10 by 20 arcsec oval aperture (L), or by

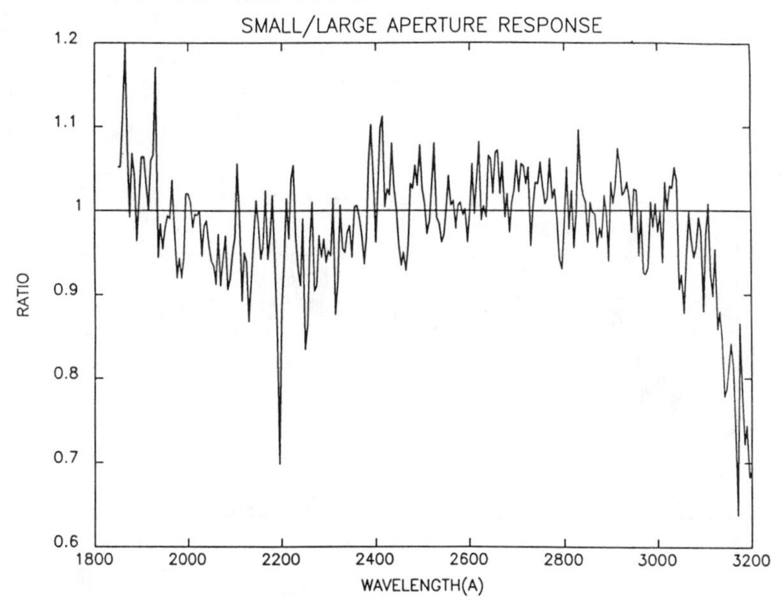

Fig. 2 Ratio of the IUE response to stars in the small aperture (S) to the large aperture (L) for LWR. Since the small aperture is not photometric but transmits about 50% of the starlight, the approximate normalization to unity is achieved by multiplying the actual ratio by 2. A total of 58 S and 77 L spectra of the 5 IUE standards are combined to define the S/L ratio.

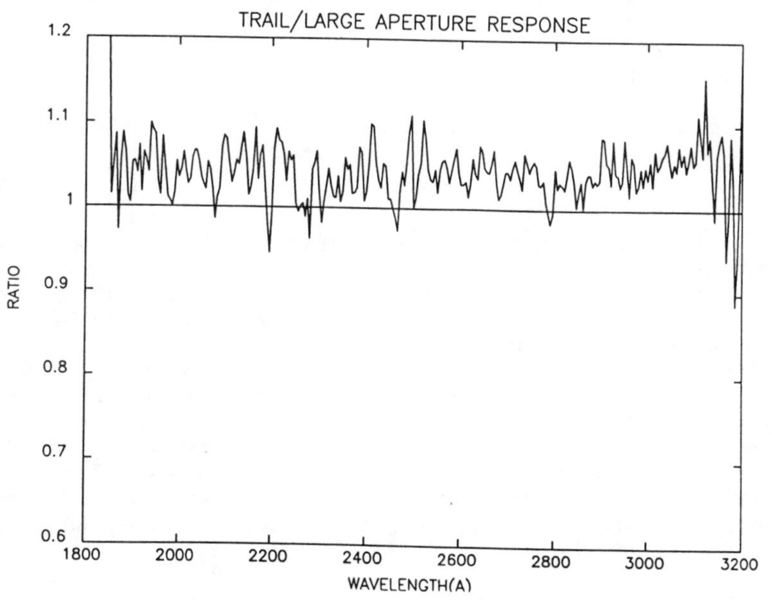

Fig. 3 Ratio of trailed (T) to L response for LWR. A total of 92 T and 272 L spectra define this T/L ratio.

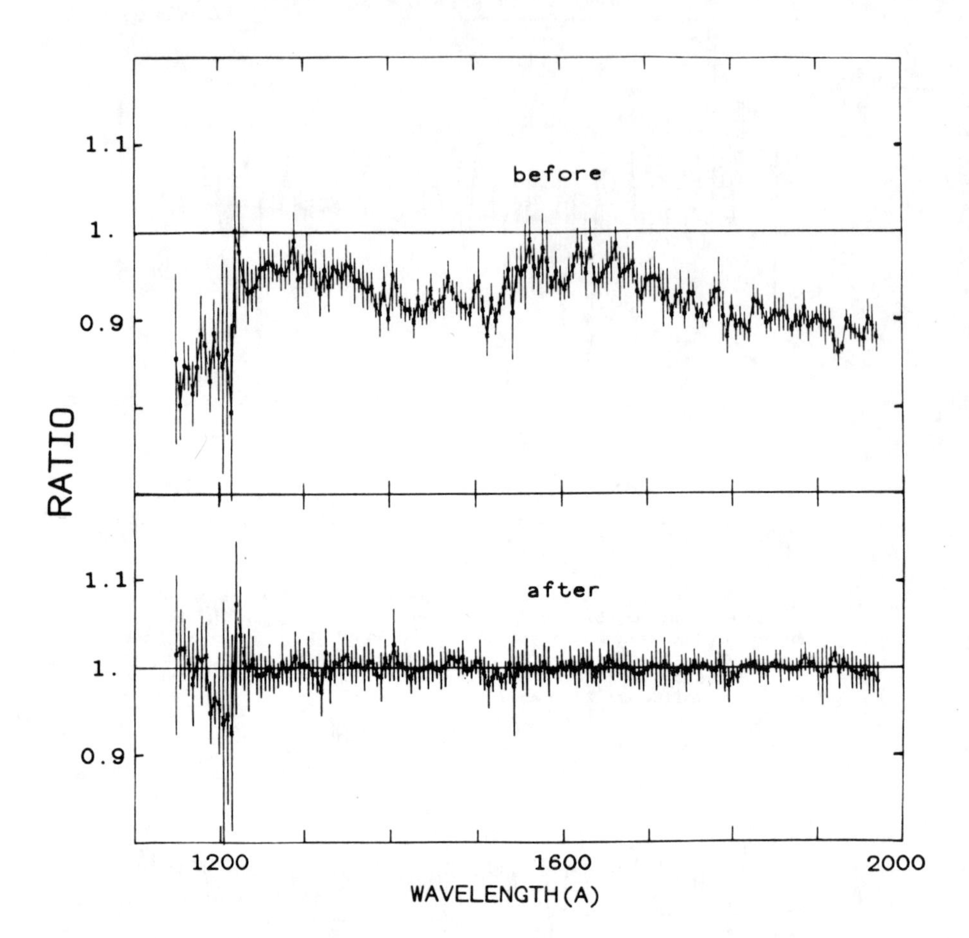

Fig. 4 Mean ratio of the five standard stars in 1985 to the baseline 1978 fluxes in five Å bins, before and after correction for the change in sensitivity. The error bars are the one-sigma scatter among the five individual ratios. The large glitch near the strong Lα absorption is caused by the division of fluxes that approach zero and by small wavelength errors that have a large effect on the five Å bins in the steep sides of Lα. The mean ratio from 1230 to 1970 Å is 0.929 before correction and 1.000 after correction. The rms scatter in the corrected average values is 0.8% and is indicative of the limiting accuracy of the technique, as applied to large samples of spectra. The number of spectra in parentheses follows each star for 1978 and 1985, respectively: HD 60753 (41, 20), BD +75°325 (28, 30), HD 93521 (34, 11), BD +33°2642 (15, 6) and BD +28°4211 (24, 19).

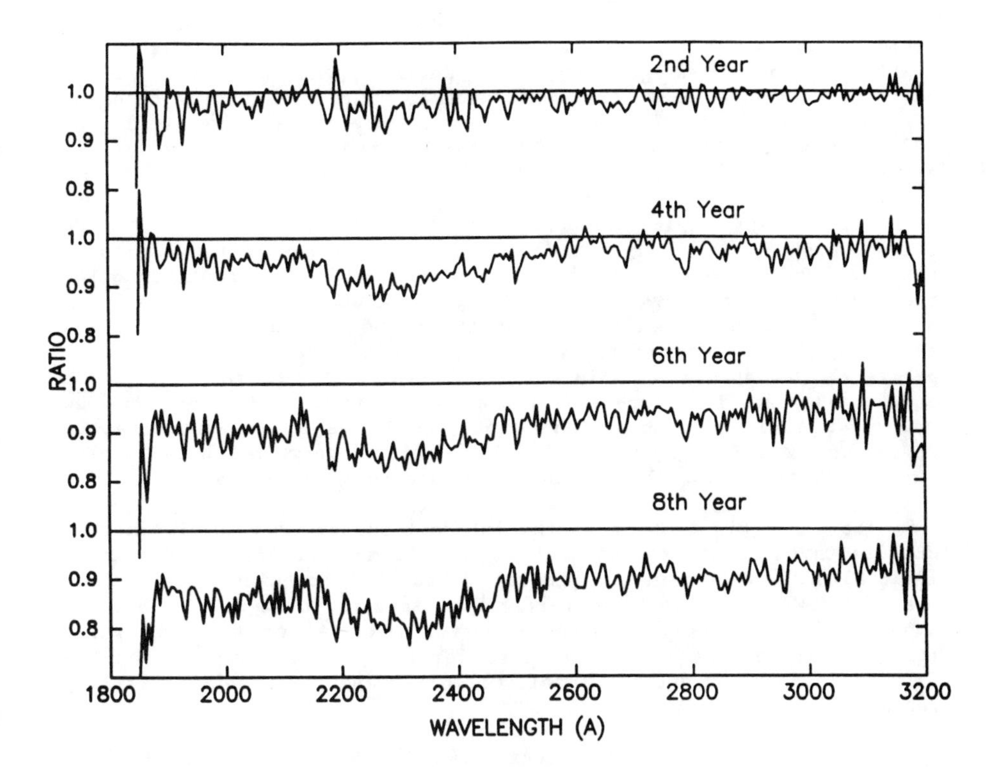

Fig. 5 Uncorrected mean ratios of the standard stars to their baseline fluxes in five Å bins from the 1978.36 - 1979.36 period. A total of 141 spectra are used to define the five baseline average spectra, while somewhat fewer spectra are available in each of the later years.

trailing the point source along the length of the large aperture (T). The early IUE calibration work used an average of spectra from both the large and the small entrance apertures with the assumption that their relative response was gray over the entire spectral range. Since recent studies have shown that the small-aperture sensitivity drops by as much as 30% at 3200 Å in the LWR camera, an error as large as 10% at 3200 Å was present in the IUE calibration when applied to a large-aperture spectrum.

Fig. 2 is a plot of the mean S/L response for LWR. The corresponding T/L plot in Fig. 3 is flat but is systematically above unity by about 4% due to the fact that Panek (1982) underestimated the true length of the large LWR aperture by about 4%. Bohlin (1986) determined the proper corrections to the S and T spectra to bring all low dispersion spectra from SWP and LWR onto the same scale as L spectra. The absolute calibration is specified for the L mode in Bohlin (1986) and is corrected for the initial error. However, one new error that crept into the Bohlin (1986) paper is an incorrect set of T/L ratios for LWR. The proper LWR T/L ratios appear in Table II. The S/L and T/L ratios for the LWP camera remain to be determined.

The other big problem in the internal photometric consistency of IUE data is the change in sensitivity with time. Bohlin and Grillmair (1988a,b) have determined a precise correction technique for five Å bins for SWP and LWR low dispersion spectrophotometry. Fig. 4 demonstrates that the precision is better than 1% for the SWP, while Figs. 5 - 6 show the similar results for LWR. Fig. 7 shows that the LWR correction technique of Clavel, Gilmozzi, and Prieto (1988) is less precise.

The spectra used to construct Figs. 4 - 6 are averages of L, S, and T spectra with the appropriate S/L and T/L correction. The combination of these aperture corrections and the corrections for the change in sensitivity with time demonstrate that a large ensemble of IUE data can be corrected to better than 1%. A more relevant test of the accuracy for the HST standard star spectra is shown in Figs. 8 - 9, where spectra taken with the same technique and observed in the same time frame as the bulk of the HST standard star data are compared to the same 1978 - 9 baseline as in Figs. 4 - 6. Even for the HST data, where heavier exposures are used to bring up the signal for regions of low sensitivity, the set of four stars is again within 1% of the baseline (Fig. 8). However, one of the four stars, BD +75°325, has a flux that is systematically high by 1.8% (Fig. 9). Thus, the limiting internal consistency of the set of HST standard star spectra may be as good as 2%.

The LWP camera shows much less loss in sensitivity than LWR; and a correction for the change in sensitivity over the relevant years of 1983 - 1988 may not be necessary to achieve the 2% internal accuracy goal (cf. Sonneborn and Garhart 1987).

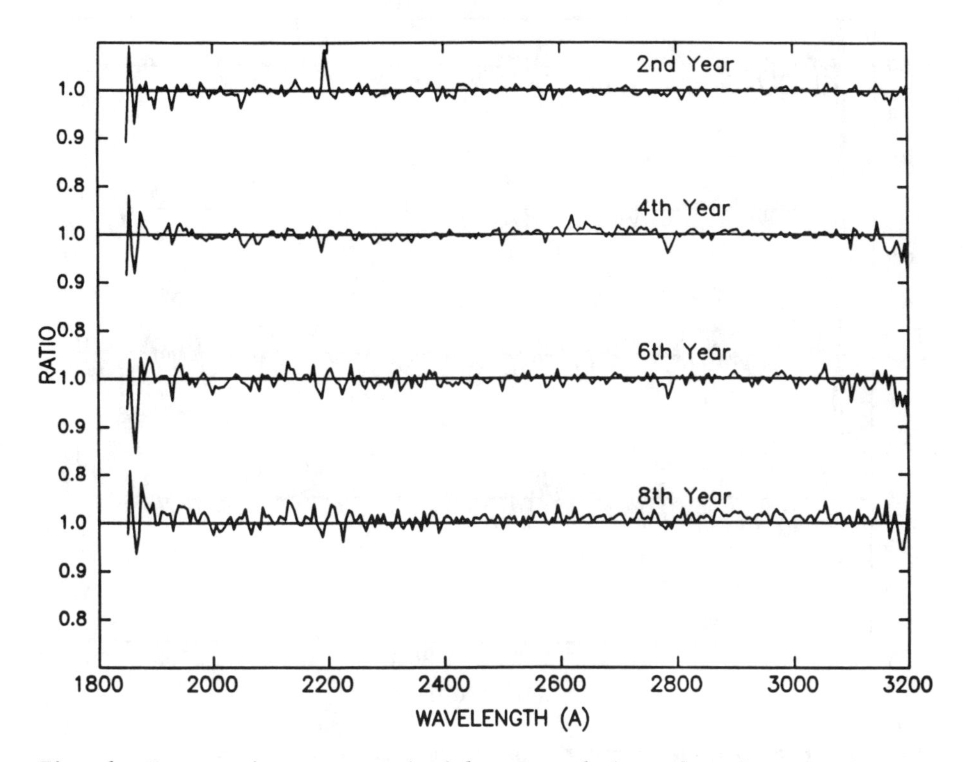

Fig. 6 Mean ratios corrected with our technique for the same data as in Fig. 5. Most of the larger glitches are caused by the bright spot near 2200 Å and the large aperture reseaux at 1855, 2055, 2580 and 2785 Å

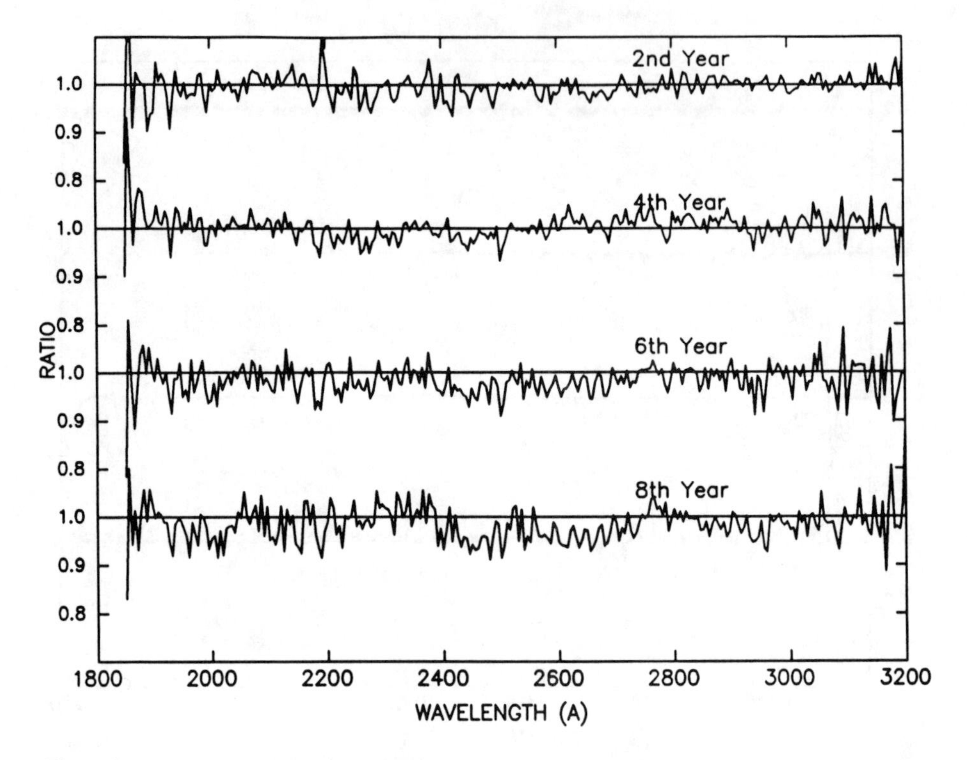

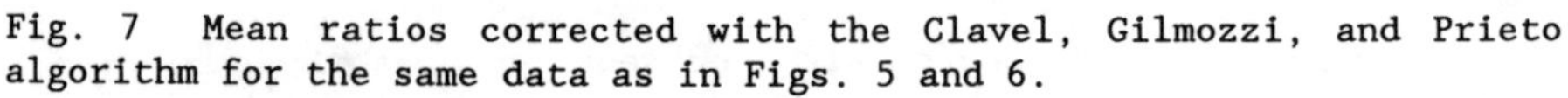
Fig. 7 Mean ratios corrected with the Clavel, Gilmozzi, and Prieto algorithm for the same data as in Figs. 5 and 6.

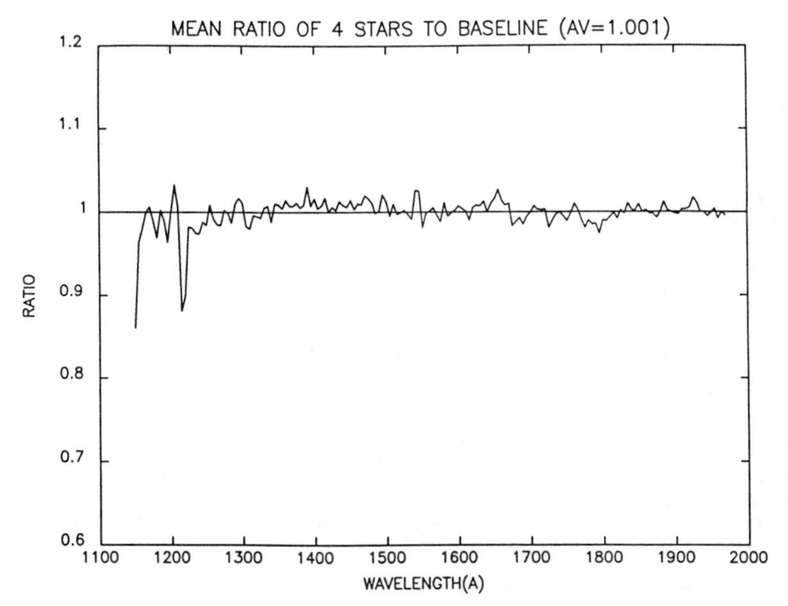

Fig. 8 Corrected mean ratios for 4 stars, HD 60753, BD +75°325, HD 93521, and BD +28°4211, to their baseline fluxes in the case where the numerator fluxes are observed with the same technique and in the same time frame as the bulk of the HST standard star observations.

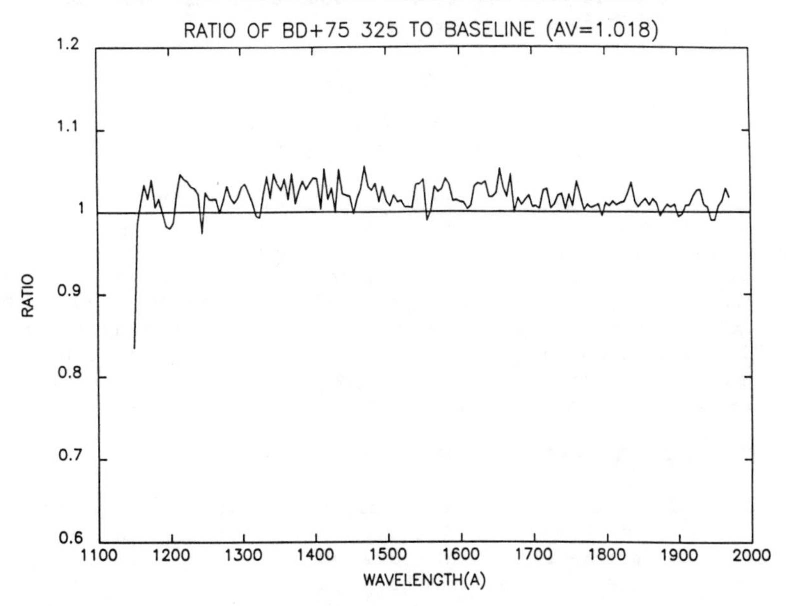

Fig. 9 Ratio of BD+75°325 to its baseline flux for the same case as Fig. 8. The 1.8% systematic error may be indicative of the errors in the internal consistency of the HST UV standard star fluxes.

TABLE II

CORRECTED[a] TRAIL SENSITIVITY FOR LWR

λ (Å)	T/L[b]	λ (Å)	T/L[b]	λ (Å)	T/L[b]
1850	1.058	2350	1.038	2850	1.043
1900	1.056	2400	1.039	2900	1.043
1950	1.053	2450	1.040	2950	1.044
2000	1.051	2500	1.040	3000	1.046
2050	1.049	2550	1.040	3050	1.053
2100	1.047	2600	1.041	3100	1.063
2150	1.045	2650	1.041	3150	1.068
2200	1.043	2700	1.041	3200	1.070
2250	1.041	2750	1.042	3250	1.070
2300	1.039	2800	1.042	3300	1.070
				3350	1.070

a. The values of T/L for LWR in Bohlin (1986) are wrong and should be replaced by the values in this Table.

b. Ratio of trailed response to point sources in the large aperture. The absolute calibrations are related by

$$S\text{-}1(\text{Trail}) = S\text{-}1(\text{LAp})/(T/L).$$

The values of T/L are for trailed exposure times computed using lengths for the large apertures of 21.4 for SWP and 20.5 for LWR (Panek 1982)

Another check on the photometric consistency is to compare ANS photometry to the mean IUE fluxes averaged over the ANS bandpasses as shown in Fig. 10. The one sigma scatter of the points about the mean are 2.4% (1550 Å), 2.5% (1800 Å), 4.0% (2200 Å), and 4.6% (2500 Å). The increase in scatter in the longer wavelength bands is probably due to a systematic difference between the LWP and LWR flux scales; all five LWR fluxes fall below the rest of the points in the 2200 Å band. Once again, the expectation of a 2% internal consistency is not unreasonable, even over the entire dynamic range of Fig. 1.

4. ACCURACY OF THE ABSOLUTE IUE FLUXES

Once the photometric repeatability of the low-dispersion IUE spectrophotometry is established, the accuracy of the IUE fluxes depends on the precision of the fluxes of the reference standard stars used to calibrate IUE and on the fidelity of the transfer of the calibration.

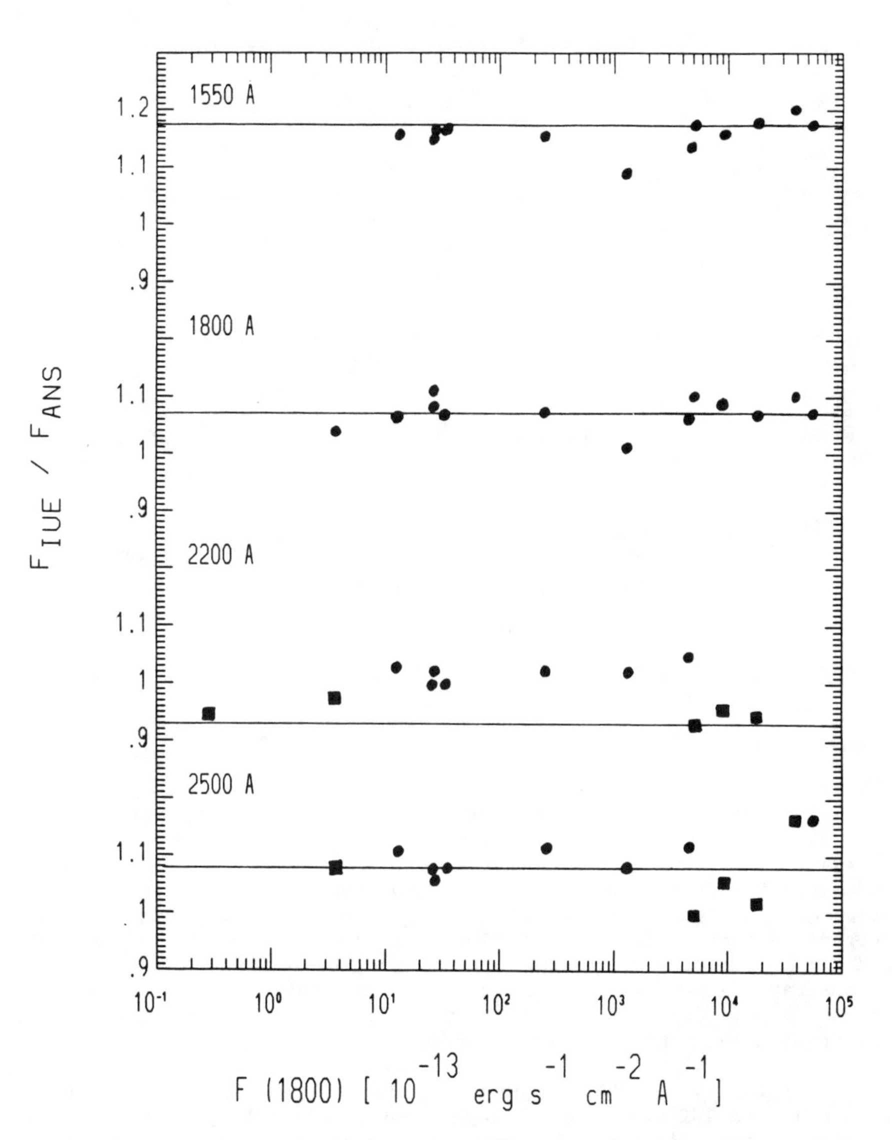

Fig. 10 Ratio of the mean IUE to ANS flux in the four ANS bands as a function of flux that ranges over five decades in brightness. The horizontal lines for each bandpass are at the value of the systematic difference in the absolute flux scales of the two satellites. The squares in the two long wavelength bands are based on LWR spectra only, while LWP fluxes dominate the other long wave length data points.

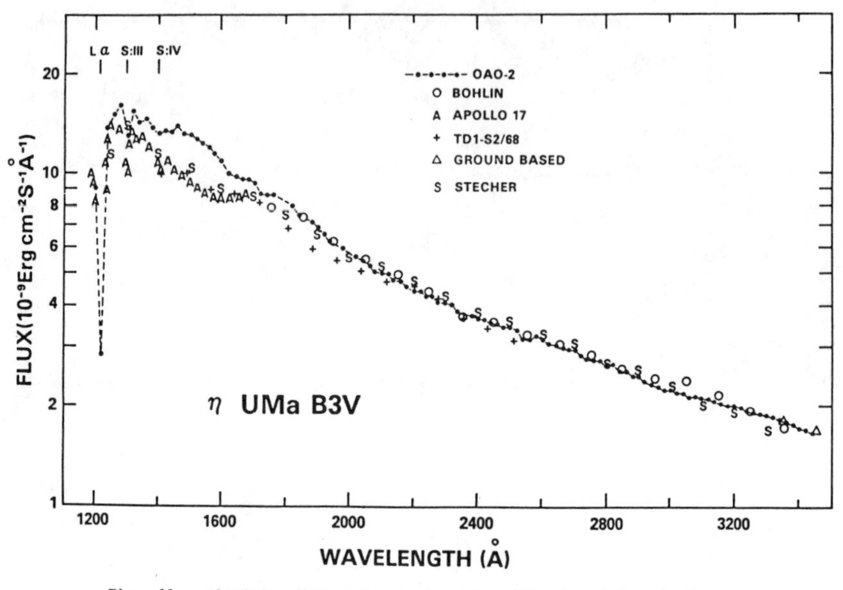

Fig. 11 Absolute flux measurements of η UMa by independently calibrated experiments.

The flux for η UMa and for the other OAO-2 and TD-1 reference standards is specified by Bohlin et al. (1980) and by Bohlin and Holm (1984) to an estimated accuracy of 10%. Fig. 11 shows all the independently determined UV flux measurements for η UMa. This plot was used to draw-in a guess for the true η UMa flux, which defined correction factors for the published flux catalogs, including the OAO-2 and TD-2 reference standards used to calibrate IUE. Bohlin (1986) demonstrated that IUE fluxes are in reasonably good agreement with ground based fluxes at the longest wavelengths and with Voyager spectrophotometry at the shortest wavelengths.

Originally in 1979 - 1980, the transfer of the reference-star flux-scale to IUE was done to an inadequate precision, because the IUE spectra of the bright OAO-2 standards were too few and of poor quality. In the 1984 - 1985 time frame, the IUE project obtained a vastly superior set of spectra of the six OAO-2 standard stars for the purpose of a comprehensive recalibration. The change in the IUE absolute flux calibration that is indicated by the upgraded set of spectra is shown in Figs. 12 - 13 for SWP and LWR. Some similar work remains to be done on the LWP absolute calibration.

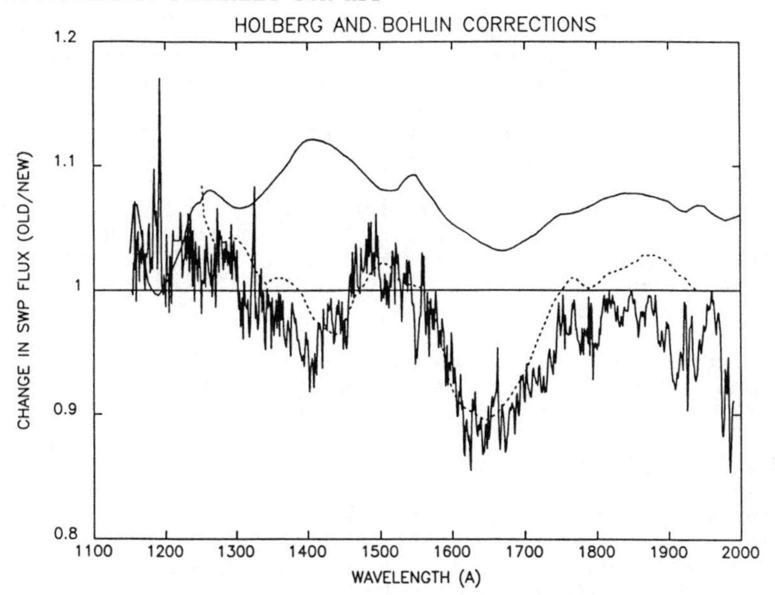

Fig. 12 Change in IUE SWP fluxes resulting from a comprehensive recalibration (upper smooth line) in comparison with the change required to fit theoretical models of the white dwarf G 191B2B (jagged line) or BL Lac objects (dashed line). The work on white dwarfs is from J. Holberg (1988), while Hackney, Hackney, and Kondo (1982) have produced the dashed line.

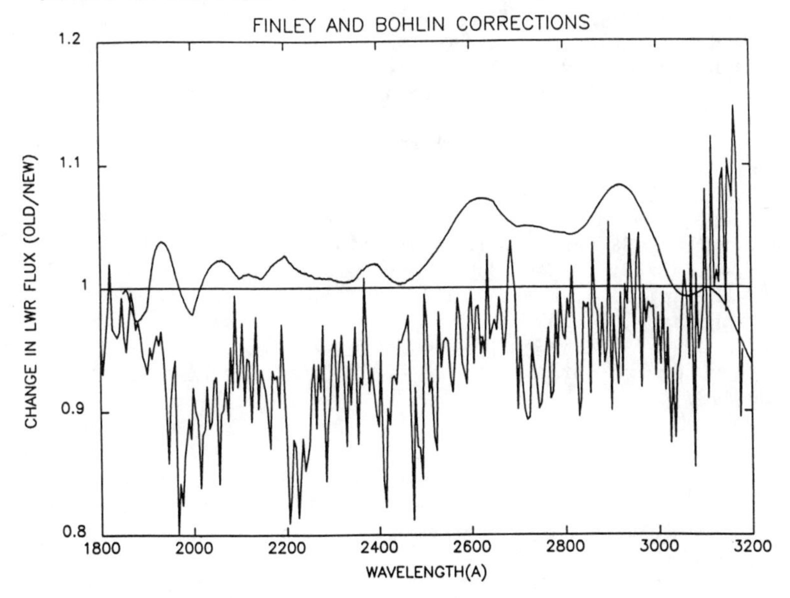

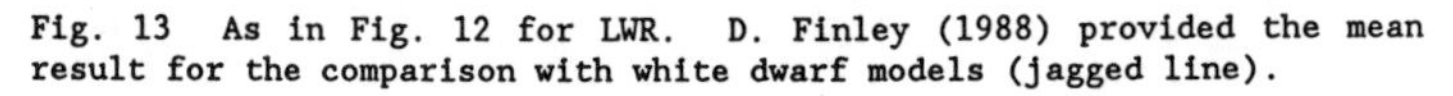
Fig. 13 As in Fig. 12 for LWR. D. Finley (1988) provided the mean result for the comparison with white dwarf models (jagged line).

Beyond the steps outlined above, further improvements in the IUE absolute flux scale could come from rocket experiments to make precise stellar flux measurements or from comparison with theoretical flux distributions of white dwarf stars. New rocket flights do not seem to be in the cards; but the preliminary results from comparisons of white dwarf models with IUE data by J. Holberg and D. Finley are shown in Figs. 12 - 13. Unfortunately, the recalibration and models do not agree, but there is some similarity in the shapes of the curves. Perhaps, the mean level of the theoretically-based calibrations would be closer to the level of the recalibration, if slightly cooler model flux distributions are used.

I thank the staff at the Goddard Space Flight Center, who made the data available and are always ready to cheerfully discuss any technical detail of IUE data.

REFERENCES

Bohlin, R. C. 1986 Astrophys. J. **308**, 1001.
Bohlin, R. C., Blades, J. C., Holm, A. V., Savage, B. D. and Turnshek, D. A. 1987 Standard Astronomical Sources for HST: 1. UV Spectrophotometric Standards, STScI.
Bohlin, R. C. and Grillmair, C. J. 1988a Astrophys. J. Suppl. **66**, 209.
Bohlin, R. C. and Grillmair, C. J. 1988b Astrophys. J. Suppl. in press.
Bohlin, R. C. and Holm, A. V. 1984 NASA IUE Newsletter **24**, 74 (=1984, ESA IUE Newsletter **20**, 22).
Bohlin, R. C., Holm, A. V., Savage, B. D., Snijders, M. A. J. and Sparks, W. M. 1980 Astron. Astrophys. **85**, 1.
Clavel, J., Gilmozzi, R. and Prieto, A. 1988 Astron. Astrophys, **191**, 392.
Finley, D. 1988, private communication.
Hackney, R. L., Hackney, K. R. H. and Kondo, Y. 1982 in Advances in UV Astronomy: Four Years of IUE Research, Y. Kondo, J. M. Meade and R. D. Chapman, eds., (NASA CP-2238), p. 335.
Holberg, J. 1988, private communication.
Oke, J. B. 1988 in New Directions in Spectrophotometry, A. G. D. Philip, D. S. Hayes and S. J. Adelman, eds., L. Davis Press, Schenectady, P. 139.
Panek, R. J. 1982 NASA IUE Newsletter **18**, 68.
Sonneborn, G. and Garhart, M. P. 1987 NASA IUE Newsletter **33**, 78.

DISCUSSION

PETERS: How confident are you that BD +75° 325 and η UMa are not variable? Are there Voyager data on these stars and if so have they been scrutinized for variability? It is possible that long-term, low-amplitude variations may exist, but these would be difficult to spot.

BOHLIN: Variability in the UV is studied by IUE in the 1150-3200 Å region. Because of the extensive set of observations of BD +75° 325 by IUE, the limit on the long term secular change is in the range of 1-2%, while the short term limit is in the 3-4% range. The limits on the variability for brighter stars is somewhat less stringent, since there are typically ~ 20 observations compared to ~ 200 for the five IUE standard stars. Of the six stars with OAO-2 data that are used to provide the absolute flux reference for IUE, only μ Col with a discrepancy of ~ 6% exceeds the typical internal consistency of ~ 3% between IUE and OAO-2.

PETERS: Recent high dispersion, high S/N (mostly Reticon) data obtained by M. Smith and others have revealed variability in the line profiles in several early B stars. One wonders if the continua in these stars are also variable. Unfortunately, most early B spectrophotometric standards have broad lines, thus rendering searches for line variability (which might serve as an indicator of continuum variability) difficult.

TAYLOR: There have been suspicious-looking but not definitive reports in the literature of a small-amplitude variation of η UMa. I mentioned those in my 1984 paper on bright standards (see Taylor, Astrophys. J. Suppl. **54**, 259).

BOHLIN: Yes, more data to monitor the stability of the bright standards is needed.

HAYES: There have been "rumors" of low-level variability of most of the spectrophotometric standards, based on evidence going back (in some cases) to the beginning of photoelectric photometry. There is no question that it would be valuable to monitor the bright spectrophotometric standards, and this would be an ideal project for an automatic photoelectric telescope. Unfortunately, the monitoring of stars for variability is hard to get funding for and doesn't (usually) give any exciting results. It is important, though.

TURNSHEK: I'd like to point out that Arlo Landolt is currently monitoring all of the HST spectrophotometric standards that are fainter than about 8th magnitude.

OPTICAL SPECTROPHOTOMETRY OF HST STANDARD STARS

J. B. Oke

Palomar Observatory, California Institute of Technology

ABSTRACT: The Double Spectrograph on the 5 m Hale Telescope has been used in slitless form to measure absolute spectral energy distributions of a selection of stars chosen to be standards for the Hubble Space Telescope. The 22 stars are mainly DC white dwarfs at magnitudes of 14 to 16 and hot subdwarfs between magnitudes 9 and 14. Essentially all objects have been observed several times in each of three observing runs. All reductions have been completed and the agreement among the observations on different runs is excellent. Eight stars need further ultraviolet observations and these will be obtained in July 1988. The observations cover the spectral range from 3200 to 9400 Å. The spectral resolution is about 4 Å in the blue and 10 Å in the red.

1. INTRODUCTION

There is an obvious need to have a series of spectrophotometric standard stars which are well calibrated in the optical and in the ultraviolet. With such standards, observations made with HST can be compared directly with observations made from the ground; the standards provide the necessary continuity near 3200 Å where the two kinds of observations overlap.

There are a number of factors which have to be taken into consideration when choosing standard stars. 1. For ground based work the standards need to have declinations so that at least some of them can be observed from both the northern and the southern hemispheres. 2. The standards should be distributed around the sky so that at least a few of them can be observed at any time during the night and at any season. 3. Some of the standards should be at magnitudes 15 to 16 so that they can be observed easily with pulse counting systems on large telescopes. 4. Some of the standards should be hot so that there is adequate flux in the UV so that accurate HST observations can be made quickly.

Using the above criteria 22 stars were chosen to be spectrophotometric standards. Seven of these have magnitudes between 14 and 16 and are all DC white dwarfs with one exception; the exception is an sdF star. The remaining 15 stars are between magnitudes 7 and 14

A. G. Davis Philip, D. S. Hayes and S. J. Adelman, (eds.)
New Directions in Spectrophotometry 139 - 144

and are mainly of spectral type sdO, DA, and DB. The stars are G 158-100, HZ 4, G 191B2B, G 193-74, BD +75°325, Feige 34, HD 93521, HZ 21, Feige 66, Feige 67, G 60-54, HZ 44, Grw 70°5824, BD +33°2642, G 193-31, G 24-9, BD +28°4211, BD +25°4655, NGC 7293, LTT 9491, Feige 110 and GD 248.

A program was begun at Palomar Observatory in 1985 to obtain absolute spectral energy distributions for these 22 stars plus a few other stars to be used as standards for high speed photometry. To obtain absolute spectral energy distributions of sufficiently high quality so they can be used as reliable standards is an enormous task for such a large number of stars, even when a large telescope is used.

The program is described below along with the results to date.

2. TECHNIQUES

It was decided that the Double Spectrograph on the 5 meter Hale telescope was the only instrument which would have suitable spectral resolution and be fast enough to carry out the program. Since this instrument uses CCD detectors it was first necessary to verify that CCDs had the necessary linearity and stability for the task. Linearity of Texas Instruments 800x800 thinned back-illuminated CCDs had been previously tested and was known to be excellent. Stability was tested by making observations of our regular cool subdwarf standard stars HD 19445, HD 84937, BD +26°2606, and BD +17°4708 (Oke and Gunn 1983) on photometric nights and with an 8 to 10 arcsec wide slit. Observations were made at various zenith distances. By removing the atmospheric extinction the instrument can be calibrated absolutely. By comparing the calibrations from different observations a measure of the stability can be obtained. This measure does include any inaccuracy in the extinction law as well as inaccuracies in the standard star fluxes themselves. The results of these tests verified that the stability was at a level of 1 percent over the spectral range from 3500 to 10000 Å. Below 3500 Å the data were often too noisy to be definitive and as will be discussed below there are in fact some problems between 3200 and 3500 Å. Above 3500 Å the data are comparable in quality to data obtained with the older Multi-channel Spectrometer (Oke 1969) which used photomultiplier tubes and pulse-counting techniques.

Texas Instruments 800x800 CCDs which have been kept in a vacuum dewar have very poor quantum efficiency below 4000 Å even after applying techniques such as uv flooding. A method for enhancing violet and ultraviolet sensitivity by over an order of magnitude was developed only about one year ago (Oke et al 1988).

The wavelength range which could be covered by the observations was governed by the spectrograph design and the size of the CCDs being used. A dichroic with a switch-over wavelength nominally at 4800 Å was used. The blue side of the spectrograph was set up to cover the wavelength range from 3150 to 4800 Å. The red side went from 4700 to

9500 Å. With a 1 arcsec diameter image the spectral resolution in the blue is about 4 Å and in the red it is nearly 10 Å. Above 8800 Å the second order at 4400 Å begins to contribute light since the dichroic transmits slightly down to 4400 Å. This contamination has not yet been removed from the observations but can be done fairly accurately.

Extinction was calculated using a standard extinction table for Palomar which is based on the recipe suggested by Hayes and Latham (1975). This table has been used for many years and appears to be quite accurate. The extinction in the A- and B-bands is done with a square root law at fairly low resolution. Water vapor in the 8200 to 9700 Å spectral region is also done with a square root law; the amount of water vapor can be adjusted to provide a "best" fit.

One of the problems with a double instrument is to decide how to combine the observations from the two halves of the spectrograph in the spectral region where the dichroic switches and the spectra overlap. In our particular case the amount of light above 4700 Å is dropping very rapidly on the blue side; the drop is so fast that even tiny wavelength errors produce very large errors in the absolute flux. In the red side the sensitivity remains satisfactory down to 4700 Å. In practice the blue side should be used up to 4700 to 4720 Å and the red side should be used above that wavelength. Actual observations in this spectral region are shown by Oke (1988).

3. OBSERVATIONS

All the observations have been made with the Double Spectrograph using an 8 or 10 arcsec wide slit. With such a wide slit the wavelengths are affected by the precise position of the star in the slit. This can lead to uncertainties of a few Ångstroms in the wavelengths. These uncertainties cause problems only where there are sharp features (see below) or where sensitivity is changing rapidly as at the red end of the blue spectrum (see above). The unknown stars are observed several times near the meridian. The cool subdwarf standard stars are observed many times during the night and at various zenith distances so the extinction can be checked. Observations are usually reduced one night at a time with the standard star calibrations being carefully compared with each other to insure consistency throughout the night. Observations were made in two runs in 1985, two in 1986, and one in 1988. A final run is scheduled for July 1988. Most of the nights were photometric with the exception of the nights in January 1988.

4. RESULTS

Absolute spectral energy distributions have been obtained for each star on 3 to 6 nights and usually in three different seasons.

Errors can be calculated at three different levels. First, one can calculate errors from the photon statistics. These must always be minimum values. Second, individual observations made on a given night, usually with only a short time interval between them, can be compared. These give errors which reflect photon noise as well as small variations in extinction. Third, observations made from night to night or from run to run can be compared. These yield errors which include those already mentioned plus any errors in the standard stars used. The second and third levels of errors should be practically the same if the observations are made under photometric conditions. The first two errors have already been calculated but the third ones have not yet been done in any detail. However, visual comparison of energy distributions determined on different runs show no systematic differences. This means that errors determined at levels two and three will very probably be consistent with each other. Overall energy distributions should be accurate to about 1 to 2 percent over most of the spectral range observed.

There are, however, a number of problems, some very minor, some fairly major, which have turned up.

1. In the neighborhood of the atmospheric A-band the energy distributions are poor. One of the reasons is that in slitless spectra there is some uncertainty in the wavelength. Thus a sharp feature such as the A-band can show a P-Cygni-like profile. There is also very little light at the band head and extinction coefficients are very large. The A-band covers at most 100 Å so that interpolation over this interval is safe. Extinction at the B-band is much smaller than in the A-band and problems are rarely seen.

2. Corrections for water-vapor absorption between 8900 and 9700 Å leave residual errors in the energy distributions which can be as large as 0.05 mag.

3. Above 8800 Å the second order spectrum comes in. This can be removed readily since the ratio of the raw counts in the first order in the blue side and in the second order in the red side of the spectrograph can be measured once and for all, and the counts in the first order in the blue side are always obtained. This procedure has not yet been carried out.

4. The CCD used for the blue observations in 1988 has a few pixel-size flaws which are not corrected with the flat fields. They are not in the spectral regions where the 1988 observations are being used (see (6) below).

5. The most serious problem encountered is one that occurs below 3500 Å. In this spectral region the CCD appears to have an oscillating sensitivity with peak-to-peak variations of about 20 percent of the signal and with the peaks spaced about 100 Å apart. This characteristic is seen in both CCDs used, even when there is an order of magnitude difference in quantum efficiency. So far all attempts to remove the oscillations by flat-field measurements have proved

unsuccessful at least in some of the observations.

6. Below 3800 Å the 1985 and 1986 data are sometimes noisy because of low detected photon counts. In the 1988 observations the photon counts in this region are larger by a factor 20. The 1988 data will be combined with the earlier data to give higher signal-to-noise results.

7. Observations of some of the DC stars show an apparent broad weak Hα line in absorption. Since it is unlikely that this line is real it may be caused by slightly incorrect absolute fluxes in the line wings of the cool sub-dwarf standard stars. This will be investigated further.

5. SUMMARY

Observations of 22 HST spectrophotometric standard stars have been made over the spectral region from 3200 to 9400 Å. Over most of this range the accuracy relative to the cool F-type subdwarf standard stars is about 1 to 2 percent. Below 3500 Å and above 8800 Å the uncertainties are somewhat larger; comparable uncertainties in these same spectral ranges still exist in the absolute calibration of α Lyrae.

REFERENCES

Hayes, D. S. and Latham, D. W. 1975 Astrophys. J. **197**, 593.
Oke, J. B. 1969 Pub. Astron. Soc. Pacific **81**, 11.
Oke, J. B. and Gunn, J. E. 1983 Astrophys. J. **266**, 713.
Oke, J. B. 1988 in Instrumentation for Ground-Based Optical Astronomy, Santa Cruz Workshop 1987.
Oke, J. B., Harris, F. H., Oke, D. C. and Wang, Delong 1988 Pub. Astron. Soc. Pacific. **100**, 116.

DISCUSSION

HAYES: Have you eliminated an extinction effect due to the ozone band as an explanation of the "waves" in the spectra at the short-wavelength end?

OKE: Yes. These bands do not coincide in wavelength with the ozone bands. The bands also do not correlate with sec z.

WEAVER: Early calibration spectrophotometry used bandpasses which were larger than the typical Paschen spacing. How did you calibrate in this region?

OKE: The F-type subdwarf standards have only very weak Paschen lines so this problem is minimal. It is necessary to define fluxes in bands which are midway between Paschen lines. The same technique is used for the Balmer line region below 3900 Å.

KURUCZ: You said that the flat field shows the waviness at about 3200 Å. But you flat-fielded, so shouldn't it have been removed?

OKE: This is true. The flat fields should have eliminated the variable response which produces the bands. The basic problem is that the flat fields do not eliminate them completely.

APPLICATION OF SYNTHETIC PHOTOMETRY TECHNIQUES TO SPACE TELESCOPE CALIBRATION

Keith Horne

Space Telescope Science Institute

ABSTRACT: The large number of novel passbands available with the Space Telescope has stimulated the development of new approaches to photometric calibration. We discuss calibration algorithms based on synthetic photometry that place photometric and spectrophotometric data on precisely the same system. This unified approach should be more accurate than transformations to a standard system with color terms to account for differences between instrumental and standard passbands.

1. INTRODUCTION

The Hubble Space Telescope (HST) is a versatile observatory comprising six Science Instruments with capabilities for spectrophotometry and photometry in wide to narrow passbands covering the optical and ultraviolet. In earlier talks Bev Oke and Ralph Bohlin presented optical and ultraviolet data on some of the spectrophotometric standard stars that will be available for calibrating HST observations. This talk will focus on calibration algorithms. Synthetic photometry plays a crucial role here in bridging the resolution gap between photometric and spectrophotometric data, and in linking the HST data to ground-based observations.

I will begin in Section 2 by showing off the Passband Generator that has been developed to keep track of the photometric calibration of all the observing modes of the HST. Section 3 will then present a calibration algorithm designed to ensure that photometric and spectrophotometric data are calibrated to precisely the same scale. The passband reconstruction problem is discussed in Section 4, which presents results of some simulation trials of passband reconstruction by the maximum entropy method.

2. GENERATING HST PASSBANDS

HST calibration begins well before launch with data taken in laboratories on the ground. For every filter, grating, mirror, window and detector that will be going into space, data are available on the transmission, reflection or detection efficiency as functions of

A. G. Davis Philip, D. S. Hayes and S. J. Adelman, (eds.)
New Directions in Spectrophotometry 145 - 158

wavelength. The spectral response of the HST in a given observing mode can be found by multiplying together the passbands of all the optical components encountered by photons on their way to detection. Fig. 1 illustrates the diverse array of available HST passbands with a selection from the Wide Field/Planetary Camera (WF/PC). Across all six science instruments there are about 1000 components, and these can be combined in over 10^9 different observing configurations.

The impracticality of storing such a large number of passbands prompted the development of the Passband Generator, a software tool that assembles the passband for any observing mode upon request. We needed a general way to represent the structure of a Science Instrument, which defines the permitted combinations of component passbands that form the valid mode passbands. Each instrument is represented in the form of a directed graph. Fig. 2 depicts the graph structure of the WF/PC, which closely parallels its optical layout. Each box represents one of the WF/PC components, a filter, grating, mirror, window or detector, and links between the boxes represent the paths that photons can follow through the instrument. Every possible WF/PC observing mode corresponds to a specific path through this WF/PC graph. The graph is stored in the computer as a list of links, each carrying a pointer to the appropriate component passband file, and an associated keyword. Any instrument composed of linear optical elements can be represented in this way.

All of this information has been assembled into a passband database that is accessed through the Passband Generator. To retrieve a particular HST passband, you would furnish the Passband Generator with a couple of keywords, for example 'WFC F336W'. The Passband Generator uses your keywords to trace a path through the graph, multiplies together the component passbands it encounters along the way, and returns to you the mode passband evaluated on your wavelength set. The keywords are mnemonic for the convenience of human users, but they can also be constructed from telemetry to generate calibration data as required for pipeline processing of the HST data. The Passband Generator is implemented as a FORTRAN subroutine and has also been installed as a task in the simulators package of the Space Telescope Science Data Analysis System (STSDAS), which runs under IRAF. The components plus graph representation of the HST instruments will permit more effective use of calibration observations, since observations in one mode can indirectly improve the calibration of other modes which share common components. This would not occur if the calibration of each mode were represented independently. We plan to extend the Passband Generator to include Johnson and other photometric systems to facilitate comparison between HST and ground-based observations.

3. A UNIFIED CALIBRATION ALGORITHM FOR PHOTOMETRY AND SPECTROPHOTOMETY

A handfull of the HST passbands were designed as approximations to Johnson or other standard photometric passbands, but the vast majority are not standard in any useful sense. The accepted practice

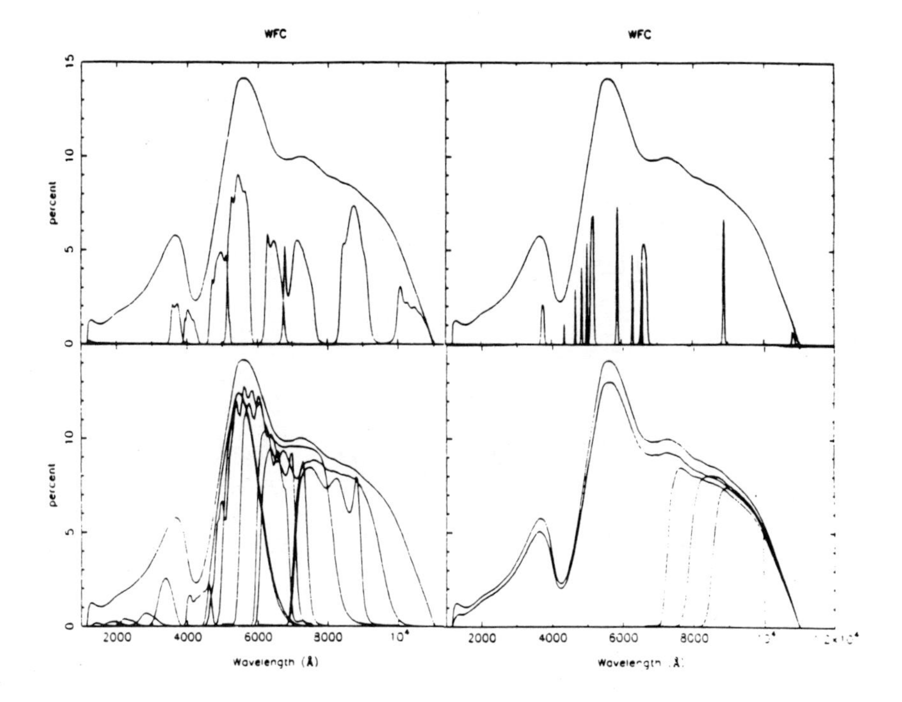

Fig 1. A selection of passbands for observing modes of the Wide Field Camera, illustrating the variety and novelty of the available photometric passbands.

WF/PC Instrument Graph

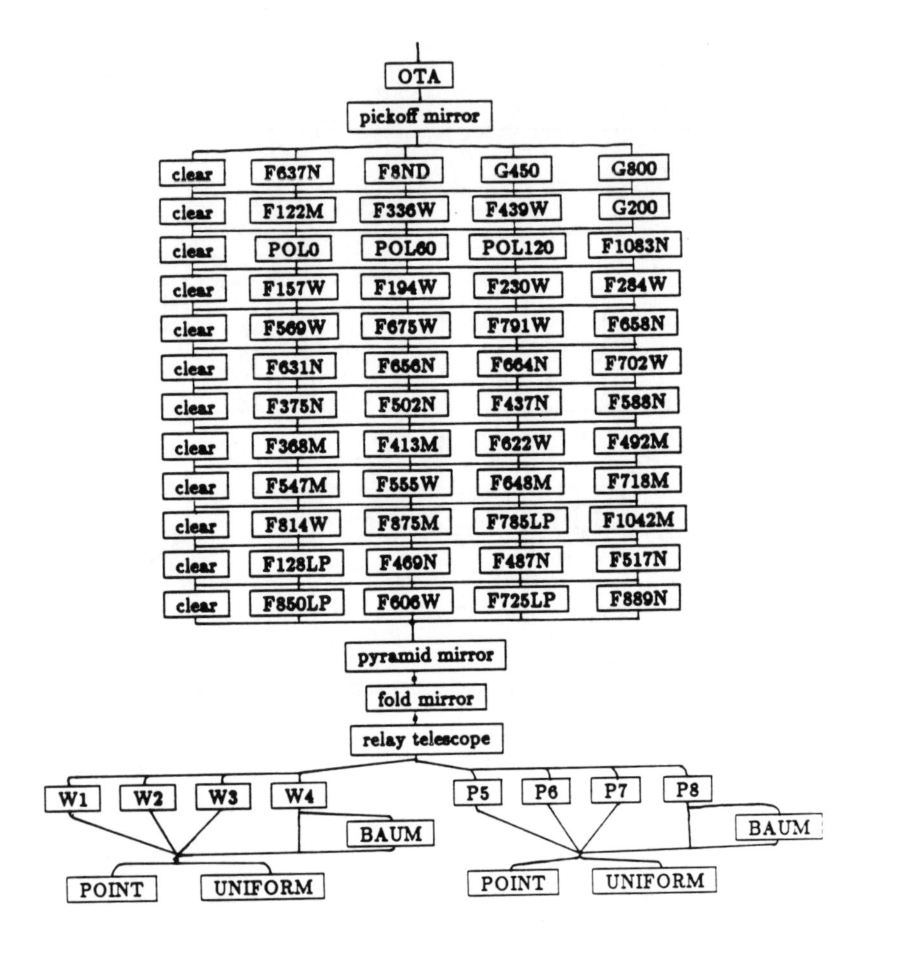

Fig. 2. A directed graph representation of the Wide Field/Planetary Camera (WF/PC). Each box represents one of the instrument's optical elements, e.g. a mirror, filter, grating or detector. The WF/PC has 12 independent 5-position filter/grating wheels, and eight CCD detectors. Every path from top to bottom through the graph represents a possible mode of observing, the passband for which is the product of the passbands for each of the components along the path. Graph representations have been developed for all six of the Science Instruments aboard the Hubble Space Telescope Observatory.

of transforming photometric data to a standard system, with color terms in the transformation to account for differences between the instrumental and standard passbands, is clearly unsuitable for most of the HST passbands. The approach to be taken is to calibrate all HST observations with respect to spectrophotometric standard stars, using synthetic photometry to bridge the gap in resolution between the spectrophotometric and photometric modes. The formulae developed in this section (see also Koornneef et al. 1986) permit precise calibration of the flux density in a passband without regard to its bandwidth. This unification of photometric and spectrophotometric calibrations ensures that the two types of data are directly intercomparable.

Spectrophotometric data taken with modern array detectors are simpler to calibrate than broadband photometry because the spectrophotometric pixels have very narrow passbands across which the detector sensitivity, atmospheric extinction, and standard star continuum are effectively constant. The count rate spectrum $C(\lambda)$ (1 count = 1 detected photon) per unit exposure time Δt and telescope area ΔA can be expressed as

$$\frac{C(\lambda)}{\Delta A \Delta t} = \frac{f_\nu(\lambda)}{U_\nu(\lambda)} = \frac{f_\lambda(\lambda)}{U_\lambda(\lambda)} ,$$

where $f_\nu(\lambda)$ (or $f_\lambda(\lambda)$) is the flux density of the target star, and $U_\nu(\lambda)$ (or $U_\lambda(\lambda)$) is the flux density required to produce a unit response of 1 count cm^{-2} s^{-1} $pixel^{-1}$. To calibrate, we want to determine the unit response curve $U(\lambda)$, which gives as a function of wavelength the factor that converts observed count rate to source flux density. $U(\lambda)$ is determined from count rate spectra observed on spectrophotometric standard stars whose flux spectra are known, and can then be used to convert count rate spectra observed on other targets to absolute fluxes.

Synthetic photometry permits a precise generalization of this procedure from the narrow passbands of spectrophotometric pixels to a passband $P(\lambda)$ of arbitrary width and shape. The analogous expression is simply

$$\frac{C(P)}{\Delta A \Delta t} = \frac{f_\nu(P)}{U_\nu(P)} = \frac{f_\lambda(P)}{U_\lambda(P)} ,$$

where $f_\nu(P)$ denotes the appropriate average of the star's flux density spectrum over the passband, and the flux conversion factor is now also a functional of the passband P. Because the photon detection rate is given by

$$\frac{C(P)}{\Delta A \Delta t} = \int P(\lambda) \frac{f_\nu(\lambda) d\nu}{h\nu} = \int P(\lambda) \frac{f_\lambda(\lambda) d\lambda}{h\nu} ,$$

the precise definitions of mean flux densities in a broad passband must be

$$f_\nu(P) \equiv \frac{\int P(\lambda) f_\nu(\lambda) d\lambda/\lambda}{\int P(\lambda) d\lambda/\lambda}$$

$$f_\lambda(P) \equiv \frac{\int P(\lambda) f_\lambda(\lambda) \lambda d\lambda}{\int P(\lambda) \lambda d\lambda} ,$$

and the count rate to flux density conversion factors are

$$U_\nu(P) = \frac{h}{\int P(\lambda) d\lambda/\lambda}$$

$$U_\lambda(P) = \frac{hc}{\int P(\lambda) \lambda d\lambda} .$$

The calibration procedure for broadband photometry is now essentially parallel to that for spectrophotometry. First calculate mean flux densities for each combination of passband and spectrophotometric standard star, using the known flux spectra and passband shapes. (Section 4 deals with uncertainties in the passband shape). Next derive the flux conversion factors from the standard stars by comparing their calculated mean flux densities against the corresponding observed count rates. Finally use the flux conversion factors to express the count rate observed on an unknown target to the precise flux density of that target.

Note that the mean flux densities depend on the spectrum and the passband shape but not on its overall normalization, while the conversion factors depend only on the integrated area of the passband. The different wavelength weighting functions used in the f_ν and f_λ averages are important for precision. A formula commonly used in synthetic photometry calculations,

$$f_\lambda(P) \equiv \frac{\int P(\lambda) f_\lambda(\lambda) d\lambda}{\int P(\lambda) d\lambda} ,$$

gives equal weight to equal photon energy rather than photon number. This subtle difference leads to calibration errors approaching 10% in the Johnson passbands, as indicated in Table I, and even more for wider passbands.

TABLE I

Comparison between photon-based and energy-based definitions of the broadband flux density f_ν.

Temperature	U	B Blackbodies	V	U	B Stars	V
2000	-.070	-.096	-.049	-.074	-.090	-.051
5000	-.026	-.027	-.010	-.023	-.039	-.008
10000	-.006	-.003	-.001	-.015	-.009	.007
20000	003	.007	-.006	.003	.001	.009
50000	.008	.012	.008	.010	.011	.010
100000	.009	.014	.008	.011	.012	.010

The advent of HST photometry may be an appropriate moment to abandon the astronomical tradition which sets the zero-point of magnitude systems on the spectrum of Vega. Spectrophotometry and now broadband photometry can be expressed with high precision directly in terms of absolute fluxes. The magnitude scale will undoubtedly remain in use, but it is no longer necessary to assign magnitude zero to a spectrum full of absorption lines. Spectrophotometric magnitude systems confer zero magnitude to a hypothetical spectrum with a constant flux density at all wavelengths. Oke's AB_ν magnitudes are based on a constant flux per unit frequency,

$$AB_\nu \equiv -2.5 \log_{10} f_\nu - 48.60 \quad ,$$

with f_ν in cgs units, and the analogous magnitude based on a constant flux density per unit wavelength is

$$ST_\lambda \equiv -2.5 \log_{10} f_\lambda + 21.10$$

with f_λ in erg cm^{-2} s^{-1} Å^{-1}. The zero points are set for convenience so that Vega has magnitude 0 in both systems for the Johnson V passband (see Fig. 3).

Our precise formulae for mean flux densities in arbitrary passbands extend the spectrophotometric magnitude systems to broadband photometry. One potential difficulty arises in the conversion between $f_\nu(P)$ and $f_l(P)$, since it is not at first obvious from their definitions that their relationship is independent of the source spectrum. Fortunately, such conversions can be made precisely by the familiar relation

$$f_\lambda(P) = f_\nu(P)\frac{c}{\lambda_P(P)^2}$$

where the passband's pivot wavelength is

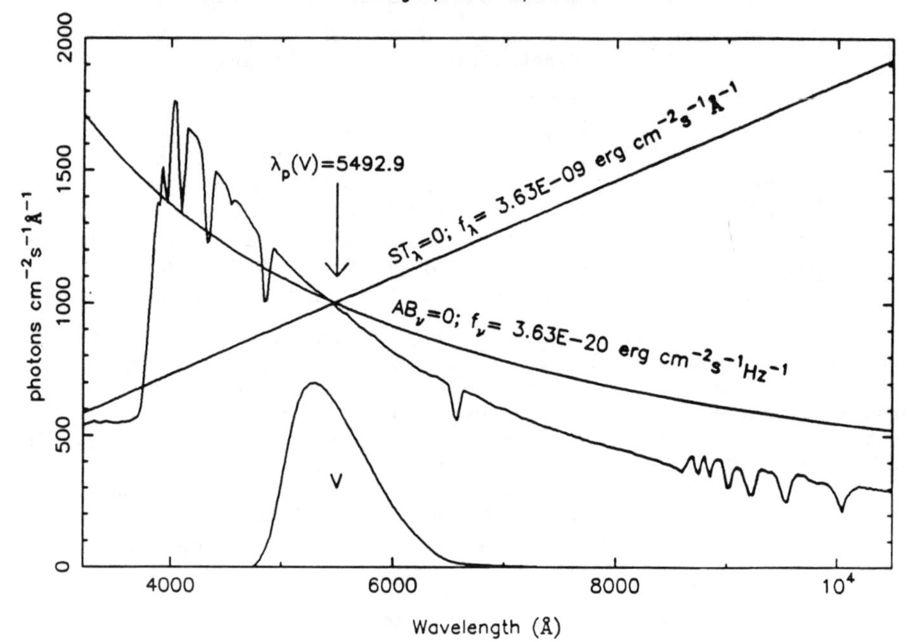

Fig. 3. Standard photometric systems generally use the spectrum of Vega to define magnitude zero. The spectrophotometric magnitudes AB_ν and ST_λ refer instead to spectra of constant f_ν and f_λ respectively. Magnitude zero in both systems is defined to be the mean flux density of Vega in the Johnson V passband. Thus all three of the spectra shown here produce the same count rate in the Johnson V passband. The pivot wavelength of Johnson V is defined to be the crossing point of the $AB_\nu = 0$ and $ST_\lambda = 0$ spectra.

$$\lambda_P(P) \equiv \sqrt{\frac{\int P(\lambda)\lambda d\lambda}{\int P(\lambda)d\lambda/\lambda}} \ .$$

4. PASSBAND RECONSTRUCTION

HST calibration begins with a database of ground-based, IUE and Voyager data for many of the planned calibration targets, plus laboratory measurements of the passbands. The pre-launch passbands undoubtedly will require some adjustments to account for the new observations from space. Our knowledge of the calibration target spectra will also improve as HST observations accumulate. Passband reconstruction techniques are being investigated as a means of fine-tuning the HST calibration based on observations in flight.

One possible approach to passband reconstruction is to make an adjustable passband by multiplying the initial default passband by a polynomial,

$$\log P(\lambda) = \log P_0(\lambda) + \Delta(\lambda) \ ,$$

where

$$\Delta(\lambda) = \Delta_0 + \Delta_1 \left(\frac{\lambda - \lambda_p}{\Delta\lambda}\right) + \Delta_2 \left(\frac{\lambda - \lambda_p}{\Delta\lambda}\right)^2 + \dots$$

Thus the polynomial coefficient Δ_0 controls the passband area, i.e. the sensitivity, Δ_1 controls the central wavelength, Δ_2 the passband width, etc. Given observations of several different stars with this passband, we can use a χ^2 statistic,

$$\chi^2 = \sum_{\text{data}} \left(\frac{\text{observed} - \text{predicted}}{\text{uncertainty}}\right)^2 \ ,$$

to determine best-fit values of the passband parameters Δ_i, and to judge the level of consistency between the parameterized passband and the data.

A second possibility is the Monte Carlo (or Las Vegas ?) approach. We can perturb the initial default passband by letting $\Delta(\lambda)$ be a random function with a correlation length somewhat less than the passband width. After each passband perturbation the χ^2 is calculated and used to assign a probability to that perturbed passband. Successive perturbations then allow the passband to take a random walk in function space, sampling the χ^2 at each step, and building up a probability density for the passband at each wavelength. The random walk must be guided to prevent the passband from wandering too far away from the χ^2 minimum. This method is computer intensive, but offers prospects for estimates of the passband uncertainty at each wavelength.

A third approach is the maximum entropy method (MEM). The entropy of the passband correction function can be defined as

$$S(\Delta) \equiv - \int d\lambda p(\lambda) \log p(\lambda)$$

where

$$p(\lambda) \equiv \frac{e^{\Delta(\lambda)}}{\int d\lambda e^{\Delta(\lambda)}} \ .$$

The entropy is a measure of the structure in the passband correction function. It takes on its maximum value when $\Delta(\lambda)$ is completely uniform. Thus MEM will simply adjust the normalization of the initial default passband, leaving its shape intact, if it is thereby possible to fit the data. The data constraints are imposed by adjusting the passband to maximize S subject to the constraint that $\chi^2 \leq C$ for some appropriate value C.

Fig. 4 illustrates the results of some initial trials of MEM passband reconstruction. We simulated observations of seven stars representing spectral types O, B, A, F, G, K and M using the passbands given by the dotted curves in the figure. Noise was added to the predicted data to simulate observational errors at the level of 1%. The dashed curves were then assumed to be the default passbands, and MEM was used to adjust the default passbands until they produced results consistent with the simulated observations of the seven stars. The solid curves are the final MEM passbands found by the general maximum entropy fitting code MEMSYS, which implements the algorithm described in Skilling and Bryan (1984).

In the upper panel, MEM has restored the areas of four passbands while retaining the default passband shapes. Passband sensitivities can thus be recovered to within the accuracy permitted by the noise in the simulated observations. In the middle panel, passband areas and central wavelengths are recovered, but the passband widths and peaks are incorrect. These tests suggest that observations of a few stars with a range of colors can provide enough information to correct a passband sensitivity error or wavelength shift, but that passband widths and higher moments may not be very well constrained by such observations.

The lower panel of Fig. 4 tests MEM's ability to discover the presence of a red leak. In this case the true passband was a Gaussian plus a long red tail. The default passband was a Gaussian plus a constant equal to 10^{-6}. The signature of the red leak in the data is an anomalously high count rate on the red stars. MEM achieved a good fit to the data partly by shifting the centroid of the Gaussian passband to the red, and partly by elevating the red tail of the passband.

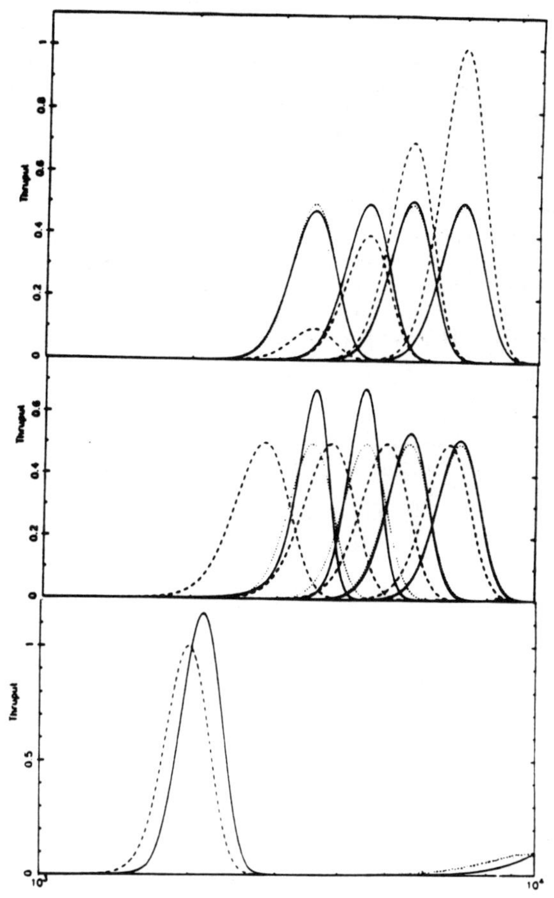

Fig. 4. Passband reconstruction techniques are being investigated as a means of making fine adjustments to the HST calibration based on observations in flight. Here we show three tests of passband reconstruction using the maximum entropy method (MEM). Simulated observations of seven stars (OBAFGKM) were generated using the dotted passbands. The dashed curves were then assumed to be the default passbands. MEM was used to adjust the default passbands until they produced results consistent with the simulated observations of the seven stars. The solid curves are the final MEM passbands. In the upper panel, MEM restores the passband areas, i.e. the passband sensitivities, to within the accuracy of the simulated observations. In the middle panel, passband areas and central wavelengths are recovered, but the passband widths are incorrect. Evidently the width and presumably also the higher moments of the passband are not very tightly constrained by the observations. Finally, in the lower panel, MEM detects a red leak.

These initial passband reconstruction trials suggest that broadband photometry can be accurately calibrated by using observations of spectrophotometric standard stars to fine-tune passbands measured in the laboratory. Because the adjusted passbands can be used to accurately predict data for targets with very non-standard spectra, this method should be more accurate than a conventional calibration based on color terms and transformation to a standard photometric system. For HST the passband shapes are generally well defined by the pre-flight measurements, and small adjustments can be made based on in-flight observations of calibration targets. Further development will be directed toward simultaneous adjustment of many different passbands and spectra to bring about a global calibration consistent with all of the available calibration observations.

REFERENCES

Koornneef, J., Bohlin, R., Buser, R., Horne, K. and Turnshek, D. 1986. in *Highlights of Astronomy*, Vol. 7, J. -P. Swings, ed., Reidel, Dordrecht, p. 833.

Skilling, J. and Bryan, R. K. 1984 *Monthly Notices Roy. Astron. Soc.* 211, 111.

DISCUSSION

HAYES: I have two related questions. Certainly, the filters in the HST instruments will change with age, and it is now several years since they were manufactured. 1) Will there be any chance to re-determine the transmission function before launch?, and 2) Will you actually have enough telescope time to apply the technique you've outlined in your paper to real observations?

HORNE: The filter transmission curves will probably not be re-measured before launch since this would require disassembly of the instruments. Changes in the detector quantum efficiency curves will be tracked by measurements with the spectrographs, or with a subset of the passbands. There will be some additional data on this from the thermal-vacuum tests prior to launch, and this will be taken into account for all passbands. The 5-10% of HST time that we expect to be available will not permit observations of several calibration stars in every mode. We will concentrate on wide and popular passbands, and also try to follow up on any anomalies as they occur.

BELL: Do you have sufficient accuracy of measurement (e.g. 0.01 mag.) to get these shifts to the necessary accuracy?

HORNE: This is a very difficult question to answer in general. It depends on what sort of corrections to the passbands you contemplate making. There is always a tradeoff between statistical noise and wavelength resolution. A 1% accuracy in the area of the passband requires 1% accuracy data. To adjust two parameters, passband area and central wavelength, you would need 1% accuracy in a red and a blue star. The question is best answered by simulation tests on the specific passbands you have in mind.

WEAVER: In the maximum entropy reconstruction technique, how do you choose the maximum χ^2?

HORNE: The maximum χ^2 value is used to control the trade-off between noise and resolution. If χ^2_{max} is set too high, the reconstruction is more uniform, but the data are not fitted very well. If χ^2_{max} is set too low, the data will be well fitted, but the reconstruction will develop short-scale noise. In calibration work it is very important to fit the data well, so χ^2_{max} is set just above χ^2_{min}.

BOHLIN: Why does the peak of the reconstructed passband fall so far off the true peak?

HORNE: The peak is higher, but the area of the passband is conserved, which is the prime and most important constraint.

KURUCZ: Can you use your reconstruction technique on ground-based photometry in your spare time?

HORNE: Yes, if I had any spare time. We have started to apply the reconstruction techniques to ground-based systems, but this is going rather slowly.

OKE: I take it that the classical way of reducing photometric observations, i.e. with color terms, etc., will not be used on HST observations. Is this correct?

HORNE: You are correct. The classical method, using color terms to transform instrumental magnitudes to magnitudes on a standard system, will not be used for HST observations. This method is flawed in principle, and we have found that errors of up to 0.3 mag. would be committed.

SCHEMPP: Because of the long wait between bandpass measurement and HST launch and the aging of filters you will essentially be making the first standard star measurements with filters of unknown bandpass. How will this be handled?

HORNE: I don't think the problem is as severe as you fear. The main change is expected to be in the area under the passband curves, and this is straightforward to calibrate. Small changes in shape will be treated by observing red and blue stars. We expect to do this for the wide passbands, and for the modes that are in the highest demand. I agree that it would be good to re-measure the filter transmission curves, but this will probably not be possible.

SPECTRAL SYNTHESIS OF POPULATION II SYSTEMS

Carla Cacciari

Space Telescope Science Institute

ABSTRACT: The application of spectrophotometry to the study of Population II stellar systems is reviewed. In particular, the following items are discussed:

1. The spectral synthesis techniques: brief description, and discussion of some of the most important limitations, which are relevant for understanding the observed spectral energy distributions (SEDs).

2. Examples of applications of the spectral synthesis method to integrated SEDs of Population II stellar systems.

3. Future needs, both on the theoretical and the observational side, for a better understanding of Population II stellar systems from the analysis of their integrated light.

1. INTRODUCTION

This review will concentrate on the use of spectrophotometry for the study of Population II stellar systems. The concept of "population" has been associated with different characteristics of a stellar sample, i.e. its dynamics, chemical abundance, or age, therefore, we first need a definition of Population II. In the following, Population II will mean old and/or metal-poor ($Z < 0.1\ Z_0$) stellar systems. This definition includes a vast range of objects such as galactic globular clusters (old metal-poor), bulges of spiral galaxies and early type galaxies (range of age and metallicity), and blue compact or H II galaxies (young metal-poor).

A stellar system can be a simple stellar population, namely a homogeneous set of stars characterized by the basic parameters age (t), chemical composition (Y, Z), and initial mass function. Examples of simple stellar populations are globular clusters and possibly some elliptical galaxies. This does not exclude the possibility that other parameters as well (rotation, [CNO/Fe], mass loss) vary from star to star and affect the character of the population in a lesser degree.

A. G. Davis Philip, D. S. Hayes and S. J. Adelman, (eds.)
New Directions in Spectrophotometry 159 - 173

Most stellar systems, however, are complex stellar populations, where one or more basic parameters (age, metallicity) vary from star to star. The study of the population content of a given stellar system therefore provides information on the formation and evolution history of the individual stellar components and of the systems themselves. Such a study is based on the statistical analysis of data on individual stars (CM diagram), for nearby systems. This, however, is possible only for a limited number of objects which are detectable at the required level of photometric accuracy and spatial resolution by the currently available technology.

Fig. 1, taken from O'Connell (1986a), shows the population types observable by the Hubble Space Telescope (HST) at the distances of selected objects, on the assumption that HST limiting magnitude is V = 27. For more distant objects only the study of their integrated light can provide information on their population content. This procedure (i.e. the spectral synthesis technique) is less precise than the analysis of the CMD, but it is potentially more powerful since in principle it can be applied to any observable stellar system in the universe.

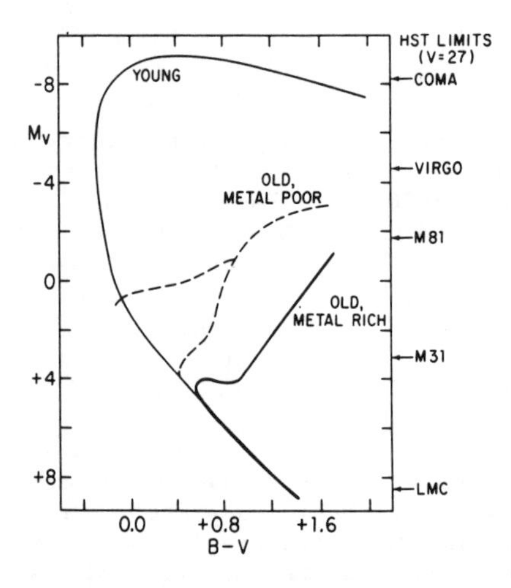

Fig. 1. A Color-Magnitude Diagram for several population types with limiting magnitudes for the Hubble Space Telescope indicated at the distances of selected objects.

Conceptually, the spectral synthesis is the combination of individual stellar spectra of given types and in given proportions to produce a model composite, which is then compared to an observed integrated SED. The foundations of the spectral synthesis techniques have been laid approximately five decades ago by Whipple (1935), but intensive applications of this procedure, supported by the additional information from the stellar evolution theory, began only more than thirty years later (Tinsley 1968, Spinrad and Taylor 1971, Faber 1972, 1973). In time this method has evolved into two different approaches, the so-called "evolutionary synthesis" and "population synthesis" (Tinsley 1980), which are described in the following section.

2. SPECTRAL SYNTHESIS TECHNIQUES

In the evolutionary synthesis approach, one starts from a given CMD, resulting from an evolutionary scenario or from empirical data, and predicts its integrated SED. This method is mathematically simple, and is generally used to explore the range of integrated spectral properties produced by the variation of the dominant population parameters. For applications and reviews of this method we refer to Tinsley (1980), Lequeux et al. (1981), Barbaro and Olivi (1986), Renzini and Buzzoni (1986), Buzzoni (1987), and references therein.

The population synthesis approach, recently called also "optimizing synthesis", starts from the other end of the problem, i.e. from an observed SED, and estimates the CMD that best fits it. This method is mathematically much more complicated (use of linear or quadratic programming methods, automatic optimizing algorithms over a very large parameter space), but allows one to evaluate the goodness of the fit and the uncertainties in the derived parameters. Evolutionary constraints may or may not be imposed when performing the fitting procedure, and the unphysical solutions (e.g. negative numbers of certain types of stars) one may find with unconstrained syntheses also provide interesting information on the population content and on the goodness of the fit. Applications and reviews of this method have been presented by Pickles (1985b), O'Connell (1986a,b), and Thuan (1986).

Both approaches are affected by a few important limitations:

A. Inaccurate or incomplete grids of stellar evolutionary tracks: the stellar evolution theory for low mass stars has been very successful, during the past decades, in explaining most of the observed features and characteristics of stars and stellar systems. Nevertheless a few phenomena, such as convection, mass loss during the RGB and AGB phases, and chemical mixing are still poorly understood, and their different treatment is probably responsible for the discrepancies among different sets of theoretical isochrones (Ciardullo and Demarque 1977; VandenBerg 1983, 1985a; VandenBerg and Bell 1985; VandenBerg and Laskarides 1987; Green, Demarque and Ciardullo 1988). Clearly a better understanding and empirical calibration of these phenomena is required, as well as the inclusion of non-solar [CNO/Fe] abundances in theoretical models (Rood and Crocker 1985; VandenBerg 1985b). Later stages of stellar evolution present even more serious problems: stellar models for Red Giant (RG) and Horizontal Branch (HB) stars are available (Sweigart and Gross 1976, 1978; Sweigart 1987; Seidel et al. 1987), but the mechanism(s) responsible for the observed spread in the morphology of horizontal branches in galactic globular clusters (the "second parameter" effect) are not known. The evolution of Asymptotic Giant Branch (AGB) and Post-AGB (Planetary Nebula, White Dwarf) stars is determined by many poorly-known phenomena, and the stellar models available for these phases (Gingold 1974, 1976; Paczynski 1971; Schönberner 1979, 1981, 1983; Wood and Faulkner 1986) should be used with caution.

For high-mass metal-poor stars, which are relevant for the study of blue compact galaxies (BCG) and of primordial galaxies that are being discovered at high redshift, the limitation is at a more basic level, since it is due to incompleteness in the grid of stellar evolutionary tracks for the required range of masses and metallicities (Maeder 1981; Hellings and Vanbeveren 1981; Brunish and Truran 1982; Eggleton, unpublished, described by Melnick, Terlevich and Eggleton 1985). These limitations affect more seriously the evolutionary synthesis approach, since in the optimizing synthesis approach the evolutionary input is treated as a set of parameters which are allowed to vary until an acceptable fit is obtained.

B. Incomplete libraries of stellar energy distributions: in constructing the integrated SED of a given stellar system, it is essential to use a complete library of individual stellar energy distributions (either theoretical model atmospheres or observed spectra), which cover the relevant range of temperatures, gravities, and metallicities, at reasonable intervals in the parameter space. Missing stellar types in the library cannot generally be replaced by "surrogates", for example by stellar energy distributions of different metallicity, since both the continuum and the line spectrum would be different at specific wavelengths and temperatures. No existing library covers all the Population II stellar types.

On the theoretical side, stellar model atmospheres have been calculated for a wide range of parameters covering the most important evolutionary phases in the HR diagram (Bell and Gustafsson 1978; Kurucz 1979a, b; Buser and Kurucz 1987), and low-mass Population II stars are fairly well represented. Pure-hydrogen model atmospheres have been calculated by Terashita and Matsushima (1969), Shipman (1971, 1977), and Wesemael et al. (1980), and a systematic collection of unpublished model atmospheres has been carried out by Buser in collaboration with Shipman (high gravity stars), Böhm-Vitense (Population II giants), Kurucz (O-F stars), and Gustafsson (late-type dwarfs). However, a huge amount of work still needs to be done, because existing models for Population II stars "have numerical errors, an unphysical treatment of convection, an inadequate or non-existent treatment of statistical equilibrium, an arbitrarily chosen microturbulent velocity, an arbitrarily chosen helium abundance, and a greatly underestimated line opacity for iron group elements" (Kurucz 1987). Moreover, metal-poor high-mass (i.e. temperature) models are not represented, since models with [m/H] = -2.0 extend only to 10,000 K. We refer to Kurucz (1988) for an update on this issue.

On the empirical side, the collection of the intrinsic stellar SEDs necessary for a complete coverage in spectral type, luminosity class, and metallicity is extremely difficult and time-consuming, and it also involves the problem of estimating the reddening, which shall not be discussed here. Moreover there is a real shortage in the solar neighborhood of stellar types that are important in some stellar populations (e.g. very low or very high metal abundance). To represent these "rare" types of stars, one may have to rely on model atmospheres, or to include in the empirical library the composite spectra of population (e.g. the M 31 metal-rich clusters), at the risk of introducing "non-pure" spectral ingredients in the study of other systems. We list in Table I the most relevant atlases of stellar spectrophotometric data available. They include also a number of Soviet catalogs of continuous energy distributions, which are described in more detail by Hayes (1986). No attempt has been made to separate these catalogs according to metal abundance, since the entire range in metallicity from the metal-poor to the super metal-rich end can be found in Population II stellar systems. For a more detailed analysis of catalogs of spectrophotometric data, see Labhardt (1988).

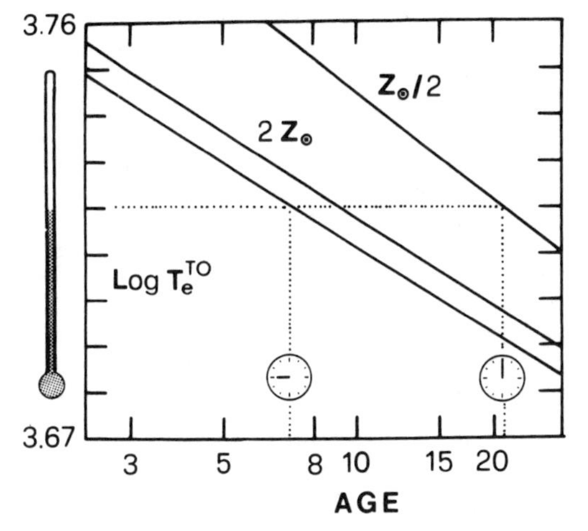

Fig. 2. The turnoff effective temperature for two values of the metallicity and Y = 0.20. The middle line refers to (Y,Z) = (0.30, 0.04).

C. Interdependence of age and metal abundance: both age and metal abundance affect the temperature of a given integrated SED, in the sense that increasing the metal abundance produces a lower temperature, which is the same effect one obtains by increasing the age of a more metal-poor population (O'Connell 1980, Renzini and Buzzoni 1986). Fig. 2, taken from Renzini and Buzzoni (1986), shows the effect of age and metallicity on the temperature of the turnoff for old stellar systems.

Therefore, even for a simple stellar population characterized by a relatively small range of age and metallicity, broad-band photometry (i.e.- energy distribution alone) cannot disentangle the contribution from these two fundamental parameters. This is, essentially, the much debated problem of non-uniqueness in synthesis analyses. However, it has been demonstrated (Pickles 1985b; O'Connell 1986b) that the use of

Table I

Atlases of Observed Stellar Energy Distributions.

Source Reference	Types	[m/H]	Res.	Wavelength
Spinrad & Taylor 1971 and Taylor 1982	MS,SG,G	SMR to MP	16,32	3300-10,700
Straižys & Sviderskienė 1972	O-M,V-I	solar	50	3000-10,000
O'Connell 1973	all	SMT to MP	20,30	3300-10,800
Oke 1974	WD		10-160	3200-10,500
Breger 1976	all	solar	10-100	3200-12,000
Burnashev 1977	?	solar	25	3200-7550
Greenstein & Buser 1977	sd F-M	MP	40-320	3100-11,000
Oke & Buser 1977	Giants	MP	80,160	3180-11,320
Alekseev et al. 1978	?	solar	25	3100-7375
Christensen 1978	SG,G,HB	solar,MP	160	3448-10,800
Kharitonov et al. 1978	?	solar	50	3225-7975
Ardeberg & Virdefors 1980	all	solar	10-100	3300-10,526
Burnashev 1980	?	solar	25	3200-7550
Cochran 1980	all	solar	200	4600-10,200
Kolotilov et al. 1980	B-M	solar	50	3225-10,825
Bartkevicius & Sviderskienė 1981	F-K	MP	50	3000-11,000
Cochran 1981	O-B	solar	200	4600-10,200
Tobin & Nordsieck 1981	B-K	solar,MP	10,20	3900-8700
Voloshina et al. 1982	B-M	solar	50	3225-10,825
Voloshina et al. 1982	all	solar	50	3225-7625
Gunn & Stryker 1983	all	SMR,solar	20,40	3130-10,800
Hayes & Philip 1983	HB	MP	160-360	3410-6840
Philip & Hayes 1983	FHB,PopI	solar,MP	40-360	3400-6790
Voloshina et al. 1983	B-M	solar	50	3225-7625
Wu et al. 1983	all	solar	5	1150-3200
Heck et al. 1984	all	solar	5	1150-3200
Huenemoerder et al. 1984	HB	MP	10-200	1170-5162
Glushneva et al. 1984	all	solar	50	3225-7625
Jacoby et al. 1984	O-M,V-I	solar	4.5	3510-7427
Taylor 1984	G V	MP,solar	32-100	3288-7000
Cacciari 1985	SG,G,HB	MP,solar	5	1150-3200
de Boer 1985	UV-bright	MP	30-150	1180-2980
Pickles 1985a	all	MR to MP	10-17	3600-10,000
Voloshina et al. 1985	B-M	solar	50	3125-10,825
ST ScI	WD		2	1150-3200

Acronyms: MS=Main Sequence, SG=Sub-Giant, G=Giant, WD=White Dwarf, HB=Horizontal Branch, FHB=Field HB, SMR=Super Metal-Rich, MP=Metal-Poor.

a long wavelength baseline (UV to IR), of high S/N data, and of a sufficiently high resolution (~ 20 - 30 Å) or alternatively of a careful selection and weighting of continuum and spectral features, can greatly restrict the solution area. The additional imposition of only few astrophysical constraints (such as non-negative solutions or mild smoothing of the derived luminosity function) generally leads to unique and quite accurate solutions, with the possible exception of integrated SEDs containing comparable contributions from a wide range of populations, where the uncertainty in the final solution will be larger than in a single population case.

Estimating the age of young stellar systems ($t < 10^9$ yrs), where blue horizontal branches have not yet formed, is a somewhat easier problem, since most of the integrated light comes from the hot stars at the turnoff, and their SEDs, particularly in the UV range, are a monotonic function of age with smaller dependence on metallicity (Cohen, Rich and Persson 1984; Barbaro and Olivi 1986). In the next section some examples of synthesis modeling of Population II stellar systems are presented.

3. SOME APPLICATIONS OF SPECTRAL SYNTHESIS TO POPULATION II STELLAR SYSTEMS

Population synthesis techniques are generally used to estimate the population content of unresolved stellar systems, however, a number of nearby resolved systems, mainly globular clusters in the Galaxy and the Magellanic Clouds, have also been analyzed, to test and calibrate the reliability and accuracy of the method. The following examples concentrate on the use of spectrophotometry in population synthesis applications, and represent only a small fraction of the hundreds of integrated light population studies which have appeared in the last 20 years.

3.1 Old Systems

Early applications of the spectral synthesis technique to galactic globular clusters (Christensen 1972; Tobin and Nordsieck 1981) were already sufficiently accurate to achieve consistency between the synthetic models and the observed CMDs. Later comparative studies of globular clusters and nuclei of spiral and elliptical galaxies (Burstein et al. 1984, 1985; Rose 1985) revealed important metal content and age differencies between these objects, which were thought until then to constitute a homogeneous family of old stellar populations. The availability of UV data from IUE opened a new series of globular cluster population studies in the ultraviolet (de Boer 1981; Nesci 1981, 1983a, b; Cacciari et al. 1987), which stressed the importance of the hot stellar component, in particular the horizontal branch morphology (second parameter effect) and the presence of UV-bright (post-AGB) stars. Early-type galaxies and bulges of spirals were extensively analyzed in the visual and infrared (O'Connell 1976a, b, 1980; Aaronson et al. 1978; Gunn, Stryker and Tinsley 1981; Bruzual

1981, 1983; Pickles 1985b; Whitford 1985; Arimoto and Yoshii 1986), and revealed the presence of a large metallicity range and super metal-rich stars in the bulges of bright galaxies, and of a significant intermediate-age population in E galaxies.

Observations with IUE, reviewed and discussed by Snjiders (1984), showed on the other hand an upturn in the UV flux at $\lambda < 2500$ Å (already detected by OAO, TD1 and ANS), which indicated the existence of a very hot stellar component, whose nature (hot HB, young stars, post-AGB, binaries) is still under debate (Wu et al. 1980; Nesci and Perola 1985; Pickles 1985b).

3.2 Young Metal-Poor Systems

Evolutionary synthesis of BCGs was used by Lequeux et al. (1981) to model IUE spectral energy distributions, while Benvenuti (1983) applied the optimizing approach, using, however, a limited stellar library. Their results have been confirmed and improved by subsequent studies (Viallefond and Thuan 1983; Huchra et al. 1983; Rosa et al. 1984; Gondhalekar 1986; Thuan 1986), which found evidence of intense, short bursts of massive star formation superposed on an underlying old population, a similar IMF to that in the solar neighborhood except perhaps at the high-mass end where it could be flatter, and a smaller extinction in the UV than deduced from optical or radio data (probably a selection effect), with the indication that the extinction law is dependent on chemical composition. Spectral synthesis of the ionizing UV in H II galaxies, using observations of the nebular emission lines, is currently being applied by Campbell (1988a, b) to study the upper part ($M > 15\ M_{\odot}$) of the IMF. Recent reviews on this topic have been presented by Thuan (1986), and Kunth and Weedman (1987).

4. FUTURE NEEDS

Synthesis techniques rest on three basic elements: a. the synthesis code; b. the stellar evolution input, and c. the stellar library input. All of them need some improvement or calibration.

a. The synthesis codes, particularly in the optimizing approach where fitting algorithms have become very complex and refined, should be tested for stability and consistency against a template population.

b. Stellar evolutionary tracks are needed to cover the missing elements in the age-metallicity plane, in particular for metal-poor high-mass stars. A better understanding and calibration against a template population are required for at least three free parameters, i.e. mixing-length treatment of convection, mass loss, and chemical mixing.

This analysis should be performed on all evolutionary phases, from the MS to the Post-AGB.

c. The most relevant element for the subject of this conference is the stellar library input. In principle, the use of a theoretical or empirical stellar library is equivalent, provided they both include accurate and complete samples of the stellar types that form the composite population. A standard library of stellar SEDs should meet the following requirements: (i) complete coverage, at reasonable intervals in the parameter space, of the full range in temperatures (spectral types), surface gravities (luminosity classes), and metal abundances observed in the CMDs; (ii) coverage of the entire wavelength range accessible to ground-based and space instruments (UV to IR), with a resolution or a selection of data points appropriate to disentangle the effects of metallicity and temperature on a given SED; (iii) consistency, for each SED in the library, between the corresponding theoretical and observed energy distributions, i.e. calibration of synthetic (spectro)-photometric colors against the observed ones. A number of projects are presently being carried on, with the aim of building a standard grid that meets the above requirements. On the theoretical side, they include the ongoing work by Kurucz (1988), and Buser, Horne and Koornneef (1986). This latter project goes along with the construction of an empirical grid covering the whole wavelength range of HST, i.e. 1200 - 12,000 Å, also for the purpose of calibrating the HST spectrophotometric system (see Bohlin, Oke, and Horne, 1988).

Other projects include the construction of an empirical library of UV intrinsic energy distributions of solar metallicity stars (Fanelli et al. 1987) using the existing IUE Spectral Atlases (Wu et al. 1983; Heck et al. 1984), and the completion of the existing grid of UV-IUE spectra (Cacciari 1985) for metal-poor field stars (Cacciari, Buser and Malagnini 1988). The comparison of observed stellar energy distributions with model atmospheres is an important step, and should be performed on all stars and atmospheric models that are to be members of the standard library, in order to: a) verify and calibrate the accuracy of the theoretical models; b) make sure that only "normal" stars representative of the desired evolutionary phases are included in the library; and c) derive the physical parameters that identify the location of the star in the HR diagram. We refer to Gustafsson and Jorgensen (1985), Gustafsson (1986) and Malagnini (1988) for a more detailed discussion of this subject.

REFERENCES

Aaronson, M., Cohen, J. G., Mould, J. and Malkan, M. 1978 Astrophys. J. 223, 824.

Alekseev, N. A., Alekseeva, G. A., Arkharov, A. A., Belyaev, Yu. A., Boyarchuk, A. A., Boyarchuk, M. E., Burnashev, V. O., Galkin, G. D., Galkin, L. S., Galkina, T. S., Demidova, A. N., Kamionko, L. A., Kulagin, E. S., Neshpor, Yu. I., Nikonov, V. B., Novidov, V. V., Novopashennyj, V. B., Pakhomov, V. P., Polozhentseva, T. A., Pronik, V. I., Ruban, E. V., Chistyadov, Yu. N., Schegolev, D. E. and Yakomo, A. A. 1978 Trudy Glavnoi

Astron. Obs. Pulkovo **83**, Ser II, 4.
Ardeberg, A. and Virdefors, B. 1980 Astron. Astrophys. Suppl. **40**, 307.
Arimoto, N. and Yoshii, Y. 1986 Astron. Astrophys. **164**, 260.
Barbaro, G. and Olivi, F. M. 1986 in Spectral Evolution of Galaxies, C. Chiosi and A. Renzini, eds., Reidel, Dordrecht, p. 283.
Bartkevicius, A. and Sviderskiene, Z. 1981 Vilnius Obs. Bull. **57**, 35.
Bell, R. A. and Gustafsson, B. 1978 Astron. Astrophys. Suppl. **34**, 229.
Benvenuti, P. 1983 in Highlights of Astronomy, R. West, ed., Reidel, Dordrecht, p. 631.
Bohlin, R. C. 1988 in New Directions in Spectrophotometry, A. G. D. Philip, D. S. Hayes and S. J. Adelman, eds., L. Davis Press, Schenectady, p. 121.
Breger, M. 1976 Astrophys J. Suppl. **32**, 7.
Brunish, W. M. and Truran, J. W. 1982 Astrophys. J. Suppl. **49**, 447.
Bruzual, A. G. 1981 Ph.D. Thesis, University of California, Berkeley.
Bruzual, A. G. 1983 Astrophys. J. **273**, 105.
Burnashev, V. I. 1977 Bull. Crimean Astrophys. Obs. **57**, 44.
Burnashev, V. I. 1980 Bull. Crimean Astrophys. Obs. **62**, 1.
Burstein, D. 1985 Pub. Astron. Soc. Pacific **97**, 89.
Burstein, D., Faber, S. M., Gaskell, C. M. and Krumm, N. 1984 Astrophys. J. **287**, 586.
Buser, R., Horne, K. and Koornneef, J. 1986 St ScI Internal Report 16/05/86.
Buser, R. and Kurucz, R. L. 1987 in The Impact of Very High S/N Spectroscopy on Stellar Physics, G. Cayrel de Strobel and M. Spite, eds., Reidel, Dordrecht, in press.
Buzzoni, A. 1987 in Towards Understanding Galaxies at Large Redshifts, R. G. Kron and A. Renzini, eds., Reidel, Dordrecht, in press.
Cacciari, C. 1985 Astron. Astrohys. Suppl. **61**, 407.
Cacciari, C., Buser, R. and Malagnini, M. L. 1988, in preparation.
Cacciari, C., Clementini, G., Fusi-Pecci, F. and Zinn, R. J. 1987 in Stellar Evolution and Dynamics in the Outer Halo of the Galaxy, M. Azzopardi and F. Matteucci, eds., ESO, p. 511.
Campbell, A. W. 1988a Astrophys. J., in press.
Campbell, A. W. 1988b, in preparation.
Christensen, C. G. 1972, Ph.D. Thesis, Cal. Inst. of Technology.
Christensen, C. G. 1978 Astron. J. **83**, 244.
Ciardullo, R. S. and Demarque, P. 1977 Trans. Yale Univ. Obs. 33.
Cochran, A. L. 1980 Pub. Astron., Univ. of Texas, **16**, 1.
Cochran, A. L. 1981 Astrophys. J. Suppl. **45**, 83.
Cohen, J. G., Rich, R. M. and Persson, S. E. 1984 Astrophys J. **285**, 595.
de Boer, K. S. 1981 in Astrophysical Parameters for Globular Clusters, A. G. D. Philip and D. S. Hayes, eds., L. Davis Press, Schenectady, p. 25.

de Boer, K. S. 1985 Astron. Astrophys. **142**, 321.
Faber, S. M. 1972 Astron. Astrophys **20**, 361.
Faber, S. M. 1973 Astrophys. J. **179**, 731.
Fanelli, M. N., O'Connell, R. W. and Thuan, T. X. 1987, preprint.
Gingold, R. A. 1974 Astrophys. J. **193**, 177.
Gingold, R. A. 1976 Astrophys. J. **204**, 116.
Glushneva, I. N., Voloshina, I. B., Doroshenko, V. T., Mossakovskaya, L. V., Ovchinnikov, S. L. and Khruzina, T. S. 1984 Trudy Gos. Astron. Inst. Shternberga **54**, 3.
Gondhalekar, P. M. 1986 in Star-Forming Dwarf Galaxies and Related Objects, D. Kunth, T. X. Thuan and J. T. T. Van, eds., Frontieres, p. 145.
Green, E. M., Demarque, P. and Ciardullo, R. 1988 Yale Univ. Trans., in preparation.
Greenstein, J. L. and Buser, R. 1977, unpublished.
Gunn, J. E. and Stryker, L. L. 1983 Astrophys. J. Suppl. **52**, 121.
Gunn, J. E., Stryker, L. L. and Tinsley, B. M. 1981 Astrophys. J. **249**, 48.
Gustafsson, B. and Jorgensen, U. G. 1985 in IAU Symposium No. 111, Calibration of Fundamental Stellar Quantities, D. S. Hayes, L. E. Pasinetti and A. G. D. Philip, Reidel, Dordrecht, p. 303.
Gustafsson, B. 1986 in Highlights of Astronomy Vol. 7, J. -P. Swings, Reidel, Dordrecht, p. 805.
Hayes, D. S. 1986 in Highlights of Astronomy, Vol. 7, J. -P. Swings, p. 819.
Hayes, D. S. and Philip, A. G. D. 1983 Astrophys. J. Suppl. **53**, 759.
Heck, A., Egret, D., Jaschek, M. and Jaschek, C. 1984 IUE Low-Dispersion Spectra Reference Atlas - Part I. Normal Stars. ESA SP-1052, Paris.
Hellings, P. and Vanbeveren, D. 1981 Astron. Astrophys. **95**, 14.
Horne, K. 1988 in New Directions in Spectrophotometry, A. G. D. Philip, D. S. Hayes and S. J. Adelman, eds., L. Davis Press, Schenectady, p. 145.
Huchra, J. P., Geller, M. J., Gallagher, J., Hunter, D., Hartmann, L., Fabbiano, G. and Aaronson, M. 1983 Astrophys. J. **274**, 125.
Huenemoerder, D. P., de Boer, K. S. and Code, A. D. 1984 Astrophys. J. **89**, 851.
Jacoby, G. H., Hunter, D. A. and Christian, C. A. 1984 Astrophys. J. Suppl. **56**, 257.
Kharitonov, A. V., Tereschenko, V. M. and Knyazeva, L. N. 1978 A Combined Stellar Spectrophotometric Catalog, Nauka, Alma-Ata.
Kolotilov, E. A., Glushneva, I. N. Voloshina, I. B., Novikova, M. F., Fetiseva, T. S., Shenavrin, V. I., Mossadovskaya, L. V. and Doroshanko, V. V. 1980 Soobshch. Gos. Astron. Inst. Im. Shternberga, No. 219.
Kunth, D. and Weedman, D. 1987 in Exploring the Universe with the IUE Satellite, Y. Kondo, ed., Reidel, Dordrecht, p. 623.
Kurucz, R. L. 1979a Astrophys. J. Suppl. **40**, 1.
Kurucz, R. L. 1979b in Problems of Calibration of Multicolor Photometric Systems, A. G. D. Philip, ed., Dudley Obs. Rep. No. 14, p. 363.

Kurucz, R. L., 1987, in IAU Colloquium No. 95, The Second Conference on Faint Blue Stars, A. G. D. Philip, D. S. Hayes and J. W. Liebert, eds., L. Davis Press, Schenectady, p. 129.
Kurucz, R. L. 1988 in New Directions in Spectrophotometry, A. G. D. Philip, D. S. Hayes, and S. J. Adelman, eds, L. Davis Press, Schenectady, p. 25.
Labhardt, L. 1988 in New Directions in Spectrophotometry, A. G. D. Philip, D. S. Hayes and S. J. Adelman, eds, L. Davis Press, Schenectady, p. 17.
Lequeux, J., Maucherat-Joubert, M., Deharveng, J. M. and Kunth, D. 1981 Astron. Astrophys. **103**, 305.
Maeder, A. 1981 Astron. Astrophys. **99**, 97.
Malagnini, M. L. 1988 in New Directions in Spectrophotometry, A. G. D. Philip, D. S. Hayes, S. J. Adelman, eds., L. Davis Press, Schenectady, p. 187.
Melnick, J., Terlevich, R. and Eggleton, P. P. 1985 Mon. Notices Roy. Astron. Soc. **216**, 255.
Nesci, R. 1981 Astron. Astrophys. **99**, 120.
Nesci, R. 1983a Astron. Astrophys. **121**, 226.
Nesci, R. 1983b Astron. Astrophys. **121** , 325.
Nesci, R. and Perola, G. C. 1985 Astron. Astrophys. **145**, 296.
O'Connell, R. W. 1973 Astron. J. **78**, 1074.
O'Connell, R. W. 1976a Astrophys. J. Letters **203**, L 1.
O'Connell, R. W. 1976b Astrophys. J. **206**, 370.
O'Connell, R. W. 1980 Astrophys. J. **236**, 430.
O'Connell, R. W. 1986a in Stellar Populations, C. A. Norman, A. Renzini and M. Tosi, eds., Cambridge Univ. Press, p. 167.
O'Connell, R. W. 1986b in Spectral Evolution of Galaxies, C. Chiosi and A. Renzini, eds., Reidel, Dordrecht, p. 321.
Oke, J. B. 1974 Astrophys. J. Suppl. **27**, 21.
Oke, J. B. 1988 in New Directions in Spectrophotometry, A. G. D. Philip, D. S. Hayes and S. J. Adelman, eds, L. Davis Press, Schenectady, p. 139.
Oke, J. B. and Buser, R. 1977, unpublished.
Paczynski, B. 1971 Acta Astron. **21**, 417
Philip, A. G. D. and Hayes, D. S. 1983 Astrophys. J. Suppl. **53**, 751.
Pickles, A. J. 1985a Astrophys. J. Suppl. **59**, 33.
Pickles, A. J. 1985b Astrophys. J. **296**, 340.
Renzini, A. and Buzzoni, A. 1986 in Spectral Evolution of Galaxies, C. Chiosi and A. Renzini, eds., Reidel, Dordrecht, p. 195.
Rood, R. T. and Crocker, D. A. 1985 in Production and Distribution of CNO Elements, J. Danziger, F. Matteucci and K. Kjar, eds., ESO, p. 61.
Rosa, M., Joubert, M. and Benvenuti, P. 1984 Astron. Astrophys. Suppl. **57**, 361.
Rose, J. A. 1985 Astron. J. **90**, 1927.
Schönberner, D. 1979 Astron. Astrophys. **79**, 108.
Schönberner, D. 1981 Astron. Astrophys. **103**, 119.
Schönberner, D. 1983 Astrophys. J. **272**, 708.

Seidel, E., Demarque, P. and Weinberg, D. 1987 Astrophys. J. Suppl. **63**, 917.
Shipman, H. L. 1971 Astrophys. J. **166**, 587.
Shipman, H. L. 1977 Astrophys. J. **213**, 138.
Snijders, M. A. J. 1984 in Fourth European IUE Conference, ESA SP-218, p. 3.
Spinrad, H. and Taylor, B. J. 1971 Astrophys. J. Suppl. **22**, 445.
Straižys, V. and Sviderskienė, Z. 1972 Vilnius Obs. Bull. **35**.
Sweigart, A. V. 1987 Astrophys J. Suppl. **65**, 95.
Sweigart, A. V. and Gross, P. G. 1976 Astrophys. J. Suppl. **32**, 367.
Sweigart, A. V. and Gross, P. G. 1978 Astrophys. J. Suppl. **36**, 405.
Taylor, B. T. 1982 Astrophys. J. Suppl. **50**, 391.
Taylor, B. T. 1984 Astrophys. J. Suppl. **54**, 167.
Terashita, Y. and Matsushima, S. 1969 Astrophys J. **156**, 203.
Thuan, T. X. 1986 in Star-Forming Dwarf Galaxies and Related Objects, D. Kunth, T. X. Thuan and J. T. T. Van, eds., editions Frontieres, p. 105.
Tinsley, B. M. 1968 Astrophys. J. **151**, 547.
Tinsley, B. M. 1980 Fund. Cosmic Phys. **5**, 287.
Tobin, W. and Nordsieck, K. H. 1981 Astron. J. **86**, 1360.
Turnrose, B. E. 1983 The IUE Ultraviolet Spectral Atlas, NASA Newsletter No. 22 (Special Ed.),
VandenBerg, D. A. 1983 Astrophys. J. Suppl. **51**, 29.
VandenBerg, D. A. 1985a Astrophys. J. Suppl. **58**, 711.
VandenBerg, D. A. 1985b in Production and Distribution of CNO Elements, J. Danziger, F. Matteucci and K. Kjar, eds., ESO, p. 73.
VandenBerg, D. A. and Bell, R. A. 1985 Astrophys. J. Suppl. **58**, 561.
VandenBerg, D. A. and Laskarides, P. G. 1987 Astrophys. J. Suppl. **64**, 103.
Viallefond, F. and Thuan, T. X. 1983 Astrophys. J. **269**, 444.
Voloshina, I. B., Glushneva, I. N. and Khruzina, T. S. 1982 Trudy Gos. Astron. Inst. Shternberga **52**, 182.
Voloshina, I. B., Glushneva, I. N., Doroshenko, V. T., Kolotilov, E. A., Mossakovskaya, L. V., Ovchinnikov, S. L. and Fetisova, T. S. 1982 Spectrophotometry of Bright Stars, Nauka, Moscow.
Voloshina, I. B., Glushneva, I. N., Doroshenko, V. T., Mossakovskaya, L. V., Ovchinnikov, S. L., and Khruzina, T. S. 1983 Trudy Gos. Astron. Inst. Shternberga **53**, 50.
Voloshina, I. B., Glushneva, I. N. and Shenavrin, V. I. 1985 Trudy Gos. Astron. Inst. Shternberga **55**, 84.
Wesemael, F., Auer, L. H., Van Horn, H. M. and Savedoff, M. P. 1980 Astrophys. J. Suppl. **43**, 159.
Whipple, F. L. 1935 Harvard Obs. Circ. No. **404**, p. 1.
Whitford, A. E. 1985 Pub. Astron. Soc. Pacific **97**, 205.
Wood, P. R. and Faulkner, D. J. 1986 Astrophys. J. **307**, 659.
Wu, C. C., Faber, S. M., Gallagher, J. S., Peck, M. and Tinsley, B. M. 1980 Astrophys. J. **237**, 290.
Wu, C. C., Ake, T. B., Bogess, A., Bohlin, R. C., Imhoff, C. L., Holm, A. V., Levay, Z. A., Panek, R. J., Schiffer, F. H. and Turnrose B. E 1983 The IUE Ultraviolet Spectral Atlas, NASA Newaletter No. 22 (Special edition).

DISCUSSION

OKE: I hope all observers will take to heart the need to observe stars of many different kinds. Part of what is needed is metal-poor stars.

CACCIARI: Yes, this is important, because you can't use surrogate stars. You have to use metal-poor stellar libraries.

BELL: Do you foresee that an optimizing synthesis method could be "intelligent" enough to note that it didn't have all the necessary stars in its library e.g. stars with very strong CN bands?

CACCIARI: In principle, yes, but it depends on a number of factors, e.g. the S/N of the data, the completeness of the library, a proper weighting of the feature to be matched, a very careful use of astrophysical constraints, etc. I guess each integrated spectral energy distribution has to be evaluated separately, in particular those that may contain some "special" population, and this involves the skill and judgment of the astronomer performing the analysis, as well as the "intelligence" of the optimizing synthesis code. Pickles (1985b) has a good discussion on this point.

HAYES: You mentioned the need for higher-S/N spectrophotometry. The present catalogs of stellar spectrophotometry in the visual covering a wide range of stellar types are not of the highest quality. Would you profit from having new catalogs with higher-quality data?

CACCIARI: Surely the better the S/N in the stellar spectrophotometry, the better the results from the spectral synthesis analysis. However, I believe a lot of the stellar data collected so far is already of sufficiently good photometric accuracy for spectral synthesis purposes. It is generally more difficult to obtain very high S/N in integrated spectral energy distributions of stellar systems, which however is essential for the reliability of the result.

ETZEL: Would you care to hazard a guess as to what the prime candidate is for the UV excess in early-type galaxies?

CACCIARI: Well, no. None of the suggested candidates, i.e. HB stars, young OB stars, post AGB stars or binaries, provides an explanation without problems. Personally, I would like some young stars as responsible for the UV upturn. But I refer you to the papers by Nesci and Perola (1985), Pickles (1985b) and Renzini and Buzzoni (1986) for a discussion of this issue.

TEAYS: In dwarf spheroidal galaxies you find anomalous Cepheids, which are overluminous, but, with one exception, are not found in galactic globular clusters. The difference is suggested to be due to the different stellar densities in the two systems, and the effect of this on binary orbits. These stars have not had their energy distributions studied due to their faintness. To what extent might their absence in the spectrum synthesis affect the results (as well as other "missing objects")?

CACCIARI: If their contribution to the total integrated light at any wavelength is not negligible, then their absence from the stellar library may affect the results. Usually, when a stellar type is missing, the algorithm tries to compensate for the missing energy at given wavelengths by adding more stars of a different type, which, however, may produce a flux excess at other wavelengths. This stresses the importance of using a stellar library which includes all the possible stellar types that can be found in the studied stellar system.

SPECTROPHOTOMETRY OF EARLY-TYPE POPULATION II STARS

A. G. Davis Philip

Van Vleck Observatory, Union College and
Institute for Space Observations

D. S. Hayes

Fairborn Observatory and
Institute for Space Observations

ABSTRACT: A set of spectrophotometric observations of early-type Population II stars is described which would be suitable as an observing program on an automatic spectrophotometric telescope (ASPT). A relatively large number of stars can be measured in a spectrophotometric system well calibrated to a set of standard stars. From the data obtained independent estimates could be made of astrophysical parameters such as temperature and surface gravity. In such a program hundreds of stars could be measured and the distribution of the points representing the astrophysical quantities of the stars will yield information concerning the nature of the various groups of Population II stars. Two of the major groups found so far are FHB stars which have evolved very little from the ZAHB and a second set of stars which have evolved past the turn (from left to right) of the evolutionary track in the theoretical HR diagram.

1. INTRODUCTION

In an article, "Spectrophotometry and Photometry of A-Type Field Horizontal-Branch Stars" (Philip 1987), a review was presented concerning measures of early-type Population II stars in the galactic halo. With the advent of automatic spectrophotometric and photometric telescopes (ASPTs and APTs) it will be possible to pursue these studies for a great number of stars. In the last few years, at the national observatories, it has been impossible to make spectrophotometric measures such as the ones made in Philip and Hayes (1983) and Hayes and Philip (1983). Classical photometric observing time is becoming increasingly difficult to obtain. The new telescopes being discussed at this meeting will help to solve the critical problem of obtaining observing time to work on projects in these areas.

In the papers mentioned above, spectrophotometric measures of

A. G. Davis Philip, D. S. Hayes and S. J. Adelman, (eds.)
New Directions in Spectrophotometry 175 - 186

horizontal-branch A-type stars were presented. Both stars in the field and in globular clusters were analyzed and it was shown that there were FHB stars which had nearly identical energy distributions as did BHB stars in globular clusters, thus giving further evidence concerning the similarity of FHB and BHB stars. If the field stars are true analogs of the cluster BHB stars then they become important probes for studies of the evolution of Population II stars. The field stars range up to six magnitudes brighter than the brightest BHB stars, allowing much more detailed studies to be made of the former group. A review of the work done on FHB stars at the galactic poles is given in Philip (1988a) and readers are referred to this paper for additional details.

In this paper we will present an observing program for an ASPT. A table is presented containing FHB stars confirmed by means of four-color measures, divided into stars with normal δc indices and stars with higher than normal δc, and ordered by y magnitude. These are the stars which should be measured first. Various suggestions are made for extending the program to other areas.

2. STARS WITH FOUR-COLOR MEASURES

A catalog of FHB stars with four-color measures is in preparation (Philip 1988b). The literature has been searched for papers in which four-color measures of FHB stars are presented. In the spectral range of A 0 to F 0 the FHB stars have very distinct features; namely a δc index > 0.2 and for stars later than A 5 a δm index a few hundredths of a magnitude smaller than a Pop I star of the same color. These two signatures make it quite simple to separate Population II early-type stars from those of Population I. Some giant stars of Population I fall in the same region of the four-color diagrams, but in regions at high galactic latitude these stars are quite rare and one does not expect to find a serious contamination of Population I stars in a group of stars selected as Population II from four-color measures of stars in a high galactic latitude region. Figures showing the distribution of FHB stars in the c_1, (b-y) and m_1, (b-y) diagrams can be found in Philip (1988a, Fig. 6).

Table I presents a list of FHB stars identified by means of four-color photometric measures. The stars in the table have been separated into two groups. The first group contains stars with δc indices between 0.15 and 0.28, the second group contains stars with δc indices > 0.28. Stars in the first group are thought to be stars on the lower evolutionary track in the theoretical HR diagram; those in the latter group are thought to be stars on the upper evolutionary track and thus more evolved than the first group. Finding charts and positions for many of the stars in Table I can be found in Philip (1984). Numbers in the table give the references for each star (the source of the four-color measures and a paper containing a finding chart).

TABLE Ia

FHB stars with δc_1 <0.28

Star	V mag.	Ref 1	Ref 2
HD 064488	7.1	1	HD
SAO 217082	8.3	8	HD
HD 083041	8.5	1	HD
HD 252940	8.7	9	HD
HD 074721	8.7	1,8,9	HD
HD 106301	9.1	8	HD
HD 117880	9.1	1,9	HD
HD 184779	9.2	10	HD
HD 008376	9.3	9	HD
HD 213468	10.9	9	HD
SS 206 II	11.0	1	3
BD -07 0230	11.1	7	BD
1 HLF 2 18 098	11.2	1	1
1 HLF 4 4 11 II	11.2	1	1
PHL 8657	11.6	6	7
4 HLF 3 16	11.7	1	1
PHL 6539	11.7	6	7
Feige 098	11.8	4	6
PHL 3275	11.9	6	7
LB 3367	11.9	6	7
1 HLF 2 17 024	11.9	1	1
LB 3126	12.0	6	7
1 HLF 5 7	12.2	1	1
SS 229 II	12.2	1	3
PHL 7209	12.4	6	7
Feige 016	12.4	4	6
4 HLF 5 1	12.8	1	1
SS 193 II	12.8	1	3
SS 179 II	13.0	1	3
LB 1707	13.0	6	7
PS 53 II	13.1	1	4
PS 52 II	13.1	1	4
PS 27 II	13.1	1	4
1 HLF 3 2 II	13.2	1	1
1 HLF 2 13 110	13.2	1	1
LB 3121	13.4	7	8
4 HLF 4 48	13.4	2	2
PS 45 II	13.5	1	4
SS 234 II	13.5	1	3
4 HLF 4 27	13.6	2	2
PS 59 II	13.7	1	4
SB 338	13.8	3	5
PS 23 II	14.0	1	4
PS 25 II	14.0	1	4
PS 21 II	14.1	1	4
4 HLF 4 36	14.3	2	2
SB 830	14.8	3	5
Feige 003	15.2	4	6

TABLE Ib

FHB stars with $\delta c_1 > 0.28$

Star	V mag.	Ref 1	Ref 2
HD 161817	7.0	1,9	HD
HD 109995	7.6	1,9,11	HD
HD 086986	8.0	1,8,9	HD
HD 130095	8.1	1,9	HD
HD 031943	8.2	9	HD
HD 139961	8.6	9	HD
HD 093329	8.8	8	HD
HD 106373	9.0	1	HD
HD 060778	9.1	1,7,9	HD
HD 012293	9.1	10	HD
HD 078913	9.1	9	HD
HD 069778	9.1	8	HD
HD 180903	9.1	9	HD
HD 202759	9.1	1	HD
BD +18 4873	9.5	1	BD
HD 087047	9.5	9	HD
HD 004850	9.6	9	HD
HD 013780	9.6	9	HD
BD +01 0514	9.8	2	BD
HD 002857	9.9	1	HD
HD 176387	9.9	1	HD
BD +25 2602	10.0	9	BD
HD 014829	10.3	1	HD
NGC 2437 003	10.3	8	9
BD +01 0548	10.7	2	BD
SS 287 I	10.8	1	3
SB 714	11.0	3	5
PHL 2408	11.0	5	7
SB 406	11.2	3	5
1 HLF 5 10	11.2	1	1
SS 194 II	11.3	1	3
4 HLF 2a 44	11.8	1	1
4 HLF 4 14	12.0	2	2
SS 210 II	12.0	1	3
LB 3211	12.0	6	7
LB 3147	12.1	6	7
Feige 068	12.1	4	6
1 HLF 2 18 021	12.3	1	1
4 HLF 4 47	12.4	2	2
Feige 097	12.4	4	6
SB 909	12.5	3	5
LB 3114	12.5	6	8

TS 149	12.7	7	10
1 HLF 1 3	12.7	1	1
4 HLF 6 2	12.7	1	1
1 HLF 3b 3	12.8	1	1
SS 202 II	12.8	1	3
4 HLF 4 19	12.8	2	2
Feige 060	12.9	4	6
4 HLF 6 7	13.0	1	1
SS 199 II	13.0	1	3
1 HLF 2 S 70	13.1	1	1
PS 35 II	13.1	1	4
4 HLF 4 31	13.1	2	2
PS 55 II	13.1	1	4
1 HLF 2 17 017	13.2	1	1
Feige 042	13.2	4	6
Feige 096	13.2	4	6
SS 222 II	13.3	1	3
4 HLF 4 49	13.4	2	2
LB 3203	13.4	6	7
SB 453	13.4	3	5
Feige 032	13.4	4	6
Feige 002	13.5	4	6
4 HLF 6 19	13.5	1	1
PS 50 II	13.5	1	4
1 HLF 2 16 111	13.5	1	1
4 HLF 4 42	13.7	2	2
SS 286 II	13.7	1	3
4 HLF 4 5	13.7	2	2
PS 06 II	13.7	1	4
1 HLF 2 17 136	13.7	1	1
PS 36 II	13.7	1	4
PS 16 II	13.8	1	4
Feige 008	13.9	4	6
SS 208 II	14.0	1	3
PS 15 II	14.1	1	4
PS 05 II	14.1	1	4
SB 107	14.2	3	5
SS 227 II	14.2	1	3
PS 01 II	14.3	1	4
4 HLF 4 30	14.5	2	2

REFERENCES IN TABLE 1a (1.x photometry), 1b (2.x chart)

1.1 Philip, A. G. D. 1988 Four-Color Catalog, unpublished.
1.2 Drilling, J. S. and Pesch, P. 1973 Astron. J. **78**, 47.
1.3 Graham, J. A. and Sletteback, A. 1973 Astron. J. **78**, 295.
1.4 Graham, J. A. 1970 Publ. Astron. Soc. Pacific **82**, 1305.
1.5 Kilkenny, D. 1984 Monthly Notices Roy. Astron. Soc. **173**, 173.
1.6 Kilkenny, D. and Hill, P. W. 1975 Monthly Notices Roy. Astron. Soc. **173**, 625.
1.7 Kilkenny, D., Hill, P. W. and Brown, A. 1977 Monthly Notices Roy. Astron. Soc. **178**, 123.
1.8 Stetson, P. 1981 Astron. J. **86**, 1500.
1.9 Stetson, P. 1988, personal communication.
1.10 Bond, H. E. and Philip, A. G. D. 1973 Publ. Astron. Soc. Pacific **85**, 332.

2.1 See Philip references in Philip (1978)
2.2 Philip, A. G. D. and Drilling, J. S. 1970 Bol. Yon. y Tac. **5**, 297.
2.3 Sletteback, A. and Stock, J. 1959 A Finding List of Stars Of Spectral Type F2 and Earlier in a North Galactic Pole Region, Hamburg-Bergedorf Observatory.
2.4 Philip, A. G. D. and Sanduleak, N. 1969 Bol. Ton. y Tac. **30**, 253.
2.5 Sletteback, A. and Brundage, R. K. 1971 Astron. J. **76**,338.
2.6 Feige, J. 1958 Astrophys. J. **128**, 267.
2.7 Haro, G. and Luyten, W. J. 1962 Bol. Ton. y Tac. f63, 37.
2.8 Luyten, W. J. 1969 A Search for Faint Blue Stars, No. 50, University of Minnesota.
2.9 Smyth, M. J. and Nandy, K. 1962 Pub. Roy. Obs. Edinburgh **3**, 23.
2.10 Cavira, E. 1958 Bol. Ton. y Tac. **2**, 31.

3. SURVEYS FOR FHB STARS

When all the stars in Table I are measured the survey can be extended to additional areas. A major spectroscopic survey has been made by Beers, Preston and Schectman (1985, 1988) using the Michigan Curtis Schmidt telescope at Cerro Tololo and the Burrell Schmidt telescope at Kitt Peak with a 4 degree prism and a narrow-band interference filter isolating the H and K lines of Ca II. 90 minute plates reach B = 16 mag. They pick stars showing weak or absent Ca II but strong Hϵ. At the present time more plates have been taken at Cerro Tololo and have southern declinations. However there are a number of fields which are reachable from latitudes of +30°. These fields are listed below in Table II. As the survey progresses areas at the North Galactic Pole will be searched for FHB candidates and this search will produce a large number of FHB candidates with northern declinations. Another important source for locating metal-poor early-type stars, all over the sky, is the spectroscopic survey being carried out by Nancy Houk at the University of Michigan. She has published a list of Weak-Lined HD Stars in the Michigan Spectral Catalogues (Houk 1986) and as new volumes of the Michigan Spectral Catalog are published additional stars will be added to this list. This list is a very good source for identifying FHB stars.

TABLE II

Areas in the Beers et al. Survey Reachable from +30°

Plate	R. A. (1950)	Dec.	Number of Stars by color classes		
			b-vb	mb-m	mf-f-vf
29521	22 58.2	+09 48	11	08	05
22965	21 58.4	-05 08	08	13	21
22967	01 18.4	-05 09	06	07	03
29512	22 08.1	-10 10	06	11	13
29517	23 52.1	-15 11	04	05	03
29506	21 24.9	-20 09	07	26	14
29528	02 19.3	-20 09	03	04	07
29527	00 34.2	-20 11	05	07	09
			50	81	75

b=bright, f=faint, v=very, m=medium.

According to Fig. 2 in Beers et al. (1988) the magnitudes of stars in the first group go down to about V = 13 mag., for the second group the limit is near V = 14 mag. and for the third group the magnitudes range from 14 to 16. They estimate that 85% of these stars are FHB stars. The remainder are stars with main-sequence gravity, A stars of near solar metallicity, high luminosity or binary stars. There are 49 additional spectral plates that have been taken at the North Galactic Pole and these plates should produce about 300 stars down to V = 13 mag., based on the data contained in Table II. Combining the stars from each source we have a potential list of 450 stars down to V = approximately 13 mag. If such a large sample of FHB stars were to be measured spectrophotometrically and analyzed to yield temperatures and surface gravities then an important test of the predictions of stellar evolutionary theory could be made. The theory predicts the distribution of stars along a typical evolutionary track and when the derived temperatures and gravities are plotted in the theoretical HR diagram there will be sufficient data points to make such an analysis.

4. THE SPECTROPHOTOMETRIC TELESCOPE

The spectrophotometer and telescope which we propose to use for the initial phase of this investigation is described by Hayes, Adelman and Genet (1988) in this Symposium, so only a brief summary will be given here. The telescope will be an 0.75-m Cassegrain reflecting telescope, fully automated so that it will run by itself without an astronomer being in attendance. The spectrophotometer will be a grating device using a CCD as the detector. The equivalent dispersion will be 4 Å/pixel in the second-order blue and 8 Å/pixel in the first-order red. During the reduction phase, the data will be binned into resolution elements of 32 Å to match our previous work and to match the spectrophotometric standards. The integration times which are necessary to reach 10^4 counts per 32 Å interval at 3500 Å for an A-star have been calculated from the information given by Hayes, Adelman and Genet (1988); they are given in Table III.

TABLE III

Integration times for stars of various magnitudes

Mag.	Time	Mag.	Time
8	2 min	11	23 min
9	4	12	58
10	9	13	144

Taking the stars in the two parts of Table I as an example, we can calculate how much observing time would be necessary to observe the stars. First, we note that at 13th mag. the integration time is nearly 2 1/2 hours, which is too long to spend in a single observation. The best strategy would be to break this into seven 20-min. integrations, each of which would be broken into four integrations of 5 min. each. The telescope would have to be recentered after each 5-min integration. Since the spectrophotometer will have two diaphragms, so that the sky can be measured simultaneously with the star, the telescope will be beam-switched in ABBA sequences with A and B denoting the two diaphragms. Thus, each ABBA sequence will take 20 min. The seven such sequences necessary to observe a 13th-mag. star are best done on seven different nights. It is probably not wise to attempt to extend this approach to fainter stars than 13th. In an initial phase, then, we will limit the observations to stars of 13th mag.; it is hoped that in a later phase a large spectrophotometric telescope might be built and the fainter stars on the list could be observed. There are 77 stars of 13th mag. or brighter in Table I. If we take the integration time for the faint end of each magnitude interval (which results in a very conservative estimate), we can sum the integration times for all 77 stars. The total is 15 nights (8 hours per night). If we add 50%, or 7.5 nights (1/3 of the total), for observing standards and extinction stars, we get a total of 22.5 nights. Acquisition will take 2-3 min per star, which accumulates to 2-3 hours for the 77 stars. Thus, about 23 nights should suffice for the program. This will quadruple the number of FHB stars with spectrophotometry, and is thus a small investment for such an increase in the statistical sample.

DISCUSSION

We have presented an observing program which is well suited to an automatic spectrophotometric telescope (ASPT). The program is modest in its total time requirements, but its completion will have a dramatic impact on the statistical sample of early-type Population II stars which can be assigned values of T_{eff} and log g and compared with evolutionary models. Further, it is possible to add more stars brighter than 13th mag. to the observing list, so with more telescope time available, a much enhanced sample could be observed.

REFERENCES

Beers, T. C., Preston, G. W. and Schectman, S. A. 1985 Astron. J. 90, 2089

Beers, T. C., Preston, G. W. and Schectman, S. A. 1988, preprint.

Hayes, D. S. and Philip, A. G. D. 1983 Astrophys. J. 53, 759.

Hayes, D. S., Adelman, S. J. and Genet, R. M. 1988 in New Directions in Spectrophotometry, A. G. D. Philip, D. S. Hayes and S. J. Adelman, eds, L. Davis Press, Schenectady, p. 311.

Houk, N. 1986 in Spectroscopic and Photometric Classification of Population II Stars, A. G. D. Philip, ed., L. Davis Press, Schenectady, p. 19.
Philip, A. G. D. 1978 in IAU Symposium No. 80, The HR Diagram, A. G. D. Philip and D. S. Hayes, eds., Reidel, Dordrecht, p. 209.
Philip, A. G. D. 1984 Contrib. Van Vleck Obs. 2, 1.
Philip, A. G. D. 1987 in New Generation Small Telescopes, D. S. Hayes, R. M. Genet and D. R. Genet, eds., Fairborn Press, Mesa, p. 165.
Philip, A. G. D. 1988a, in preparation.
Philip, A. G. D. 1988b, in preparation.
Philip, A. G. D. and Hayes, D. S. 1983 Astrophys. J. 53, 751.

DISCUSSION

BELL: Do you get M_V for the horizontal-branch stars in the globular clusters which you have studied?

PHILIP: No, at present there are not sufficient standards set up in the CCD frames of the globular cluster regions to derive an independent V or y magnitude. Rather I am using the published UBV V magnitudes to set the y magnitude in the four-color system. So I am working in an "instrumental" four-color system but this is sufficient to map out the morphology of the BHB stars in each cluster. I hope to do this calibration work at some future time.

WARREN: How do the techniques of transformation to the standard four-color system, relative to filter observations of the classical method, differ when reducing CCD data?

PHILIP: Transforming to the standard four-color system is a problem. For each of the CCD four-color observing requests I have pointed out the need to make conventional four-color observations of a few standard stars in each frame. When the observing time is awarded the CCD time is granted but the photoelectric time in the single-channel mode usually is not.

In the globular clusters M 4 and M 92 there are a few stars that have been measured on the 60 inch telescope at CTIO and the 90 inch at Steward Observatory and frames have been selected which contain some of these stars. What I do is to measure three frames per color with the program DAOPHOT at the Dominion Astrophysical Observatory and then compute an average instrumental magnitude in each filter. By comparing the magnitudes computed on each of the three frames the internal errors of observation can be found. These are very good for stars at V magnitudes of 13.6 and 15.6. In most cases I have found internal probable errors of from ±0.003 to ±0.008 which is much better than the internal probable errors found in the earlier photoelectric photometry. These probable errors were ±0.05 in the c_1 index and the only way that I could improve on the statistics was to measure some of the more important stars ten times or more. The CCD measures have been made with 0.9 m telescopes so one can see the greatly improved precision of the new method.

To compute the instrumental four-color indices I reduce each set of frames and find the average y, b, v and u magnitudes. Then I compute (b-y), c_1 and m_1 by performing the proper subtractions of the various magnitudes. The sets of three frames are all taken at nearly the same time so their airmass corrections are quite similar. For the longer u frames I try to take a series of exposures at the same time on consecutive nights, so in these cases the airmass corrections are similar also. For clusters that have standards, the instrumental colors can be transformed to the standard four-color system. For clusters that have no standards as yet, it is possible to study the morphology of the horizontal branch in the instrumental plane.

PHYSICAL PARAMETERS FROM STELLAR SPECTROPHOTOMETRY

M. L. Malagnini

University of Trieste

C. Morossi

Astronomical Observatory, Trieste

ABSTRACT: Spectrophotometric measurements and theoretical atmosphere models have been used to derive the effective temperatures, radii and metallicities for Population I and II non-supergiant stars. The aim of this project is to find the actual positions of stars in the L_{bol} - log T_{eff} diagram through direct determination of these quantities, thus providing a statistical database, as large as possible, for calibration purposes.

1. INTRODUCTION

In the framework of the stellar population synthesis approach for interpreting evolutionary phenomena in galaxies, it is highly desirable to start with a fully tested and exhaustive library of spectral energy distributions representative of each constituent star. Since, in most cases, only integrated spectra from galaxies are available, suitable stellar population mixtures are sought for reproducing the observed integrated light. This approach involves the solutions of different classes of problems, both on the theoretical plane and from the observational point of view, referring to reliable and calibrated stellar evolutionary sequences and theoretical models for the spectral evolution of stellar populations (Renzini and Buzzoni, 1986, and references therein). Theoretical models can be successfully used only when accurate tests prove their ability to reproduce real stars and stellar systems. On the observational side, this possibility is related to the availability of a complete reference library of stellar SED's (Spectral Energy Distributions). On the other side, empirical SED's can be replaced in the reference library by model atmosphere predictions only after the comparison between observed and computed fluxes has been proved to satisfy criteria of consistency within acceptable limits on errors.

In principle, one must provide complete coverage of each

A. G. Davis Philip, D. S. Hayes and S. J. Adelman, (eds.)
New Directions in Spectrophotometry 187 - 196

non-void box in the HR diagram, and correspondingly a complete link between observable or empirical quantities and the physical parameters needed to compare stars and models in the theoretical HR diagram. At present no existing library covers all stellar types nor do SED's cover all the desirable wavelengths.

These considerations have motivated our line of research: in particular, we focused on the derivation of fundamental physical parameters for stars, starting from available observational data for large samples of different stellar types, and from atmospheric models. Optical spectrophotometric measurements, and, when available, ultraviolet data from space instrumentation, have been used in connection with model atmospheres for deriving effective temperatures, gravities, and metallicities for non-supergiant stars of Populations I and II.

In this paper we will present a review of the results we have obtained so far for about 300 solar chemical composition stars representative of solar vicinity objects of Population I, and a discussion of the relationship between T_{eff} and its estimate based on the analysis of the energy distributions only in the optical region, T_{opt}, together with new results concerning the derivation of radii for stars with accurate spectrophotometry and reliable trigonometric parallaxes.

2. DETERMINATION OF T_{eff} FOR MAIN-SEQUENCE STARS

Ideally, only when the absolute energy distribution for the stellar surface of a star over the whole wavelength range is known, can its T_{eff} be computed. Knowledge of the stellar radius will permit the derivation of the bolometric luminosity, and detailed spectral features will provide information about chemical composition.

Many methods have been scrutinized and successfully used to arrive at the derivation of the stellar parameters (T_{eff}, R, chemical composition) by resorting to different approaches to overcome the limits due to partial spectral information. Along this line, we concentrated at first on the study of solar vicinity "normal" stars, i.e. stars having a chemical composition equal to that of the Sun.

The method we followed is that of comparison between spectrophotometric measurements, which sample the stellar SED's, and the fluxes theoretically predicted by means of model atmosphere codes. This approach permitted us to i. test the consistency of existing models with observations, ii. derive temperature and gravity information for large samples of stars, iii. obtain reliable estimates of stellar angular diameters, and iv. determine a good empirical scale for the bolometric correction versus T_{eff}. The first step is extremely important in many aspects, and preliminary for the correct use of theoretical SED's in stellar population synthesis applications.

Indeed, since observations cannot be obtained over the whole spectral range, model fluxes must be used instead, but the substitution is feasible if and only if they provide a good description of real stars in all the observable parts of the spectrum for all the different classes of stars; moreover, only "realistic" models would permit the derivation of reliable chemical abundances.

For the analysis, we refer to the most complete grid of LTE model atmospheres (Kurucz 1979) and to homogeneous collections of spectrophotometric measurements, in the UV and/or in the optical spectral regions, for different samples of stellar types (the following catalogues have been used: the Ultraviolet Bright-Star Catalogue, hereafter S2/68 Catalogue, by Jamar et al. 1976, the IUE Ultraviolet Spectral Atlas, NASA Catalogue, by Wu et al. 1983, the IUE Low Dispersion Spectra Reference Atlas, part I: Normal Stars, ESA Catalogue, by Heck et al. 1984 and the Catalog of Spectrophotometric Scans of Stars, Breger Catalogue, by Breger 1976). Details about the numerical procedure we developed can be found in Malagnini and Morossi (1983), and detailed results in Malagnini et al. (1982, 1983, Paper I and II, respectively) and Morossi and Malagnini (1985, Paper III). We report here the main conclusions referring to non-supergiant normal stars, whose observed fluxes were homogeneously normalized and dereddened in accordance with classical methods.

For each star of the complete set of 129 A and F non-supergiant stars observed by the S2/68 experiment, the results are represented by the pair of values [T_{UV}, log g_{UV}] of the interpolated model which gives the best description of the observed UV data according to the LMS error criterion. These results have been compared with analogous determinations obtained by different authors from the analysis of the optical region. The main conclusion of the comparison was that practically the same model accounts for the observed stellar SED's from the optical to the UV region.

More insight has been gained when analyzing both the UV and the optical regions, referring to the 48 stars of spectral type B 5 - A 0 of luminosity classes from V to III, included in the S2/68 Catalogue and in the ample collection of uniformly calibrated optical data by Breger. Particular care has been taken in the analysis of the region between the last UV and the first optical data, and in the comparison of our results with similar determinations from different authors. The main conclusions are that Kurucz models reproduce fairly well, both in shape and in absolute level, the SED's observed in the optical region, while the residuals between computed and observed fluxes tend to be fairly large in the UV region (the largest discrepancies appear for the data referring to the 2740 Å band). Even though observational uncertainties might be corrected for, it still appears clear that in the temperature range 10,000 - 16,000 K no reliable fit can be reached by using UV information only, while results of a quality equivalent to that obtained from optical data are reached by limiting the degrees of freedom of the fit. For instance, a useful constraint is to fix the apparent angular diameter to the value obtained from the best fit in the optical region. Instrumental effects for S2/68 data, in particular at 2740 Å, are likely to have been responsible for lack of continuity in the 2500 - 3200 Å region. Stellar angular diameter determinations have been compared with photometric estimates, which prove to be highly correlated to each other, with linear correlation slope close to 1.0; direct comparison with interferometric measurements, possible in three cases, proved that our results were within observational uncertainties.

The above conclusions suggested the possibility of extending our analysis to all normal stars having optical spectrophotometry reduced by Breger, without the constraint of the availability of UV space observations, foreseeing good accuracy for the two main parameters, temperature and apparent angular diameter. After excluding supergiant and/or peculiar stars, the Breger Catalogue provides 435 sets of measurements, referring to 302 stars in the spectral type range O 9 - G 8. We fix, a priori, the log g value assigned to each star: the criterion was the correspondence between luminosity class and gravity from the calibration reported in Allen (1976). In any case, computations have been made for each possible gravity value, while the solution [T_{opt}, θ_{opt}] reported in Morossi and Malagnini (1985) always refers to this choice for log g; in most but not all cases the solution is also the best fit one. In general, the accuracy of the comparison between observed and predicted optical fluxes is on the same order as the uncertainty of the observations (~0.03 mag); the results in T_{opt} and in θ_{opt} are estimated to be accurate to within 5% and 10%, respectively. The average relationship between T_{opt} and $(B-V)_o$ for main-sequence stars shows a good agreement with previous results, although individual stars may significantly deviate from it. Therefore, when precise information on individual stars is required, individual analysis is better suited than the use of average relationships; on the other hand, the larger the number of stars used for calibration purposes, the better the quality and reliability of the calibration itself. This is clearly the case of the empirical determination of the bolometric scale versus T_{eff}.

The fundamental work by Code et al. (1976) rests on the data measured for 32 stars in the spectral type range O 5 to F 8: we improved that scale by almost doubling the number of stars used for the calibration, since our method permits us to overcome the limitation, due to observational constraints, of direct measurements of stellar angular diameters (Malagnini et al., 1986).

The main result of the direct determination of T_{eff} through integration of the total flux energy distributions, scaled by taking into account the stellar angular diameter, rests on the availability of the complete set of measurements from UV to the infrared part of the spectrum, and on the reliability of the models used for adding the still unobservable parts of the SED's. Fig. 1 shows the good agreement between the apparent angular diameters we derive and the photometric or

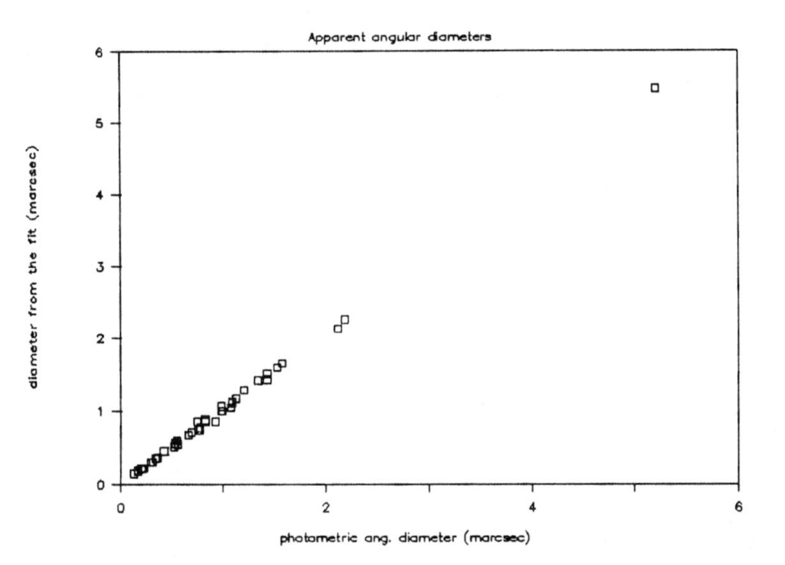

Fig. 1. The apparent angular diameters obtained through the fit are compared with the photometric angular diameters or with the interferometric determinations, if available.

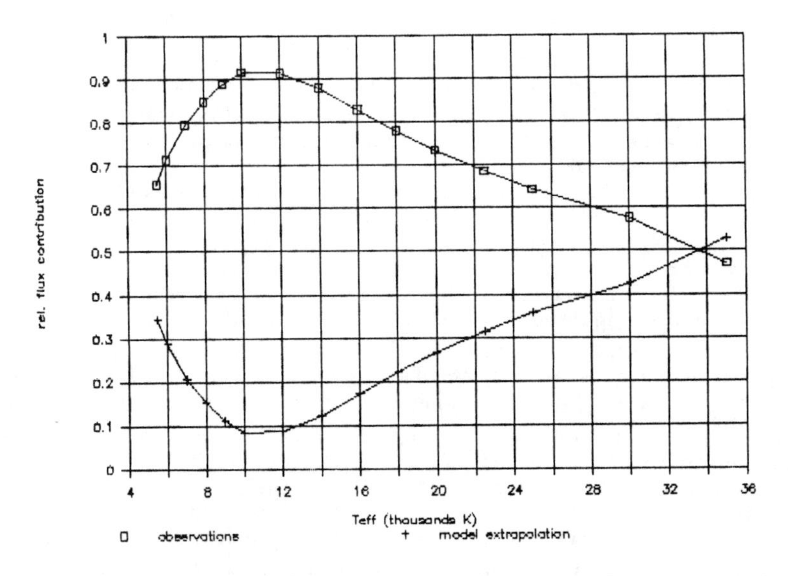

Fig. 2. The relative contribution to the total flux expected from observations in the spectral range 1250 - 10,000 Å, and that expected from model extrapolations (shortward of 1250 Å and longward of 10,000 Å) are plotted for different values of T_{eff}, at fixed model surface gravity (log g = 4.0).

interferometric determinations. Fig. 2 illustrates, as an example, the relative contribution to the total flux due to the observations (1250 - 10,000 Å) and to the models (shortward of 1250 Å and longward of 10,000 Å) as a function of T_{eff}. To give an idea of the importance of the uncertainties in different wavelength ranges on the total flux used to derive T_{eff}, Table I lists the actual data for the four stars having measured angular diameters. In general, the uncertainty affecting T_{eff} is on the order of 5% and that affecting the angular diameter is on the order of 8%. In particular, in the T_{eff} range 7000 - 18,000 K the results are largely independent of the accuracy of the models.

TABLE I

Relative uncertainties in the integrated stellar fluxes

HD	UV	IUE	VIS	IR	TOT
61421	0.0	0.6	1.7	2.1	4.4
87901	0.8	5.0	1.1	0.3	7.2
159561	0.0	1.3	1.8	1.1	4.3
216956	0.0	1.9	1.6	0.6	4.1

3. DETERMINATION OF R/R_{Sun}

The location in the L_{bol} - log T_{eff} diagram of the stars for which T_{eff} and B.C. have been obtained is conditioned by the existence of reliable distance estimates. In particular, parallaxes accurate to within 25% are available only for 23 out of 42 stars. To improve the significance of the comparison with theoretical stellar evolution predictions, it is highly desirable to increase the number of objects having complete SED's from observations and good parallaxes. Since a greater number of accurate trigonometric parallaxes are available but for stars with incomplete SED's, new accurate spectrophotometric observations are required, in the UV or in the optical range. Meantime other methods have to be used for the determination of T_{eff} to increase the number of objects in the theoretical HRD. Our approach is to make use of T_{opt} as a reliable estimator of T_{eff}: in Fig. 3 T_{opt} is plotted versus T_{eff} for the 40 stars analyzed both in Morossi and Malagnini (1985) and in Malagnini et al. (1986), together with the 45° line. From the linear regression analysis it results that the intercept and the slope are not significantly different from zero, and from one, respectively. The same results have been achieved for the estimates of the apparent angular diameters. As a consequence, by using the T_{opt} and θ_{opt} determinations for the 302 O 9 - G 8 stars analyzed in Morossi and Malagnini (1985) and the parallaxes given in the new General Catalogue of Trigonometric Stellar Parallaxes (van Altena 1986), we are

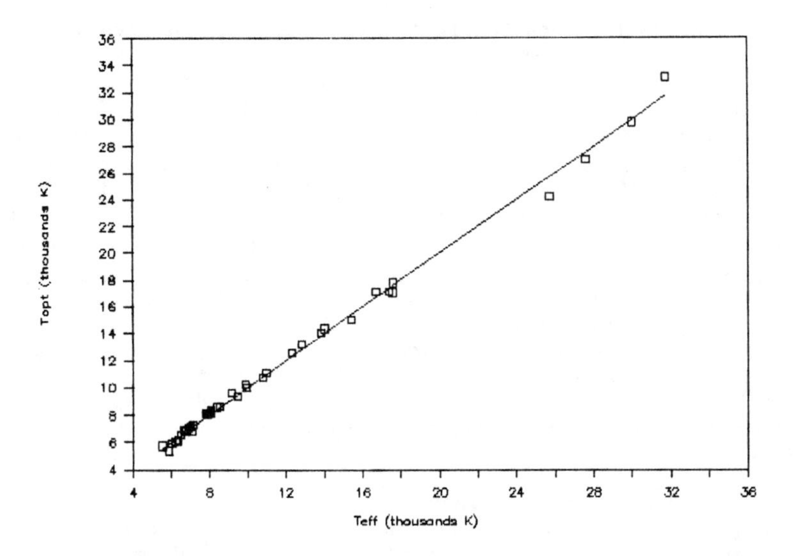

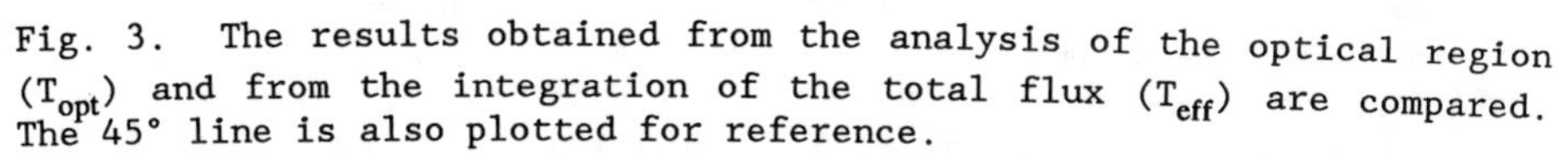
Fig. 3. The results obtained from the analysis of the optical region (T_{opt}) and from the integration of the total flux (T_{eff}) are compared. The 45° line is also plotted for reference.

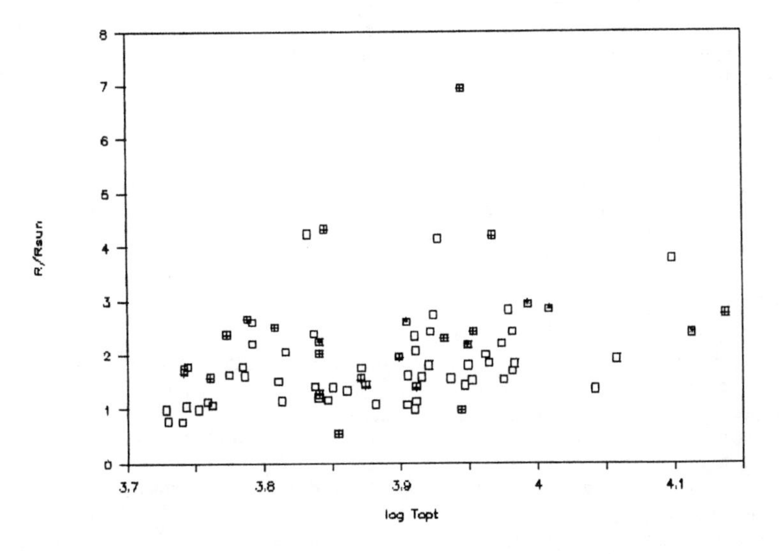

Fig. 4. Radii (R/R_{Sun}) and temperatures ($\log T_{opt}$) are plotted for the program stars (luminosity class V is represented by empty boxes, luminosity classes III, III-IV or IV are represented by marked boxes).

able to derive the location in the L_{bol} - log T_{eff} diagram for other 54 stars with relative parallax errors less than 25%. As a byproduct, Fig. 4 shows the values of R/R_{Sun} versus log T_{opt}. The different luminosity classes, V (empty boxes) and III, III-IV or IV (marked boxes), are mixed together without any strong correlation with the R/R_{Sun} values. This behavior may be due to the fact that the luminosity classification is not intended to provide quantitative information on the radius of the star or to misclassification. This last problem may also affect the values of T_{opt} and of θ_{opt}, since they have been obtained by fixing a priori the log g_{opt} values according to luminosity class. We are planning to obtain gravity estimates for each star from the analysis of gravity-dependent spectral features. Another bias is introduced by the initial assumption of solar chemical composition, which can not be checked because of the lack of the spectral energy distribution in the UV. In fact, only by looking closely at that wavelength range can one derive indications of the [M/H] values (Cacciari et al. 1987).

In conclusion, for increasing the statistical significance of any comparison between observations and theories, in particular for the safe use of the stellar population synthesis method, more observations are required to achieve, for a much larger sample of stars, i. complete SED's, ii. refined surface gravity information, iii. accurate trigonometric parallaxes. Obviously the accuracy of the results will be greatly improved, on one hand, from higher accuracy in the observations and, on the other hand, in particular for the log g and [M/H] determinations, from more refined and physically reliable atmosphere models.

ACKNOWLEDGMENTS

Partial support from a bilateral CNR grant and MPI 40% grants are acknowledged. We are grateful to Dr. W. van Altena for providing us with the magnetic tape version of the new General Catalogue of Trigonometric Stellar Parallaxes.

REFERENCES

Allen, W. C. 1976 in Astrophysical Quantities, The Athlone Press, London, p. 213.

Breger, M. 1976 Astrophys. J. Suppl. **32**, 7.

Cacciari, C., Malagnini, M. L., Morossi, C. and Rossi, L. 1987 Astron. Astrophys. **183**, 314.

Code, A. D., Davis, J., Bless, R. C. and Hanbury Brown, R. 1976 Astrophys. J. **203**, 417.

Heck A., Egret, D., Jaschek, M. and Jaschek, C. 1984 IUE Low-Dispersion Spectra Reference Atlas, Part 1: Normal Stars, ESA SP-1052.

Jamar, C., Macau-Hercot, D., Monfils, A., Thompson, G. I., Houziaux, L. and Wilson, R. 1976 Ultraviolet Bright-Star

Spectrophotometric Catalogue. ESA SR-27.
Kurucz, R. L. 1979 Astrophys. J. Suppl. **40**, 1.
Malagnini, M. L., Faraggiana, R., Mcrossi, C. and Crivellari, L. 1982 Astron. Astrophys. **114**, 170. (Pap. I)
Malagnini, M. L., Faraggiana, R. and Morossi, C. 1983 Astron. Astrophys. **128**, 375. (Pap. II)
Malagnini, M. L. and Morossi, C. 1983 in Statistical Methods in Astronomy, ESA SP-201, p. 27.
Malagnini, M. L., Morossi, C., Rossi, L. and Kurucz, R. L. 1986 Astron. Astrophys., **162**, 140.
Morossi, C. and Malagnini, M. L. 1985 Astron. Astrophys. Suppl. **60**, 365. (Pap. III)
Renzini, A. and Buzzoni, A. 1986 in Spectral Evolution of Galaxies, C. Chiosi and A. Renzini, eds., Reidel, Dordrecht p. 195
van Altena, W. 1986, private communication
Wu, C. -C., Ake, T. B., Boggess, A., Bohlin, R. C., Imhoff, C. L., Holm, A.V., Levay, Z. G., Panek, R. J., Schiffer III, F. H. and Turnrose, B. E. 1983 The IUE Ultraviolet Spectral Atlas, NASA IUE Newsletter 22.

DISCUSSION

LODEN: You said that you apply this to "normal" stars only. What is your definition of "normal" stars?

MALAGNINI: They're "normal" just in the sense that no anomalies or peculiarities are reported in the literature (see Malagnini et al. 1983 for references).

ETZEL: Were your tests demonstrating the relative lack of sensitivity to the selection of log g made only with optical data?

MALAGNINI: Yes, the figure shown here refers to the analysis of the optical region alone: in that case, log g's are fixed according to LC's, and the comment is in order.

ETZEL: If so, small systematic errors in temperature determination may result when you fit the combined optical and FUV data. For example, assuming log g = 4.0 for a B 9 V rather than something greater (e.g. 4.2) will result in a temperature estimate that will be too high as a consequence of matching the shorter wavelengths of the Balmer continuum.

MALAGNINI: The answer is negative when optical and ultraviolet data are referred to. Indeed, in this case no constraints on log g's are applied, so the best fit model determines simultaneously both the atmospheric parameters T_{eff} and log g, and the apparent angular diameter.

SYNTHESIS OF PHOTOMETRIC SYSTEMS FROM SPECTROPHOTOMETRIC DATA

David L. Crawford

Kitt Peak National Observatory*

ABSTRACT: With linear detectors, all of spectrophotometry is really "synthetic photometry," in the sense that one is doing photometry on the incoming flux, with some spectral resolution. One can and often will produce useful "photometric indices" from the spectrophotometry. This paper reviews the basic principles and gives a few elementary examples to illustrate some of the issues.

1. INTRODUCTION

For the purpose of the discussion in this paper, I will define "spectrophotometry" as a measure of I(λ), with some spectral resolution defined by the instrumentation used. "I" is the flux measured and "λ" is the wavelength. I will also assume here that we are measuring one star at a time, rather than doing "array" photometry.

The flux emitted by the star (or other object of interest) is affected by many things on the way to our reduction program: interstellar matter, the Earth's atmosphere, the telescope and the instrument's optics, the filters used (or other method used to obtain wavelength discrimination), the detector's response, and any other such factors. In some cases, we are attempting to measure the factor; in other cases, we need to remove or correct for the factor. The "filters" could be actual filters or slots or just the resolution elements in a spectrum.

---

*Operated by AURA Inc. under contract with the National Science Foundation.

A. G. Davis Philip, D. S. Hayes and S. J. Adelman, (eds.)
New Directions in Spectrophotometry 197 - 212

Let us write the concept in the following equation (I will drop the subscript "lambda", but remember that we have an equation for each spectral resolution element at each lambda):

$$I(observed) = \int I(star) * I(IM) * I(EA) * I(TL) * I(F) * I(D) * I(O) d\lambda$$

In words, the flux measured is the convolution of the flux emitted with all of the intervening effects, some controllable (or chosen), some out of our control. Sometimes we use the I(star) as a probe to measure one of the effects, such as of (IM), for example.

This type equation, and issues concerning it, are nicely discussed by Buser (1978, 1986), Buser and Kurucz (1985), Gustafsson (1986) and others. See, for example, the many references cited in these papers.

The I(star) could also be produced by theoretical models, of course, and such fluxes have been used in many synthetic photometry studies. Lester, Gray, and Kurucz (1986) have presented an excellent study of theoretical uvby and H-beta indices, for example.

We should also be aware, in understanding photometry, that we need an equation such as:

$$Mag(I) = -2.5 * \log[Flux * D^2 * Q * t * BW * SNR^{-2}]$$

where D is the telescope aperture, Q is the "quantum efficiency" of the detector, t is the time observed, BW is the bandwidth of the filter (or resolution element, for spectrophotometry), SNR is the signal-to-noise ratio obtained.

A goal of synthetic photometry is to analyze spectrophotometric data, either that already published or new data, so as to produce useful photometric indices. The above two equations are the ones to be used for such analysis.

Another goal is to use existing information on fluxes, detector sensitivity, filter transmission curves, and such to MODEL photometric systems, so as to understand the effect of the terms in the above equations, over reasonable assumptions about the make-up of the terms. There are many useful applications, such as understanding an existing photometric system or designing a new one.

There are, for example, undoubtedly several different interstellar reddening laws, observatories exist at different observing sites, and at different altitudes above sea level (different amount of atmosphere above the telescopes, and different characteristic of the atmosphere), aluminum coatings age with time and their reflective properties change (and not all reflective coatings are the same anyway), and detectors differ greatly in their sensitivity with respect to wavelength, even detectors that have the same "model number."

Reductions procedures also differ greatly. The most common thing one looks at with synthetic photometry, however, is probably to study the effects of using different filters, in an effort to reproduce standard photometric systems.

2. FILTER SYSTEMS

We will here illustrate the issues by taking one set of published stellar fluxes and convoluting them with a number of different filters. We will ignore, in this paper, all other factors. Some of these factors are critically important, and any careful investigation must do a first class job of including them, of course.

We choose the stellar fluxes published by Straižys and Sviderskienė (1972), hereafter called SS, because they are simple to enter and use on a PC, and because others have also used such data for similar studies, including SS themselves.

We will use a number of different filter curves, varying in location and in bandwidth and in shape, so as to illustrate some things that can be investigated. The filter curves used are described in Table 1. The range of types is sufficient to show some of the things that can be looked at.

Future papers will use more detailed flux curves and many more filter types, as well as including the other factors ignored in this present paper. As I noted, some of these factors are usually important to include and are of great interest in their own right. Studies of interstellar reddening, for example, should be done with care, in such a way as to minimize other factors that may confuse the issue under study. One must understand one's photometric system well before one can draw accurate conclusions.

Here, and in future papers, we will choose filters that are similar to UBV and uvby filters (at least in wavelength), so that one can have a feeling for the results. Future papers will include filter curves designed to investigate other photometric systems as well. In all such investigations, it is important to remember some general principles about photometric systems and photometric data reductions. There are many such discussions in the literature; see Hayes and Crawford (1986) for example and for other references contained there.

In this paper, I will not discuss issues such as signal-to- noise ratios vs. limiting magnitude, or any other such interrelations between the terms in the above equations. Such interrelations are an important part of understanding and using photometric systems, and future papers will discuss such issues.

Let me only remark that the interrelation between bandwidth and some of the other factors is really the difference between "photometry" and "spectrophotometry" with the new, linear detectors. Narrower bandwidth would go in the direction of spectrophotometry, of course. Often, but not always, this is also in the direction of increased "information resolution." One is measuring more parameters and with higher spectral resolution. If that is what is needed for the research problem, then that is the information resolution needed as well. For many problems, such information resolution is not needed, and "photometry" is the preferred choice. The wider bandwidths used mean one can work fainter, the other factors, such as telescope size, being the same.

Let us define, in fact, WIDE BAND (WB) photometry as that with wavelength resolution (band widths) between 500 and 1000 Å Å, MEDIUM BAND (MB) photometry as that with band widths between 100 and 300 Å, and NARROW BAND (NB) photometry as that with band widths between 10 and 50 Å, approximately.

The goal in using synthetic photometry to investigate photometric systems is to have a better understanding of the interaction of all the important factors. The result will be better matched systems, better designed for the scientific job, more homogeneous data, and better astronomy therefore.

3. SOME CALCULATIONS

For the purpose of this paper, as noted above, we will only look here at the interplay of one specific set of stellar fluxes, those published by SS, and an artificial set of filters. All other factors are ignored in this paper. We will look at several photometric systems (UBV and uvby), in the sense that the artificial filters are located approximately at the wavelengths of the filters used in the actual photometry. We will vary the wavelength somewhat and will also vary the bandwidths and shapes of several of the "filters," so as to see the results on the "observed" photometric indices. We'll also calculate the "effective wavelength" for several of the filters.

We make no pretense here at carefully matching of actual filters, but only at illustrating some of the issues and potentials of synthetic photometry. (Of course, there are many other issues and features that we will not discuss at all here.)

We have entered flux data from SS for thirteen of the spectral types given, main sequence (class V) stars from O to G5, and giants (class III) of A5, F0, and G8. For each, the flux is given at 50 Å intervals, and we entered the values from 3000 Å to 6000 Å. We then either entered a filter curve or calculated one, for the "filters" described in Table I. A simple numerical integration (a Simpson integration over 20 points for each filter) and some arithmetic then gives us a "magnitude" as "observed" through each filter for each "star." For more accurate and more realistic work, better integration techniques should be used. For the present applications, the simple approach used here should be all right.

We have located filters at the approximate wavelengths of each of the UBV and uvby filters, and at several wavelengths shifted to the blue and red of the normal wavelengths. We have also created filters of different bandwidths, centered at each of the wavelengths.

Tables IIa-f illustrate some of the output from this modeling exercise. For output from two close but different "photometric systems," such as where one of the filters has been shifted slightly in wavelength, we can easily see the effects of such mismatches. Some of the results of such comparisons are shown in Table III. We can also calculate "effective wavelengths" for each filter with each flux output; these are shown in the tables. Some discussion of the data will be given in the next section, as an illustration of a few of the issues.

4. DISCUSSION

There are many interesting things that can be discovered by careful inspection of the tables (and from similar data not reproduced here). I will only mention a few such items.

Table III illustrates the type of transformation coefficients we may expect when attempting to transform between two similar systems. As the standard errors of the transformations show, some such transformations are quite accurate, while others are not too good. Much depends on what spectral features are contained, or not contained, within the band passes of the filters.

Clearly, filter systems must be well-matched to reproduce standard photometric systems. Filters for some indices must be matched much closer than others, of course. We don't show the data here, but an example of how rather dissimilar indices can be related with accuracy is the V magnitude of the UBV system, which can be derived with high accuracy from the y measurements of the uvby system, where the bandwidths are much narrower.

We also see the advantage of using narrower bandwidths, when we can. Many potential problems become less, as "color terms" are much smaller. But some problems increase: for those filters where a strong feature is located in or at the edge of a filter transmission curve. Each photometric system must be investigated on its own. And for such a detailed study, the modeling should be as realistic as possible (not as in this current paper!).

Quite often there is a natural band width for a certain index. An example is a UV band, where one would like to have the filter band be located all above the region where atmospheric absorption is dominate, and all below the region of the Balmer discontinuity. Such a filter is the u of the uvby system.

Note that UBV as a wide band system is not bad at all: that is one of the major reasons that it has been and is so frequently used. The three filters nicely cover the region from 3000 Å to 6000 Å (the region between the atmospheric cutoff and the decline of the 1P21 sensitivity). In the same sense, the uvby system is a natural system, the filters being located with care by Strömgren to offer maximum information resolution with a intermediate band system (here called a "medium band" system). Careful consideration of I(star) is necessary to locate such optimum locations.

Let me leave it to the reader to further digest some of the information content in the table, and in similar data. There is lots to be learned. I will conclude this section by listing some of the obvious "Yet To Do" items:

Include higher resolution I(star) data.
Include other spectral types, including peculiar ones.
Test out many more filter curves.
Investigate the other terms in the basic equation, such as
    interstellar reddening effects
    atmospheric extinction effects
Look at effects of radial velocity shifts, such as on NB
    indices for stars in the Magellanic Cloud complexes.
With higher resolution data, one can investigate in detail
    some interesting questions about NBP.
Compare the calculations to observations.
Apply the calculations to specific problems and questions.

Gustafsson (1986), in particular, discusses some of the interesting application of synthetic photometry to astronomy. There is much that can be done. Computer processing of spectrophotometry data, whether gotten at the telescope or at the computer terminal, offers much of value in understanding systems, and flux output from a wide variety of astronomical objects.

5. SUMMARY

Synthetic photometry can and should be used to investigate in detail characteristics of photometric systems. Much insight can be gained by such synthetic photometric investigations. This approach will help us greatly in understanding not only our systems, but the results gotten with the system. It will be especially valuable in studies of non-normal objects, whatever they may be. It will also help us in coping with the increasing use of new filters, and even more so with new detectors. As the mis-match (sometimes even a slight one) of filters can cause problems, so can the mis-match of detectors. Not all CCDs are the same, and they are not the same as other detectors. Many things combine to cause us problems. Effective use of synthetic photometry many well save us from ourselves and our systems.

Synthetic photometry can also be of great value in designing a new photometric system, to a attack specific research problem. With the increasing ease of doing synthetic photometry, I predict that increasing use will be made of this specific advantage.

Perhaps one of the major uses, however, will be in deriving photometric indices from available spectrophotometric data, including the great volumes that will be coming as a result of using the new, sensitive, linear detectors, on all sizes of telescopes, to obtain spectrophotometric data.

As mentioned above, and elsewhere, I believe that the distinction between spectroscopy and photometry will disappear in the near future. It will all be spectrophotometry. When done with care and with understanding (it's not just applying a recipe), we will all be much ahead in doing accurate and efficient observational astronomy.

When working at low information resolution (such as getting a visual magnitude, or a color index, or a H-beta line strength), we will be doing wide bandwidth spectrophotometry, when working at high information resolution (such as on faint lines, or on many lines, ...), we will be doing narrow-bandwidth spectrophotometry. For the faintest stars, or on small telescopes, it will be more of the former; on brighter stars and with larger telescopes, it will be more of the latter. There is a need and a place for different size telescopes and different information resolutions.

REFERENCES

Buser, R. 1986 in Highlights of Astronomy, Vol 7, J. -P. Swings, ed., Reidel, Dordrecht, p. 799.

Buser, R. 1978 Astron. Astrophys. **62**, 411.

Buser, R. and Kurucz, R. L. 1985 in IAU Symposium No. 111, Calibration of Fundamental Stellar Quantities, D. S. Hayes, L. E. Pasinetti and A. G. D. Philip, eds., Reidel, Dordrecht, p. 513.

Gustafsson, B. 1986 in Highlights of Astronomy, Vol 7, J. -P. Swings, ed., Reidel, Dordrecht, p. 805.

Hayes, D. S. and Crawford, D. L. 1986 in Automatic Photoelectric Telescopes, D. S. Hall, R. M. Genet and D. Genet, eds., Fairborn Press, p. 87.

Lester, J. B., Gray, R. O. and Kurucz, R. L. 1986 Astrophys. J. **61**, 509.

Straižys, V. and Sviderskienė, Z. 1972 Bull. Vilnius Astron. Obs. **35**.

TABLE I

List of "Filters" Used in the Analysis

Square Band Passes:		
	Band Width	1000 Å
		500 Å
		300 Å
		200 Å
		100 Å
		50 Å = One resolution element is 50 Å
Bell Shaped:		
	Band Width	800 Å
		200 Å
Gaussian Shaped:		
	Band Width	300 Å
Line Filters:		
	Narrow	50 Å at the line wavelength
	Wide	100 Å or 200 Å at the line wavelength
	Side	50 Å element at blue and at red side of line

These "filters" were chosen to illustrate some of the possible ranges that can be used in a photometric system. The square ones are similiar to ones used in some photometers, those where the "filter" is a combination of spectral dispersion and slots to define the chosen wavelength. They are similiar to ones that would be used in synthetic photometry to produce indices from spectrophotometric data.

TABLE IIa

Sample of Outputs from the Calculations

Part 1: Calculation for the "Standard Systems"

The uvby System. 200 Å Wide Bell Shaped Filters.

3500 4100 4700 5500 are the peak wavelengths

Spectral Type	Effective Wavelength				Calculated photometric indices (b-y)	(u-b)	c_1	m_1
O V	3495	4096	4696	5496	-0.677	-1.056	-0.046	0.171
B3V	3496	4098	4696	5497	-0.587	-0.543	0.323	0.154
B5V	3497	4098	4696	5497	-0.536	-0.278	0.564	0.115
B8V	3498	4098	4696	5497	-0.507	0.194	0.940	0.134
AOV	3499	4100	4695	5497	-0.475	0.624	1.258	0.158
A5V	3500	4102	4696	5498	-0.416	0.742	1.150	0.212
FOV	3501	4102	4698	5499	-0.286	0.647	0.773	0.223
F5V	3500	4102	4698	5499	-0.214	0.572	0.572	0.214
GOV	3503	4103	4700	5499	-0.111	0.708	0.438	0.246
G5V	3502	4103	4700	5500	-0.071	0.857	0.453	0.273
gA5	3499	4100	4696	5498	-0.386	0.769	1.223	0.149
gF0	3501	4101	4697	5499	-0.345	0.801	0.919	0.286
gG8	3505	4105	4702	5500	0.093	1.422	0.416	0.410

The UBV system. 800 Å Wide Bell Shaped Filters.

3500 4500 5500 are the peak wavelengths

Spectral Type	Eff. Wavelength			(B-V)	(U-B)
O V	3458	4464	5465	-0.834	-0.903
B3V	3478	4467	5472	-0.728	-0.419
B5V	3492	4469	5474	-0.665	-0.191
B8V	3516	4473	5475	-0.612	0.190
AOV	3527	4474	5475	-0.561	0.586
A5V	3529	4481	5480	-0.473	0.686
FOV	3522	4491	5487	-0.305	0.592
F5V	3521	4497	5491	-0.213	0.515
GOV	3523	4508	5495	-0.073	0.644
G5V	3520	4514	5499	-0.006	0.767
gA5	3535	4482	5481	-0.447	0.675
gF0	3531	4492	5486	-0.351	0.717
gG8	3539	4531	5505	0.219	1.314

TABLE IIb

Sample of Outputs from the Calculations (continued)

Part 2: Calculation for Non-Standard UBV Filters.

The UBV system. 800 Å Wide Bell Shaped Filters.

3500 4500 5500 are the "standard" peak wavelengths

Spectral Type	Eff. Wavelength			(B-V)	(U-B)
B band shifted to 4400 Å:					
O V	3458	4364	5465	-0.920	-0.818
B3V	3478	4369	5472	-0.803	-0.344
B5V	3492	4370	5474	-0.740	-0.117
B8V	3516	4374	5475	-0.680	0.259
A0V	3527	4378	5475	-0.625	0.650
A5V	3529	4386	5480	-0.518	0.730
F0V	3522	4396	5487	-0.326	0.612
F5V	3521	4401	5491	-0.221	0.522
G0V	3523	4412	5495	-0.056	0.626
G5V	3520	4417	5499	0.023	0.739
U Band Shifted to 3600 Å and V Band Shifted to 5400 Å:					
O V	3559	4464	5364	-0.751	-0.807
B3V	3585	4467	5371	-0.660	-0.380
B5V	3605	4469	5374	-0.604	-0.195
B8V	3636	4473	5374	-0.550	0.119
A0V	3654	4474	5375	-0.502	0.466
A5V	3652	4481	5380	-0.425	0.573
F0V	3636	4491	5387	-0.272	0.513
F5V	3626	4497	5391	-0.191	0.455
G0V	3623	4508	5395	-0.060	0.589
G5V	3620	4514	5400	-0.003	0.716

TABLE IIc

Sample of Outputs from the Calculations (continued)

Part 3: Calculation for 300 Å Wide UBV Filters.

3500 4500 5500 are the "standard" peak wavelengths

The UBV system. 300 Å Wide Square Shaped Filters.

Spectral Type	Eff. Wavelength			(B-V)	(U-B)
O V	3489	4493	5492	-0.835	-0.897
B3V	3492	4496	5493	-0.721	-0.406
B5V	3494	4497	5494	-0.654	-0.157
B8V	3496	4499	5494	-0.602	0.289
AOV	3497	4504	5494	-0.546	0.695
A5V	3500	4503	5495	-0.474	0.796
FOV	3502	4503	5497	-0.301	0.655
F5V	3501	4503	5497	-0.216	0.565
GOV	3506	4504	5499	-0.088	0.670
G5V	3506	4505	5499	-0.025	0.797

The UBV system. 300 Å Wide Gaussian Shaped Filters.

Spectral Type	Eff. Wavelength			(B-V)	(U-B)
O V	3484	4488	5487	-0.837	-0.897
B3V	3490	4490	5490	-0.734	-0.397
B5V	3493	4491	5491	-0.668	-0.151
B8V	3499	4493	5491	-0.619	0.286
AOV	3502	4495	5491	-0.576	0.704
A5V	3504	4497	5493	-0.493	0.797
FOV	3504	4499	5496	-0.313	0.658
F5V	3504	4501	5496	-0.221	0.561
GOV	3509	4505	5498	-0.090	0.671
G5V	3508	4507	5499	-0.020	0.788

TABLE IId

Sample of Outputs from the Calculations (continued)

Part 4: Calculation for Different Band Widths

uvby System:

3500 4100 4700 5500 are the peak wavelengths

Spectral Type	Effective Wavelength				Calculated photometric indices (b-y)	(u-b)	c_1	m_1
500 Å Wide Square Shaped Filters:								
O V	3475	4077	4679	5480	-0.671	-1.062	-0.042	0.161
B3V	3482	4086	4679	5484	-0.577	-0.555	0.277	0.161
B5V	3488	4085	4681	5485	-0.522	-0.299	0.517	0.114
B8V	3499	4088	4681	5486	-0.493	0.141	0.807	0.160
AOV	3501	4097	4678	5486	-0.451	0.566	1.036	0.216
FOV	3506	4109	4689	5493	-0.261	0.597	0.629	0.245
GOV	3514	4122	4698	5497	-0.108	0.678	0.212	0.341
200 Å Wide Square Shaped Filters:								
O V	3496	4097	4697	5497	-0.680	-1.053	-0.047	0.177
B3V	3497	4098	4697	5497	-0.591	-0.540	0.324	0.159
B5V	3498	4098	4697	5498	-0.540	-0.275	0.573	0.116
B8V	3498	4097	4698	5498	-0.512	0.200	0.952	0.136
AOV	3499	4099	4697	5498	-0.483	0.633	1.271	0.164
FOV	3501	4101	4699	5499	-0.293	0.654	0.780	0.230
GOV	3502	4101	4700	5499	-0.118	0.720	0.464	0.246
100 Å Wide Square Shaped Filters:								
O V	3498	4098	4698	5498	-0.698	-1.055	-0.041	0.171
B3V	3499	4099	4699	5499	-0.590	-0.541	0.319	0.160
B5V	3499	4099	4699	5499	-0.540	-0.278	0.546	0.129
B8V	3499	4098	4699	5499	-0.512	0.201	0.933	0.145
AOV	3499	4099	4699	5499	-0.485	0.635	1.253	0.176
FOV	3500	4100	4699	5500	-0.294	0.659	0.785	0.231
GOV	3502	4100	4700	5500	-0.108	0.708	0.484	0.220

TABLE IIe

Sample of Outputs from the Calculations (continued)

Part 5: Calculation for a Modified uvby System.

A Modified uvby System. Widths 300 100 200 200 for uvby.

3500 4200 4700 5500 are the peak wavelengths

Spectral Type	Effective Wavelength				Calculated photometric indices (b-y)	(u-b)	c_1	m_1
O V	3489	4198	4697	5497	-0.680	-1.545	-1.439	0.627
B3V	3492	4199	4697	5497	-0.591	-1.031	-0.965	0.558
B5V	3494	4199	4697	5498	-0.540	-0.765	-0.729	0.522
B8V	3496	4198	4698	5498	-0.512	-0.291	-0.349	0.541
A0V	3497	4199	4697	5498	-0.483	0.142	0.078	0.515
A5V	3500	4199	4698	5498	-0.424	0.255	0.009	0.547
F0V	3502	4200	4699	5499	-0.293	0.160	-0.366	0.566
F5V	3501	4200	4699	5499	-0.219	0.080	-0.606	0.562
G0V	3506	4199	4700	5499	-0.118	0.209	-0.759	0.602
G5V	3506	4200	4701	5500	-0.079	0.359	-0.739	0.633

Part 6: uvby System. Wavelength Shift of the b Filter.

Peak wavelength of the b filter is shifted to 4600 Å.
Both b and y are 200 Å wide square filters.

Spectral Type	Effective Wavelength		(b-y) Index
O V	4597	5497	-0.758
B3V	4597	5497	-0.671
B5V	4597	5498	-0.613
B8V	4597	5498	-0.575
A0V	4597	5498	-0.552
A5V	4598	5498	-0.479
F0V	4600	5499	-0.312
F5V	4600	5499	-0.236
G0V	4599	5499	-0.122
G5V	4599	5500	-0.068

TABLE IIf

Sample of Outputs from the Calculations (continued)

Part 7: Calculation for a Narrow Band H-delta Index:

The central wavelength for the bands is 4100 Å

Spectral Type	Calculated photometric indices 50 Å / 100 Å	50 Å / (50 Å to blue + 50 Å to red)
O V	-1.784	1.551
B3V	-1.862	1.647
B5V	-1.842	1.662
B8V	-1.953	1.757
A0V	-2.113	1.945
A5V	-2.035	1.855
F0V	-1.934	1.734
F5V	-1.839	1.619
G0V	-1.774	1.538
G5V	-1.747	1.505
gA5	-1.900	1.692
gF0	-1.916	1.712
gG8	-1.719	1.712

TABLE III

Some Results of Wavelength Mis-matches on a Photometric System

Part 1: Using a 800 Å wide bell shaped filter.

Wavelength of the two filters	Scale Term Coefficient	Standard Error
(B-V)		
4500 / 5500	This is the "Standard System"	
4450 / 5500	0.928	0.005
4400 / 5500	0.874	.005
4500 / 5400	1.101	.003
(U-B)		
3500 / 4500	This is the "Standard System"	
3500 / 4400	1.054	0.031
3500 / 4450	1.026	.018
3600 / 4500	1.118	.040

Part 2: Using a 200 Å wide bell shaped filter.

Wavelength of the two filters	Scale Term Coefficient	Standard Error
4500 / 5500	This is the "Standard System"	
4400 / 5500	0.864	0.024
4600 / 5500	1.164	.005
3500 / 4500	This is the "Standard System"	
3500 / 4400	1.084	0.016
3600 / 4500	1.100	.035

One can see the effect of compressing or expanding the "scale." In some cases, the transformation accuracy is quite adequate. Some of the larger dispersions are caused by effects of H-delta being in one filter system and not in the other. Another cause of increased dispersion is due how much of the region of the Balmer Discontinuity is in or not in a filter.

DISCUSSION

UPGREN: I want to emphasize your plea for a determination of a broad-band magnitude along with whatever other observations that may be sought. Many problems of stellar distances and kinematics are complicated by a lack of a V magnitude or the equivalent. Basic stellar distance methods even as taught in the most introductory courses involve either the use of a standard candle to calibrate distances, or the use of primary distance methods to calibrate the standard candles. Sometimes it can be forgotten that the third variable, the apparent magnitude, is important in all of this. If it is poorly known, much information is needlessly lost.

CRAWFORD: I agree, of course. Wide band magnitudes are still very important, including V. If a wide-band measure will do the scientific job (it often does), we should not use narrower bands, which have the effect of lowering the SNR. We need information resolution at all bandwidths.

SPECTROPHOTOMETRY OF LATE M GIANTS WITH IUE

Hollis R. Johnson and Joel A. Eaton

Indiana University

ABSTRACT: We have obtained ultraviolet spectrophotometry of several semi-regular, variable M giants with IUE at low dispersion including W Cyg (M 5), NU Pav (M 6) and θ Aps (M 7). Observations have been obtained every two weeks for several months, so as to span the characteristic variability timescales. All stars are variable in Mg II line strength, but the pattern of variability differs widely: W Cyg changed by a factor of ten; NU Pav is constant to within 10%; θ Aps changed by a factor of two: it suddenly brightened, stayed at roughly the same level for two months, then decayed abruptly to the original level.

1. INTRODUCTION

Almost all red-giant stars are photometrically variable. Yet for only a small fraction have sufficient observations been made to permit a complete specification of the type of variation, and many more observations are needed. More specifically, spectrophotometry is needed to investigate possible variation in spectral lines or bands. We have, therefore, begun ultraviolet spectrophotometry with IUE of several cool giant stars. Although such stars are generally too faint for high-resolution spectra, spectrophotometry with a resolution of six Å can be obtained.

Chromospheric emission in cool stars might change either because of changes in the filling factor of active regions or by changes in the active regions themselves. A change in chromospheric heating may result from the passage of shock waves through the atmosphere of a pulsating star or modification of magnetic field structures. Alternatively, emission-line strengths can change due to changes in the absorption of circumstellar material.

K giants are normally assumed to have roughly constant ultraviolet emission although there is some evidence for variability at the 10% level (McClintock et al. 1978, Mullan and Stencel 1972, Baliunas et al. 1981, Brozius et al. 1986). The chromospheric emission of cooler giants can be more variable. Carpenter (1986) has recently reported changes in the profiles of chromospheric Fe II lines in the

A. G. Davis Philip, D. S. Hayes and S. J. Adelman, (eds.)
New Directions in Spectrophotometry 213 - 218

M 3 giant γ Cru and Dupree et al. (1984, 1987) have found the supergiant α Ori to be chromospherically variable, probably as the result of pulsations. Extensive series of spectra of the cool carbon stars TW Hor (Querci and Querci 1985) and TX Psc (Johnson et al. 1987) show variations in chromospheric Mg II emission of up to an order of magnitude. The Mira variables show even greater variability, which is associated with the passage of shocks through the cool ambient chromosphere (cf. Brugel, Willson and Cadmus 1986, Willson and Bowen 1986).

Since almost all cool M giants are variable, it is possible that pulsation-driven shocks are as important in heating their outer layers as are the shocks in Mira variables. To investigate this possibility, we have monitored several semi-regularly variable red-giant stars with IUE. Observations were obtained every two weeks over a period of four months, which spans the characteristic timescales of the SRb cycles of most of these stars. Program stars are shown in Table I.

TABLE I

Primary Targets

Star	HD	Spectrum	Variable?	Range	Period
ψ Peg	224427	M 3 III	...	4.66	...
42 Her	150450	M 3 III	Var?	4.90	...
δ^2 Lyr	175588	M 4 II	SRc	4.22 - 4.33	...
R Lyr	175865	M 5 III	SRb	3.88 - 5.0	46 d
W Cyg	205730	M 5 III	SRb	6.8 - 8.9	131
NU Pav	189124	M 6 III	SRb	4.91 - 5.26	80
θ Aps	122250	M 7 III	SRb	6.4 - 8.6	119

Note: Data listed are from BSC and GCVS

2. OBSERVATIONS

All stars in our sample are found to be somewhat variable in Mg II h and k line strength (the lines are unresolved at this resolution). Spectrophotometric light curves (Mg II strength in a 30 Å bin on a magnitude scale) against time of three stars are given in Fig. 1.

Of the stars illustrated, NU Pav is constant to within about 10% during this time. It also had a relatively large Mg II flux for its V magnitude. Other stars were variable, but in perplexing ways. W Cyg (= HR 8262) showed organized changes by a factor of ten: it decayed smoothly from an undetected maximum, only to recover abruptly, as if by shock heating, whence it started to decay again. Changes in ultraviolet flux were associated with changes in the optical region.

The very late M star θ Aps, an SRb variable discovered by Payne-Gaposchkin (1952) and found to have a 120-day period, showed changes of order 50%. These however, were not the well-defined decay of luminosity seen in W Cyg. Rather, the ultraviolet luminosity of the star increased suddenly, stayed at roughly the same level for two months, then fell abruptly to the original level (Eaton and Johnson 1987).

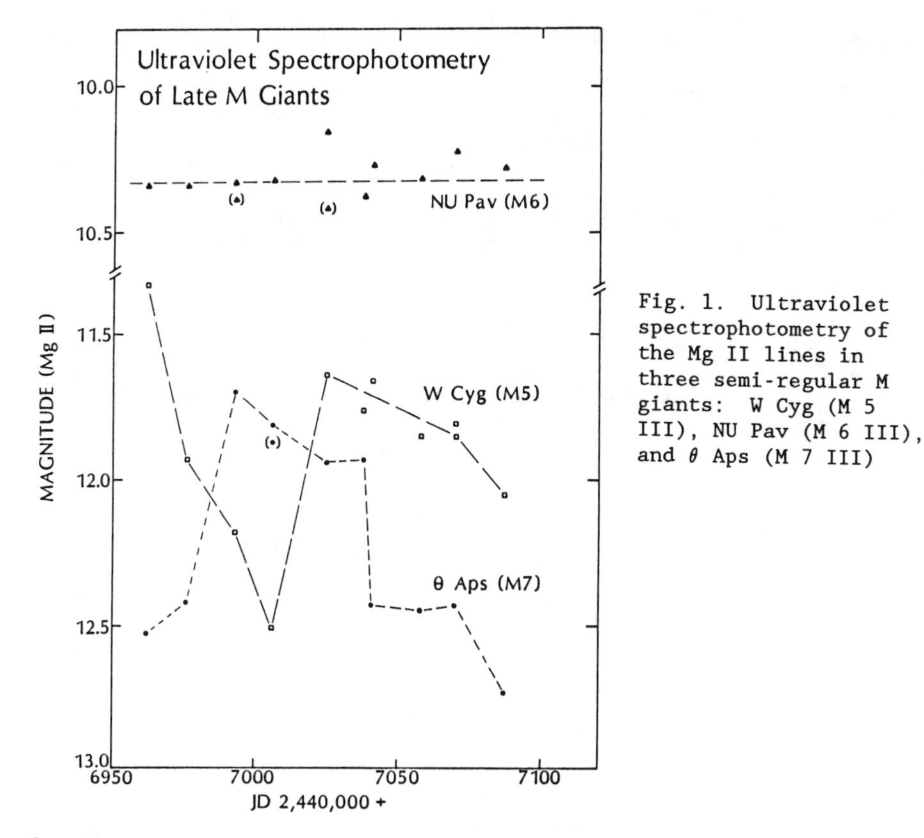

Fig. 1. Ultraviolet spectrophotometry of the Mg II lines in three semi-regular M giants: W Cyg (M 5 III), NU Pav (M 6 III), and θ Aps (M 7 III)

3. DISCUSSION

In chromospherically-variable M giants, what is the relation between optical and ultraviolet variability? As shown in Fig. 2, ultraviolet spectrophotometry of the Mg II lines for θ Aps is clearly correlated, during this time interval, with continuous flux in the IUE region (integrated flux from 2600 - 3200 Å exclusive of Mg II).

This correlation contrasts sharply with that in the N-type carbon stars, where the emission-line fluxes vary by factors of up to ten but the ultraviolet continuum appears to change only slightly. It should be noted that ultraviolet spectrophotometric fluxes for θ Aps are also well correlated in a positive sense with the FES magnitude, which varies by $\Delta m \approx 0.5$ over this interval and with the visual magnitudes measured for θ Aps by Mr. Colin Henshaw (an amateur astronomer in Zimbabwe), which vary by $\Delta m \approx 0.8$. The mechanism of variation remains uncertain, however. While changes in chromospheric flux over the pulsation-cycle time of the star were detected in two stars, it is not at all certain that these changes were produced by pulsations. To clarify the situation, one needs further observations that will allow us to determine whether these effects repeat with each cycle of the variable, or whether they are a random phenomenon, perhaps related to stochastic heating or coverage of the stellar disk with active regions. another quèstion to be addressed is whether the stars that are most variable are also the ones with the weakest level of emission, as might be expected from the data in Fig. 1. Only very preliminary modeling of chromospheres for these cool giants has been attempted (cf. Judge 1986, Eaton and Johnson 1988), but we expect this to be a fruitful field of endeavor in the near future.

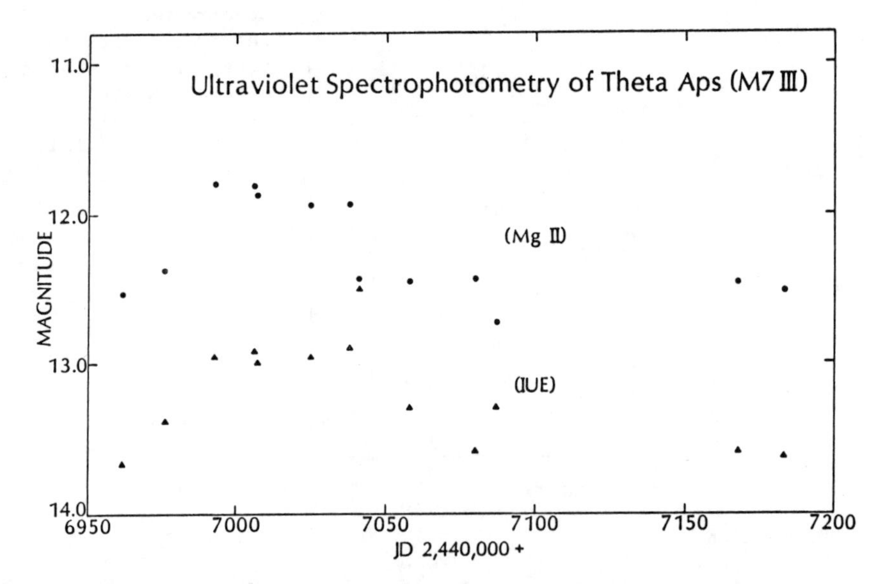

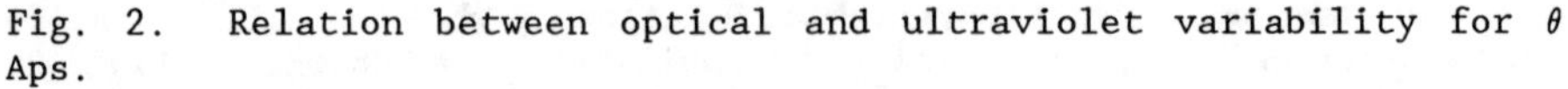
Fig. 2. Relation between optical and ultraviolet variability for θ Aps.

This research is supported by NASA grant NAG 5-192. We are grateful for Mr. Colin Henshaw's willingness to share his visual photometric observations with us.

REFERENCES

Baliunas, S. L., Hartmann, L., Vaughan, A. H., Liller, W. and Dupree, A. K. 1981 Astrophys. J., **246**, 473.
Brozius, J. W., Mullen, D. J. and Stencel, R. E. 1966 Astrophys. J. **288**, 310.
Brugel, E. W., Willson, L. A. and Cadmus, R. 1986 in New Insights in Astrophysics, (ESA SP-263), p. 213.
Carpenter, K. G. 1986 in New Insights in Astrophysics, (ESA SP-263), p. 213.
Dupree, A. K., Baliunas, S. I., Guinan, E. F., Hartmann, L. and Hayes, D. P. 1984 in Future of of Ultraviolet Astronomy Based on Six Years of IUE Research, J. Mead, R. Chapman and Y. Kondo, eds., (NASA SP-2349), p. 462.
Eaton, J. A. and Johnson, H. R. 1987 Inf. Bull. Var. Stars, No. 2974.
Eaton, J. A. and Johnson, H. R. 1988 Astrophys. J. **326**, in press.
Johnson, H. R., Baumert, J. H., Querci, F. and Querci, M. 1987 Astrophys. J. **311**, 920.
Judge, P. G. 1986 Monthly Notices Roy. Astron. Soc. **223**, 239.
McClintock, W., Moos, H. W., Henry, R. C., Linsky, J. L. and Baker, E. S. 1978 Astrophys. J. Suppl. **37**, 223.
Mullan, D. J. and Stencel, R. E. 1982 Astrophys. J. **253**, 716.
Payne-Gaposchkin, C. 1952 Harvard Ann. **1**, 1.
Querci, M and Querci, F. 1985 Astron. Astrophys. **147**, 121.
Willson, I. A. and Bowen, G. H. 1986 in Proceedings Fourth Cambridge Workshop on Cool Stars, Stellar Systems and the Sun, M. Zeilek and D. M. Gibson, eds., Springer, Berlin, p. 385.

DISCUSSION

WING: It's too bad that the monitoring program for semiregulars by Bob Cadmus includes only B and V, since it would be interesting to see how the light curve in the Johnson U filter compares with your IUE integrated flux curve, which must be dominated by light between 2900 and 3100 Å. One can't transform U to the standard system from a low-altitude observatory, but in this case one only needs to determine the sense of the changes.

JOHNSON: You're quite correct regarding the U magnitude, and I've spoken exactly the same words to Bob. Adding observations In U would mean deleting something else, however. In any case we're all grateful for the excellent B and V photometry Bob is doing.

QUERCI: Hollis, have you looked at the Fe II UV1 lines and their correlation with the Mg II UV1 primary lines (primary in the sense of shock waves)?

JOHNSON: Possibly excepting the fluorescent lines, all emission-line strengths tend to follow the Mg II lines in both carbon stars and late M giants. Lines of Fe II (UV1) lie at the longest wavelengths accessible to IUE where sensitivity is very low, and their strengths are uncertain.

THE IUE EXPERIENCE: INSIGHTS FOR FUTURE AUTOMATIC SPECTROPHOTOMETRIC TELESCOPES

C. A. Grady

Computer Sciences Corporation and
IUE Observatory, Goddard Space Flight Center

ABSTRACT: Space Observatories provide useful models for automatic ground-based observatories. As the space observatory providing absolute spectrophotometry over a wide wavelength range for the past decade, the International Ultraviolet Explorer (IUE) is the space telescope which most closely approximates the planned use of a simple automatic spectrophotometric telescope. The features of the IUE instrument, the way observations are handled, and the subsequent processing and archiving of the data which are relevant to optical facilities are discussed.

1. INTRODUCTION

An automatic spectrophotometric telescope can provide high quality optical data to support and complement a variety of multispectral observing programs. It can also provide small amounts of data obtained during simultaneous observations with other facilities, such as the Hubble Space Telescope (HST), which have much less flexibility in their scheduling. If equipped with stable instrumentation, the telescope can provide data for long-term studies. Long-duration observing programs are difficult for multiuser and multiple instrument facilities, which emphasize short observing runs and frequent instrument changes. It is also possible for an automatic observatory, if maintained for a long time, to generate a large database supporting survey or large archival research programs, provided that the data have been handled consistently and archived. In this mode an automatic spectrophotometric telescope, whether located on the ground or in orbit, reaches its maximum potential as a research tool not only for the original observers, but for other members of the astronomical community. A further benefit of a large and consistently obtained and reduced archive is that calibration data can be used to identify and monitor trends in the instrument performance, and to produce improved calibrations.

Space observatories have addressed many of the same decisions facing those planning for an automatic spectrophotometric telescope and

A. G. Davis Philip, D. S. Hayes and S. J. Adelman, (eds.)
New Directions in Spectrophotometry 219 - 226

can provide useful models for such a facility. The International Ultraviolet Explorer (IUE), as a simple spectrophotometric telescope with a long lifetime, a digital archive of more than 64,500 spectral images, and a large user community, is probably the closest to the envisioned facility. The remainder of this paper will review some of the aspects of the IUE which may be relevant to a new telescope in the 1990s.

2. ARCHIVING PHILOSOPHY

A major decision facing those planning for a new observatory with an archive is the nature of the archive and the volume of data to be stored in the archive. This decision affects the calibration philosophy for the facility, and the hardware and software planning for the observatory and individual users. As noted by the IUE Users' Committee and by the IUE Project (Willson 1985), the maximum scientific return can be achieved if the data are archived in an immediately usable form. The availability and ease of access to the IUE archive have been major factors in the success of the IUE as both an ongoing observatory and as an archival research facility. How fully reduced "immediately usable" data are will depend upon the complexity of the data reduction process, the equipment and software available to the observatory users, and the volume of data they are interested in analyzing. The simplest approach to archiving is to archive only the raw data. In this case the individual investigators would need to assemble the appropriate calibration observations, maintain their own data reduction software, and perform their own data reduction. The data reduction hardware and software needed in this case will be similar to those required for reducing data obtained at the major ground-based facilities. This approach may be particularly appropriate for data obtained with infrequently used observing modes, or for specific scientific problems where customized reduction is required. As a further plus, several astronomical software packages are available allowing users to reduce data from common instrument types, and may eliminate or greatly reduce the need to produce custom software.

This approach has some serious drawbacks, however. A raw data archive may lack a centralized observation log with sufficient information for users to assess the quality of particular observations. Since the data are archived in raw form, investigators often must fully reduce the data to determine whether they are suitable for a particular investigation. Also, different users of the same data may obtain different results as a consequence of using either different calibration data or different reduction algorithms. Comparison of results by different users of the facility may be difficult. A more serious problem to a facility serving a widely distributed user community, with many of the users at smaller institutions, is that all of the facility users would need their own data reduction software and hardware. The availability of such equipment is increasing, but may still be beyond the budgets of isolated researchers. Also software maintenance takes time and effort, which can detract from the scientific productivity of the users.

An alternative approach is to centralize the data processing at one facility. Logically, this facility might also be responsible for maintaining the instrumental calibration, generating the archive, and producing a catalog. This is the approach which has been taken by the IUE and is also planned for the HST. This philosophy has several advantages. First, the spectral data are archived in a common format, and have been reduced using well-documented algorithms and calibration data. This enables users to inspect data for suitability without themselves needing to do extensive data processing. If properly designed to take the scientific interests of the users into account, the majority of observing programs should not need additional data reduction. The availability of "fully" or largely reduced spectral data increases the scientific efficiency of the users, who can concentrate on the analysis and interpretation of the data, rather than routine reduction. Also spectra obtained for different users, or separated by months or years, can be intercompared. The individual users, who may not have the resources to manage their own data reduction facilities, do not need to have such equipment. This may make the facility attractive to astronomers needing only small amounts of data from such a telescope, as well as reducing the effort required to analyze large numbers of spectra for survey or long-term studies. This "pipeline" form of data reduction may not produce optimum data quality for all programs. This is simply because any such reduction system may not take advantage of dedicated calibration observations. A routine data reduction system also could not use assumptions about the nature of the astronomical sources which are valid for some scientific programs and not for others merely to simplify the data reduction. Provided that the raw observations are archived along with any processed data, users of such data would still be able to perform customized reductions.

A compromise between the two approaches, which permits routine programs to minimize data reduction while allowing other science programs to take full advantage of the instrument's performance, is to archive both the raw data and a series of intermediate reduction steps ranging from relatively standard steps such as removal of any detector background and flat fielding, to progressively more specialized steps such as spectral extraction, extinction correction, and absolute calibration. The number of such intermediate archiving stages would depend upon the project budget, the availability of data processing systems to the investigators, the complexity of the individual reduction steps, and the overall interests of the user community. It is the availability of a limited number of intermediate datasets that has made optimum extraction techniques feasible with IUE data, to the considerable benefit of the astronomical community.

3. CALIBRATION

Major optical observatories, such as NOAO, require that

individual observers acquire all calibration data to be used in the reduction of their observations. This philosophy, which is practical for a facility with no permanent archiving of data and frequent instrument changes, is not necessarily the most efficient way to manage a stable, long-duration, and automatic instrument, with the goals of supporting multi-year observing programs and supporting a large archive. If a goal of an automatic spectrophotometric telescope is to produce an archive containing data reduced in a standard fashion, standard calibrations are required.

The calibrations that would be used routinely for a stable, dedicated instrument can be separated into calibrations obtained on a regular basis, and those which should be obtained in conjunction with specific science observations. The number and nature of the calibrations will, of course, depend upon the choice of detector, and the overall instrument design. Regularly obtained calibration data might include detector flat field, dark background, and wavelength calibration observations. These observations can, in principle, be obtained when science observations are not possible, thus maximizing observing efficiency. This is a major advantage over space-based observatories, which are capable of observing at any time, and are over-subscribed. Thus there is considerable pressure to minimize calibration time. The absolute calibration observations, and measurements of atmospheric extinction are best done in conjunction with science observations.

4. SPECTRAL DATA PROCESSING AND CATALOG GENERATION

The nature of any data processing will depend closely on the archiving philosophy chosen for the facility. Even if only the raw observations are archived some processing will be required. This processing can be divided into processing needed to store the data in the archive and any reformatting of the data before distribution to the users. As part of storing the data in the archive, some reformatting may be needed to structure the data as compactly as possible. At this time information on the observation, and the condition of the instrument at the time of observation should be stored with the data. A catalog entry would need to be created or updated at this time and would contain, in addition to the source and observation information, the location of the data in the archive. The catalog contents are discussed in greater detail in Section 5 below. Since a format which is convenient for archiving the data may not be particularly convenient for transmitting the data across a network, or putting the data on some physical medium for mailing to an observer, some processing may be needed whenever particular observations are requested. If the decision is made to archive reduced data as well as the raw observations, additional software will be needed, as well as additional information to be stored with the data in the archive. This would include information on how the data were reduced, and identification of the calibration data used in the reduction.

Since, over the course of years, mistakes in processing do occur, and improvements to the reduction algorithms and software are made, it is vital that a log of major changes in the software and data processing be kept. Experience with the IUE has shown that major changes to calibration data used to derive the final spectra can have a significant impact on the accuracy and quality of the data. As a result, changes in calibration datasets should be cataloged and stored with the archived data. This kind of catalog can be used to identify spectra needing reprocessing, as well as those which can be used by the community without further processing. If the decision is made to process the spectral data to some more reduced state, these steps would be needed, as would others. Some form of "standard" data processing steps such as removal of instrumental backgrounds, corrections for or flagging of blemishes, flat-fielding, geometric corrections, and wavelength calibration would be needed. If the decision were made to archive "fully" reduced spectra, some standardized data reduction would be needed, along with sky subtraction, extinction correction, and absolute calibration. Any intermediate output products which were to be archived would also need to be formatted in a commonly readable structure. More specific steps will depend upon choice of detector and the research interests and equipment of the user community.

5. THE OBSERVATION CATALOG

Regardless of the contents of the archive, a catalog entry will need to be generated for each observation. The observation and archive catalog for an automatic facility of this kind will be essential, not only to retrieve data, but also as a research tool in its own right. The catalog can be used to plan upcoming or proposed observations, identify data for augmentation of an on-going observing program, or used to support a purely archival study.

Experience with the IUE Merged Log has shown that information on the quality of the data is useful in selecting particular observations for analysis. Data quality information might include indications of missing data, estimates of the exposure level, signal-to-noise ratio, and some indication of the nature of the source (e.g. emission line source, continuum spectrum, etc.). An indication of whether the data are suitable for spectrophotometry, or merely usable for spectrometry would be desirable, since an automatic facility would observe under a range of atmospheric conditions.

5.1 Catalog Consistency

With ten years of experience with the IUE merged log, it has become clear that the catalog for a long-duration facility should be as internally consistent as possible. At present, many of the data entries in the IUE merged log, such as object identification, coordinates, V magnitude (GSFC observations only), and object class are specified by each observer. This has resulted in well-observed sources frequently being classified in more than one object class, specified with different coordinates, and with V magnitudes which may bear little or no relation to the brightness of the object at the time of observation. The inhomogeneity of the IUE merged log currently limits the ease of searching the log, and can be a major obstacle to inexperienced or casual users making productive use of the archive. Work is underway to define, as part of the IUE Observatory's coordination with the ESA and SERC groups responsible for the VILSPA facility, a set of homogeneous object identifiers and to assign objects to homogeneous classifications. These additional classifications will be used to supplement the data provided by the observer. The major drawback of this approach is that the individual(s) reclassifying the catalog data may mis-classify specific objects. The primary advantage is that once the data are classified, a single search of the catalog should identify all observations of a given target which meet the other search criteria.

5.2 Catalog Distribution and Accessibility

In addition to containing the relevant information and being internally consistent, a catalog for a long-duration facility should be reasonably up-to-date, should be available to all of the interested users, and should be easy to use. Depending upon the size and complexity of the catalog these may be mutually incompatible goals, and some compromise may be necessary. The "low technology" approach of producing a printed (or microfiche) listing of catalog contents sorted by various criteria has been used by IUE and also by the STScI for the planned guaranteed time observations. This approach is very straightforward to use if one is interested in a limited number of specific sources. It does not lend itself to large surveys, large numbers of targets, or to using multiple search criteria. It can be widely distributed, and does not require access to computing or network resources. By their nature, such catalog listings are out of date, inflexible, and tend to be infrequently updated.

The "high technology" approach, which is planned for the HST catalog and archive, is to place the catalog in a relational database on a dedicated computer. The IUE RDAFs provide similar services with a simple, and not very flexible, catalog search program. Access to such a catalog is by phone line, local terminals, or an electronic network. This approach has the advantage that only one version of the catalog need be maintained, and that any catalog searches are up to date. The primary drawback is that access is only as good as your electronic connections and can be expensive. Since access to computer networks is rapidly increasing in the astronomical community, this approach is likely to become more feasible in the future, and may be the preferred approach.

An intermediate approach, which may be appropriate for a facility such as we are considering, would be to implement the catalog under a commercially available database which can be run either on reasonably cheap microcomputers or on minicomputers which are already available to the user community. The catalog available to each user could then be updated by distribution of appropriate diskettes, or by downloading the appropriate files over a network. This approach would give the user more flexible search software, at the modest cost of a less frequently updated database.

6. NETWORK AND REMOTE ACCESS

In the past few years remote access to the IUE archive and RDAFs have become available to IUE users. The desirability of remote access was recognized in the early 1980s, but it is the advent of widely distributed computer networks which has made the concept feasible. The major advantage of tying the archive, catalog, and, if needed, data reduction facility into at least one astronomical network are that users will not need to travel to a centralized facility to get or analyze their data. This allows scientists to tackle longer term, larger-scale, or more involved projects than are possible when the entire data analysis is limited to a few days. Students at all levels can be more-easily involved in projects, and can use an archival facility to "test out" ideas for theses. The ability to download data from the archive to the user's computer also means that some spectra can be made available to the scientists quite rapidly. The benefits of network access to the archive, and facility managing the observatory are considerable. Coupled with similar plans for upcoming instruments such as HST, it is clearly highly desirable for any future automatic telescope to include network access in the planning phases of the facility.

7. CONCLUSIONS

The maximum scientific return from an automatic spectrophotometric facility can be achieved if the spectral data, together with calibration observations and catalog information are archived in an easily accessible format. The maximum scientific return can be achieved if the data are archived in both raw and reduced form, with as many intermediate steps in the reduction as are feasible.

The catalog for any such facility should be implemented as part of a readily-available database system, allowing users to identify suitable observations by source descriptors, or by observation criteria. The catalog should contain all of the information needed by an investigator to determine whether particular data are of interest, as well as all engineering, environmental, and calibration source identification data to fully reduce the spectrum.

Rapid access to the archived data, together with the catalog information is a real asset in performing archival research, and should be included in any plans for upcoming facilities. The data should be able to be distributed in a widely supported data format, such as FITS.

REFERENCES

Willson, L. A. 1985 Astrophysics Data System Workshop, Workshop Report 1987 III-1 to III-3.

Draft Report to the IUE Users' Committee, September 1985 CSC/TM-85 /6103.

DISCUSSION

GENET: Based on our experience with automated photometry, all the points you raise are relevant. Is IUE automated? Perhaps NASA might learn something also from low-cost, primarily unmanned automatic ground-based telescopes.

GRADY: IUE is not a fully-automated instrument; the telescope is remotely operated. The calibration approach, handling of archived data and use of remote and service observing do have parallels with any planned automated telescope. On your last point, I agree, and this may be particularly useful for upcoming Explorer-class projects, such as Lyman.

GRAYZECK: What does the IUE spectral archive of 64,500 images convert to in bytes as a database?

GRADY: Offhand, I'm not sure, but multiple gigabytes at least. Keep in mind that we are still taking data, so the archive is still steadily growing.

BOHLIN: The extracted spectra on-line archive in the IBM mass storage device is about 30 gigabytes. Wayne Warren says the total IUE archive that includes the raw and photometrically corrected images is about 150 gigabytes on 1000 tapes.

WARREN: The figure given to Ralph Bohlin does not include a separate set for extracted spectrum tapes (about 230 in number at the present time), since the extracted spectra are also included in the full archive.

WEAVER: What are your thoughts about long-term data storage media?

GRADY: A number of projects have been exploring optical disk systems. The technology appears to be promising, but is evolving rapidly, and should be re-evaluated at the time the design for a new telescope is being considered. Choice of distribution format will depend on data volume, expense of the media, and the preferences of the users. The media for the permanent archive need to be particularly robust. HST is planning an optical disk archive, and has been taking a lead in evaluating the technology.

CONSIDERATIONS FOR THE FORMATTING, ARCHIVING AND DISSEMINATION OF SPECTROPHOTOMETRIC DATA

Wayne H. Warren Jr.

Astronomical Data Center (ADC)
National Space Science Data Center (NSSDC)
NASA Goddard Space Flight Center

ABSTRACT: The ADC has received and archived a number of atlases and libraries of spectrophotometric data over the last decade. Unlike classical catalog data, spectrophotometric observations generally consist of arrays of integer numbers pertaining to the observation of a single object, usually with relevant header records containing basic information about the object. These data require special considerations for archiving and dissemination to secondary users, since the standard formatting of a single logical record per object, as generally used for catalog data, cannot be used effectively here. These considerations are discussed in this paper. Examples of some spectrophotometric observations that have been successfully archived, documented, and distributed to the scientific community are given.

1. INTRODUCTION

An increasing number of high-quality spectrophotometric data can be expected to become available in the future, since the higher resolution of these data makes them more valuable than conventional photometry for the analysis of physical conditions in stellar atmospheres, and because increasingly-sensitive modern detectors such as CCD arrays allow observations of fainter objects to be obtained in reasonable observing times.

The proper archiving and documentation of data are very important considerations if the data are to be preserved and used by other scientists for future studies. We have seen this clearly at the NSSDC, where data for some space missions that were submitted in specialized formats, e.g., in binary form suitable for processing only on a particular computer, or only partially reduced, just sit in the archives, while observations submitted and archived in readily usable form (e.g., character coded) and well documented are frequently requested and contribute much more science to investigations beyond the initial studies for which they were obtained.

A. G. Davis Philip, D. S. Hayes and S. J. Adelman, (eds.)
New Directions in Spectrophotometry 227 - 233

Considerations for the proper preparation and documentation for archival spectrophotometric data are especially important, since these data are necessarily more complex than simple catalog data. Suggestions for the preparation of spectrophotometric data, based on past experience with several collections of data that have been successfully documented and distributed, are presented in this paper, and examples of spectrophotometric observations currently being disseminated from the ADC are given.

2. SPECTROPHOTOMETRIC DATA FORMATS

Although spectrophotometric data can be prepared in a variety of formats, their basic structure is determined by the nature of the observations themselves; this structure cannot deviate very much from an established "standard" if the data are to be readily processable by computer. The following guidelines, as illustrated by examples from existing spectrophotometric data in the ADC archives, should serve to demonstrate the best ways to submit data to be archived and distributed:

A. Each observation should be contained within a group of logical records (lines) of uniform length and equal number, if possible. If the observations contain varying numbers of data points, then the numbers of points should be declared in a header record.

B. Each observation should include a uniform number of header records to identify the object and give basic data relating to it and the observation. For example, header records may contain object identification; basic data relating to the object: photometry, spectral type, etc.; a starting wavelength, λ_o and wavelength increment, λ (if λ points are not spaced equally, it may be necessary to begin a file with an array of wavelengths).

   The header records are followed by the array of flux or magnitude values in a uniform format containing an equal number of data points per logical record (line). Each element of the array can be appended by flags to indicate missing data, points of lower accuracy, etc. Flags should be placed into separate fields so that a single numerical format specification can apply to the data themselves while the flags can be read simultaneously and checked before full processing begins.

C. All data should be in character-coded form (ASCII or EBCDIC) not in binary, and record structures should be kept as simple as possible so that any computer can be used to process the data. If it is necessary to archive the data in binary form, the FITS standard should be used in order that the data can be processed with existing software that is widely available in the astronomical community.

D. The observations should be contained in a separate file, with remarks, references, and other text in different files. Multifile data sets are easy to process, but if free text is contained within a data file, machine processing becomes more complicated than it needs to be.

E. An algorithm to read a complete observation, including the header records, should be provided in the documentation. This is especially important if any scaling of the data values is necessary. Not only does this save time for each user of the data, but it serves to avoid errors and possible misinterpretation of the observations.

Some of the points above are now illustrated by the examples given below. These examples are taken from existing spectrophotometric data sets currently archived in and disseminated from the NSSDC and the ADC.

3. EXAMPLES OF ARCHIVED SPECTROPHOTOMETRIC DATA

Several examples of archived data are shown in order to illustrate data of various characteristics. Only examples are given here, so the data are incomplete and the records may not be of the same lengths as in the original data. In most cases, the data were not originally submitted in this form, but were modified and documented at the ADC. In cases where the data were submitted in final form, much time was saved in preparing them for distribution; thus, we were able to prepare documentation and send the data out to the community on a much shorter time scale.

3.1 The Absolute Calibration of Stellar Spectrophotometry (Johnson 1980)

The absolute flux data are equal in number for all observations, but the λ increments vary in certain regions throughout the spectral range (4045 - 10220 Å). Therefore, an array of wavelengths at the beginning of the data file is processed before any of the observations are read:

```
4045 4145 4195 4245 4395 4445 4495 4545 4595 4645 4745 4795 4945 ...
5040 5090 5140 5190 5240 5290 5340 5390 5440 5490 5540 5556 5590 ...
 .    .    .    .    .    .    .    .    .    .    .    .    .    .
 .    .    .    .    .    .    .    .    .    .    .    .    .    .
9850 99501015010220
```

A header record is then read, followed by the flux data:

```
ALF DEL  HR 7906HD 196867
1.624E-211.553E-211.574E-211.526E-211.423E-211.404E-211.391E-21...
    .        .        .        .        .        .        .    ...
    .        .        .        .        .        .        .    ...
7.024E-226.757E-226.454E-226.384E-22
```

3.2 0.2 Å Resolution Far-Ultraviolet Stellar Spectra Measured with Copernicus (Snow and Jenkins 1977)

The data points are equally spaced at 0.2 Å intervals in the range 1000 Å < λ < 1450 Å; thus, each observation consists of 2250 data points. However, each data point has a number of observations associated with it (zero indicating that the data point should not be used) and some data fields are blank because they were deleted from the original observations due to an obvious flaw attributable to guidance errors or to over- or undercorrection for particle background. Thus, the numbers of observations must be read simultaneously with the data themselves and weighted or removed prior to processing. Each observation consists of a header record and an array of signal points.

```
   Zeta    Pup
   197/0  1020/1  1579/1  2023/1  2688/1  3124/1  3222/1  3574/1...
     .      .      .      .      .      .      .      .      .   ...
     .      .      .      .      .      .      .      .      .   ...
 44564/1 42417/1 41259/1 40728/1 39671/1 37988/1 36054/1 34737/1...
```

3.3 Atlas of Ultraviolet Stellar Spectra and a Second Atlas of Ultraviolet Stellar Spectra from OAO 2 Observations (Code and Meade 1979; Meade and Code 1980)

These atlases contain observations in different wavelength regions and with varying wavelength steps; thus, they are contained in separate files. However, the individual files are similar in structure and differ only in initial wavelength, λ_o, and wavelength interval, $\Delta\lambda$. Data points can be blank (no measurement) or can contain an asterisk (*) to denote that a σ/F value is $\geq$ 20% for the measurement.

(NOTE: The presence of asterisks in the data fields complicates processing because the arrays must be read initially with an A format or the data must be read into a buffer and tested before converting to real numbers. A better practice is to place the asterisks into a flag field that follows each flux value. This would be quite convenient here because flag fields are present anyway.)

Spectrophotometric data in the first file of the atlas can be either from spectrometer 1 or spectrometer 2; thus, a "+" sign is used to indicate where spectrometer 1 data begin when spectrometer 2 data are present.

The examples below present a mixture of the files to show how the data described above appear in the machine version. Each observation begins with a simple header record.

```
ALF SCO
2.63E-10   3.90E-11   4.62E-10   5.42E-10   5.60E-10   5.20E-10   ...
    .          .          .          .          .          .      ...
           2.50E-10+  2.27E-10   2.25E-10   2.15E-10   1.98E-10   ...
    .          .          .          .          .          .      ...
ALF2LIB
    .          .          .          .          .          .      ...
    *          *          *          *      8.03E-11   7.92E-11   ...
9.07E-11   1.14E-10   1.18E-10   1.36E-10   1.43E-10   1.54E-10   ...
    .          .          .          .          .          .      ...
LAM CEP
           1.13E-10   1.12E-10   1.02E-10   9.34E-11   8.74E-11   ...
    .          .          .          .          .          .      ...
```

3.4 Stellar Spectrophotometric Atlas, 3130 Å $\leq \lambda \leq$ 10800 Å (Gunn and Stryker 1983)

The atlas consists of groups of observations in uniform format, each group beginning with two header records followed by 140 data records. However, the data records contain three contiguous arrays of data of unequal sizes because the authors wished to include a flux error estimate at each of the 509 wavelengths, plus miscellaneous data related to the observations. This complex format makes it important to include a sample READ statement and wavelength-assignment algorithm in the documentation, the latter because the λ increment changes from the blue to the red. A more detailed description of the format is given by Warren (1983). The header records contain sequential numbers of the spectra in the atlas, object identifications, UBVRI data, airmass, equatorial coordinates, and other miscellaneous information. The example contains a group of flux records followed by one of flux errors and one of miscellaneous data.

```
  19 SGR     05  AV = 0.88 AP=14.0 G,R,B = 1 2 1    41G 24      4.794

-1.150 -0.337 -0.397 -0.437  1.873  3.000  6.187  0.000  4.723 -0.425

0.000E+00 0.000E+00 0.000E+00 2.568E+00 2.756E+00 2.761E+00     ...
    .         .         .         .         .         .         ...
1.157E-02 1.184E-02 6.917E-03 6.977E-03 6.874E-03 8.575E-03     ...
    .         .         .         .         .         .         ...
............................ 1.900E+02 7.770E+02 0.000E+00     ...
    .         .         .         .         .         .         ...
```

3.5 A Library of Stellar Spectra (Jacoby, Hunter, and Christian 1984)

The format is similar to the Stellar Spectrophotometric Atlas described in 3.4, but the complex contiguous arrays are not present. In this case, the second header record contains the wavelength of the first pixel and the wavelength increment per pixel (always 1.40 Å for this library).

```
HD 242908 O5    V              .60     -.67      .29     -1.11

     -.72      .28   3510.00      1.40

 2.851E-11 2.872E-11 2.899E-11 2.903E-11 2.890E-11 2.884E-11 ..
     .         .         .         .         .         .     ..
 1.513E-12 1.486E-12 1.471E-12 1.483E-12 1.444E-12
```

4. SUMMARY AND CONCLUSIONS

The foregoing examples of successfully archived and readily processable spectrophotometric data are presented as a guide for future workers who wish to archive their observations in a form that will make them most useful to other researchers. If the above precepts are followed, one can be reasonably certain that his/her data will present a minimum of difficulties when sent to other facilities. ADC personnel enthusiastically invite astronomers having data to archive to contact us for any assistance they may need to prepare the data for archiving and distribution. We would also like to work with astronomers on providing adequate documentation for archived data, and we will pass all ADC prepared documentation through data providers before it is finalized for dissemination with the data. Our primary purpose is to make data of the highest quality and usefulness available to future researchers and, toward this end, we wish to establish a close collaboration with as many members of the scientific community as possible.

ACKNOWLEDGMENTS

Appreciation is expressed to all those astronomers who have submitted data to the ADC and who have worked closely with us to make the data and documentation of high quality.

REFERENCES

Code, A. D. and Meade, M. R. 1979 Astrophys. J. Suppl. **39**, 195.
Gunn, J. E. and Stryker, L. L. 1983 Astrophys. J. Suppl. **52**, 121.
Jacoby, G. H., Hunter, D. A. and Christian, C. A. 1984 Astrophys. J. Suppl. **56**, 278.
Johnson, H. L. 1980 Rev. Mex. Astron. Astrofis. **5**, 25.
Meade, M. R. and Code, A. D. 1980 Astrophys. J. Suppl. **42**, 283.
Snow, T. P. Jr. and Jenkins, E. B. 1977 Astrophys. J. Suppl. **33**, 269.
Warren, W. H. Jr. 1983 Documentation for the Machine-Readable Version of the Stellar Photometric Atlas, 3130 Å ≤ λ ≤ 10800 Å of Gunn and Stryker (1983), NSSDC/WDC-A-R&S 83-02 (Revision 1, November, 1984).

DISCUSSION

GENET: Is the data center going to help out the people with PC's in the future?

WARREN: We have decided not to distribute data on floppies for the present because the large catalogs require so much storage and because we do not have the resources to service the large number of non-professional persons who, as indicated by recent experience, would request astronomical data on diskettes. Since most professional astronomers and serious amateurs have access to tape processing facilities, we do not consider this a major problem for them. We are, however, disseminating data via the computer networks (SPAN and BITnet) and we are in the process of preparing a pilot CD ROM containing a number of our most popular catalogs.

A HOLOGRAPHIC FOURIER TRANSFORM SPECTROPHOTOMETER

W. V. Schempp

Photometrics Ltd.

Since Stroke and Funkhouser (1965) first demonstrated that a photographic plate could directly record the interference pattern from a compensated Michelson-Twyman-Green interferometer and that proper re-illumination with coherent optics could then directly display the spectrum of the source, many authors have explored the operation of the holographic Fourier transform spectrometer (HFTS). Recently, Schempp and Smith (1988) have discussed its use for astronomical applications with a CCD camera as the detector. The common path or Sagnac interferometer is shown schematically in Fig. 1. Displacement of one of the mirrors by a distance d from its conjugate position produces the source doubling, the virtual sources being separated by a distance $l = d\sqrt{2}$. It is important to note that with the detector and virtual sources at conjugate focal points of the spherical lens, the field of view is not constrained by a luminosity-resolution product meaning that the shape and size of the entrance aperture is arbitrary, unlike a grating spectrograph which requires a narrow slit to maintain resolution.

Fig. 2 is a sketch of how the HFTS might be incorporated into a spectrophotometer. The major elements are the HFTS, some source for calibration, and the CCD detector. The detector (outlined in Fig. 3) is a key element here since a many pixel detector can be readily constructed from smaller, state of the art CCDs. This is possible since it is not necessary to sample the interferogram continuously or even regularly in order to perform the inversion needed to get the spectrum. While it is true that there will be a computational penalty in computing a periodogram rather than using the fast Fourier transform (FFT) the small number of points involved (a few thousand) and the low cost of modern computer time make this a matter of small concern.

What are the advantages and disadvantages of the holographic Fourier transform spectrophotometer? When compared to a diffraction grating based instrument, it has the usual advantages of a Fourier transform based instrument. The instrumental profile is optimal and in particular there is much less scattered light in the Fourier transform device. Scattered light arises mainly from optical imperfections in the system (i.e. ruling errors in the grating) though these can be

A. G. Davis Philip, D. S. Hayes and S. J. Adelman, (eds.)
New Directions in Spectrophotometry 235 - 240

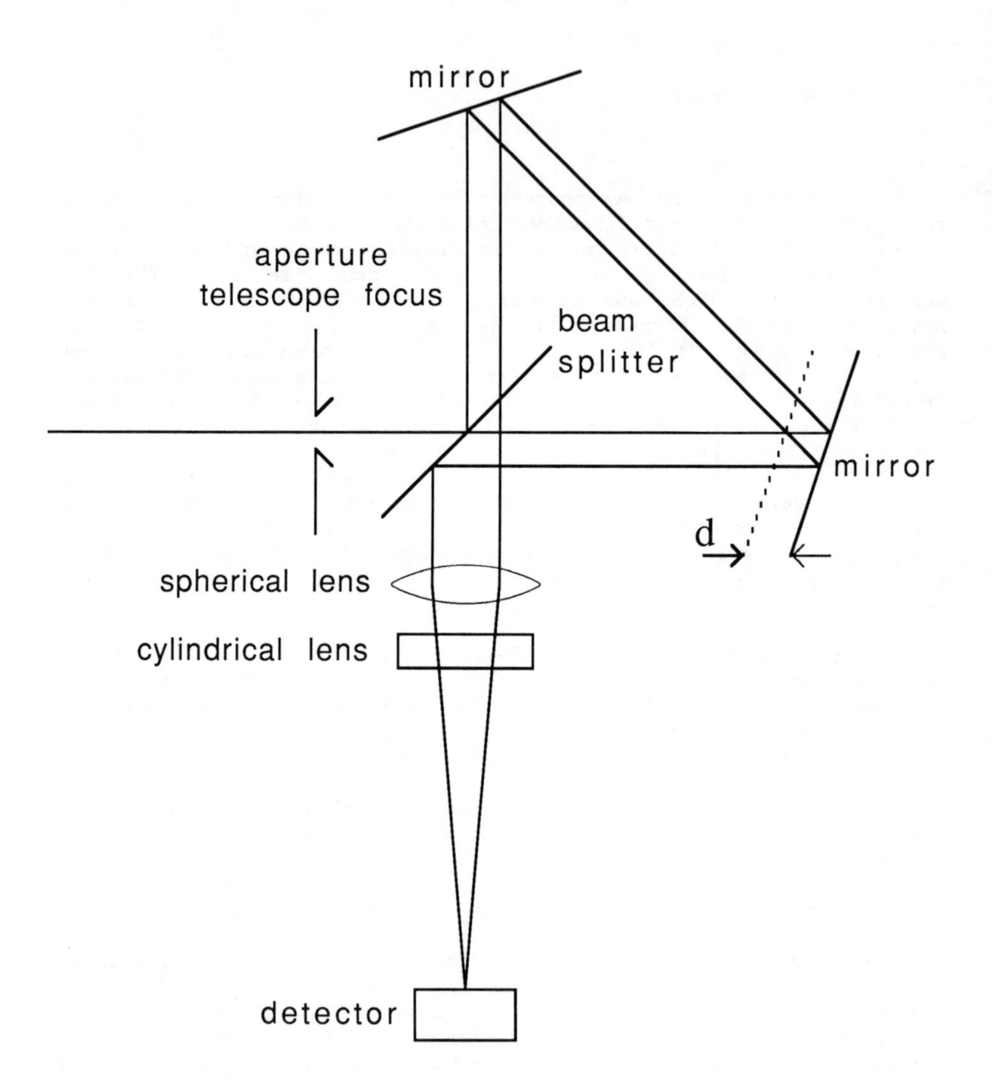

Fig. 1. A schematic diagram of the holographic Fourier transform spectrometer. The axis of the cylindrical lens is oriented perpendicular to the fringes of the interferogram.

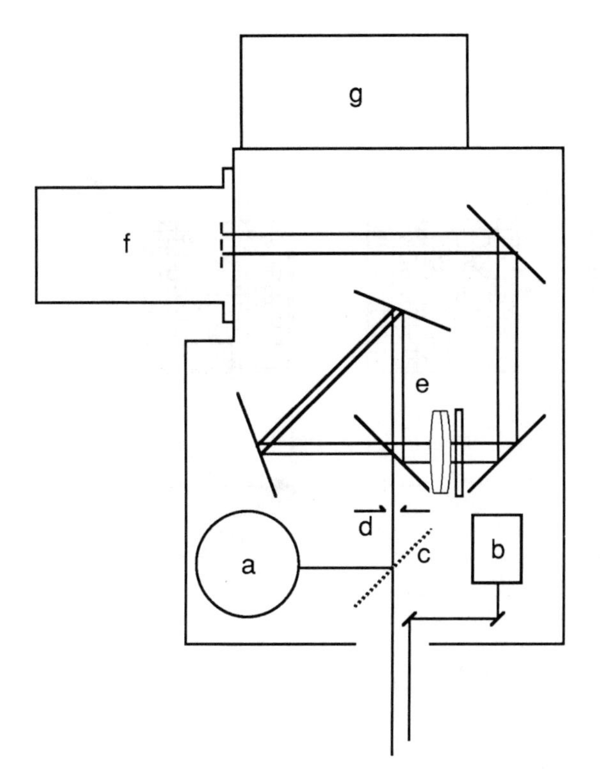

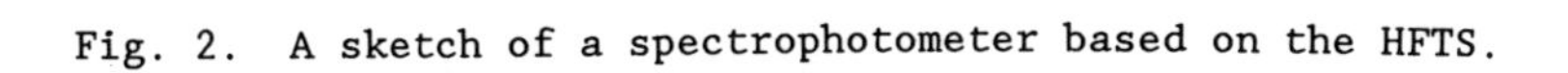
Fig. 2. A sketch of a spectrophotometer based on the HFTS.

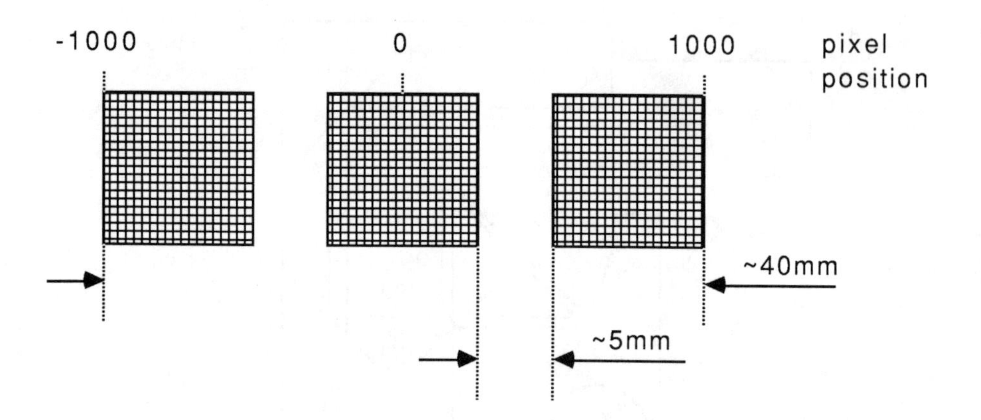

Fig. 3. The detector for the holographic Fourier transform spectrophotometer. Three PM512 CCDs form the imaging surface of the detector, which samples the interferogram over an equivalent range of 2000 pixels.

substantially reduced in holographic gratings compared to the older diamond ruled gratings. The spectral response from a beamsplitter can be made much flatter than the spectral response from a grating, in fact, a grating will require two observations to cover the visible region of the spectrum whereas the HFTS will require but a single observation. Also, the field of view of the HFTS is arbitrary and could even be tailored to fit the object under observation (e.g. comets and planets) while the field of view of the grating device is fixed by the required entrance slit. This, of course, implies that the throughput of the HFTS is much better than that of the grating. And lastly, a practical matter. While there is little doubt that state of the art CCDs of the size necessary for a grating spectrophotometer will be readily available within a few years, they are not available now, whereas the type of detector needed for the HFTS, while not an off the shelf item, can be put together today.

The HFTS is not the perfect spectrophotometer. Because noise tends to be distributed fairly uniformly throughout the spectrum, it can be difficult to detect a very faint emission line in the presence of a very strong emission line. Gratings suffer from the same problem, though for a different reason. Ruling errors lead to ghost lines in the spectrum which can be confused with faint (but real) lines, though again, holographic gratings can substantially reduce this difficulty. More importantly, however, the HFTS requires a detector with a very large dynamic range to properly encode the interferogram for absorption line spectroscopy, whereas the grating instrument requires only a limited dynamic range. It should be noted that exactly the reverse is true for emission line sources, the grating instrument requires the large dynamic range detector, whereas the HFTS has only modest requirements.

The single greatest drawback to the HFTS is the large dynamic range requirement placed on the detector for absorption line spectrophotometry. A potential fix to this problem (not as yet tested to my knowledge) is to install an objective prism type predisperser in front of the system so that instead of forming an image of the object at the entrance aperture, an image of the spectrum is formed instead. In Fig. 2 the dispersion of the spectrum would be out of the page, that is, perpendicular to the fringes of the interferogram formed by the HFTS. The cylindrical lens shown in Fig. 1 would then spatially resolve the spectrum at the CCD (as shown by Schempp and Smith) so that each column of the CCD would contain an interferogram corresponding to a small wavelength slice of the spectrum of the object. Such an instrument would greatly reduce the dynamic range requirements of the detector, but whether the many pieces of spectrum so obtained can be patched together for useful spectrophotometry remains to be seen.

REFERENCES

Stroke, G. W. and Funkhouser, A. T. 1965 Phys. Letters **16**, 272.
Schempp, W. V. and Smith, Wm. Hayden 1988, submitted to Applied Optics.

DISCUSSION

KEEL: It has been suggested that an FTS might improve sky subtraction in the region of the strong OH bands by redistributing the detector and sky noise. Can you comment on whether this would really be expected?

SCHEMPP: The point here is that at the cores of strong absorption lines the spectrum may be detector noise limited rather than photon noise limited with a grating spectrograph. With an FTS the noise is rather uniformly distributed throughout the spectrum, independent of flux level. If the uniformly distributed noise is less than the detector noise then there would be a gain. I don't know the answer to the question, but I would guess that the read noise on a CCD is so low that there would be no gain.

KURUCZ: I heard a talk by Larry Mertz about 20 years ago in which he described a similar spectrograph that was solid. Could yours be made solid?

SCHEMPP: Yes, it can. But then you lose the advantage of no luminosity-resolution product. For stellar observations that's probably not important, but for extended objects it will be.

WARREN: What technique is used to align the instrument and how stable is that alignment?

SCHEMPP: You simply change the tilt of the mirrors with some kind of screw mounting and take a picture with the CCD. This procedure fits into a loop until the fringes are properly aligned and centered on the CCD. The stability on a wooden bench was quite satisfactory. I would imagine a real optical bench would be much better. More importantly, it is a trivial matter to check and readjust the alignment if necessary.

WARREN: Can you roughly estimate how much it will cost to construct the instrument on a production basis?

SCHEMPP: Without having put a device onto a telescope that's hard to say. I should think that a serviceable device could be built for $10,000 (not including the detector) and that a cadillac system could be built for $200,000. Somewhere in between is probably what you would actually need for regular use at the telescope.

FABRY-PEROTS IN SPECTROPHOTOMETRY

Jonathan Bland and R. Brent Tully

Institute for Astronomy, University of Hawaii

ABSTRACT: Since the turn of the century, the Fabry-Perot interferometer has found wide-ranging applications from ultraviolet to millimeter wavelengths owing to its versatility and ease of operation. With the advent of areal detectors, the power of the imaging Fabry-Perot interferometer (IFPI) has been realized through the Jacquinot relation, $\mathbf{R}_\lambda \Omega = 2\pi$. For a spectral resolving power $\mathbf{R}_\lambda$, the throughput of the IFPI through solid angle Ω far exceeds that of prism or grating spectrographs, thus making it a highly efficient tool for obtaining spectral information over a wide field. The IFPI may be designed for 1 meter to 10 meter class telescopes with $10^2 < \mathbf{R}_\lambda < 10^6$ and inter-order spacing, $10^{-4}\lambda < \Delta\lambda < 0.1\lambda$. For a spectral resolution $\Lambda\lambda$, the choice of $\mathbf{R}_\lambda$ $(= \lambda/\Lambda\lambda)$ fixes the etalon finesse $(= \Delta\lambda/\Lambda\lambda)$ through a reflective coating.

The preferred method of imaging depends on the detector read-out noise and the degree of atmospheric stability above the observing site. The particular merit of photon-counting detectors (PCD) is the ability to scan rapidly in wavelength. However, the high quantum efficiency, linearity and geometric stability of the CCD prove to be essential to a large number of programs. At Mauna Kea, we have conducted optical studies using CCD-based Fabry-Perot systems on the UH 2.2 m and CFH 3.6 m telescopes. This paper addresses the application of these systems to imaging spectrophotometry. We begin by providing a context for Fabry-Perot interferometers before giving a general description of the IFPI system. A number of criteria are derived for selecting etalons and detectors. The operation and method of data acquisition is outlined followed by a discussion of the calibration and reduction procedures.

1. CONTEXT

At the heart of the Fabry-Perot is the etalon which comprises, classically, two plates of glass kept parallel over a small separation, l, where the inner surfaces are mirror coated with reflectivity, R. The response of the etalon to a monochromatic source, λ, is given by the well-known Airy function (e.g. Jenkins and White 1976),

A. G. Davis Philip, D. S. Hayes and S. J. Adelman, (eds.)
New Directions in Spectrophotometry 241 - 256

$$A = [\ 1 + 4R(1-R)^{-2} \sin^2(2\pi\mu l\cos\theta/\lambda)\ ]^{-1} \quad (1)$$

where θ is the off-axis angle of the incoming ray and μ is the refractive index of the etalon gap. The peaks in transmission occur at

$$n\lambda = 2\mu l\cos\theta \quad (2)$$

where n is the order of constructive coherence. It is evident that λ may be scanned physically through a change in θ (angle scanning), μ (pressure scanning), or l (gap scanning). Some of the early IFPI systems obtained kinematic data over a wide field by moving the telescope across the field-of-view or by tilting the etalon with respect to the optical axis (Courtes 1960; de Vaucouleurs and Pence 1980). At each telescope or etalon orientation, an interferogram was observed. Radial velocity perturbations cause wiggles in the interference rings which can, in principle, be used to measure a velocity at each pixel position crossed by a ring. This interferogram method has a number of limitations; in particular, a bright H II region at the edge of a ring distorts the least-squares fit to the ring cross-section (Pence 1981). Pressure or refractive index scanning is limited by the dynamic range in available gas pressure which generally means that the scanning range is less than 20Å (Atherton et al. 1981). For both these methods of scanning, the etalon gap remains fixed.

The success of TAURUS systems (Taylor and Atherton 1980) owes much to the development of piezo-electric, gap-scanning etalons designed originally at Imperial College, London. These optically-contacted etalons, now marketed by Queensgate Instruments Ltd. (QI), use servo-controlled, capacitance micrometry to maintain plate parallelism to better than $\lambda/100$ over a continuous range in gap spacing (Hicks et al. 1984). This allows for the ring pattern to be scanned continuously over the detector area such that line profiles can be synthesized at each pixel position. The TAURUS systems employ the Image Photon Counting System (Boksenberg and Burgess 1973; Jenkins 1987) at the image plane which enables each wavelength increment to be observed many times in a cyclical fashion. Rapid scanning allows for variations in atmospheric transparency to be averaged out during the course of a full integration. While TAURUS performs well in obtaining line profiles with high velocity resolution over a wide field, its combination of filter bandwidths and etalon free-spectral-ranges makes reliable photometry of the lines hazardous (Atherton et al. 1982). Recent developments have been possible through the use of CCDs and high finesse, wide free-spectral-range etalons which we now describe.

2. INSTRUMENT

The classical design of the IFPI is illustrated in Fig. 1. Light from the observed source is imaged by the telescope through a narrowband filter (a) onto the entrance aperture (b) of the optical system. The field lens (c) ensures that the extreme off-axis rays reach the collimator (d). The filtered light is collimated onto the Fabry-Perot etalon (e) before being reimaged by the camera lens (f) onto the detector (h). The focal lengths of the collimator and camera lenses are given by f_{coll} and f_{cam}. For our configuration, the detector is a CCD which is cooled within a dewar housing (g). The gap-scanned etalon is used on-axis to ensure the largest possible field-of-view for a given instrumental resolution. For applications involving single element detectors, the outer surfaces of the etalon were originally wedge-shaped to ensure that ghost reflections were deflected out of the beam. In the Fabry-Perot train, a blocking interference filter is placed either at the entrance aperture or in the collimated beam which confines the incoming wave to a narrow bandwidth. This is essential as equation (1) shows that the instrumental response of the Fabry-Perot, called the *Airy profile*, is periodic in wavelength. For applications involving a wide dynamic range in signal, it is necessary to keep all surfaces of the etalon, blocking filter and detector window as parallel to each other as possible. Otherwise, ghosts due to internal reflections will be imaged at the detector. Generally, we apply an anti-reflective coating to these optical surfaces but this is unlikely to exorcise all ghosts. By tilting the etalon through a small angle with respect to the optical axis, it is possible to deflect the first-order ghosts to the edge of the field. However, tilting reduces the effective solid angle of the etalon and degrades the resolution limit in equation (5).

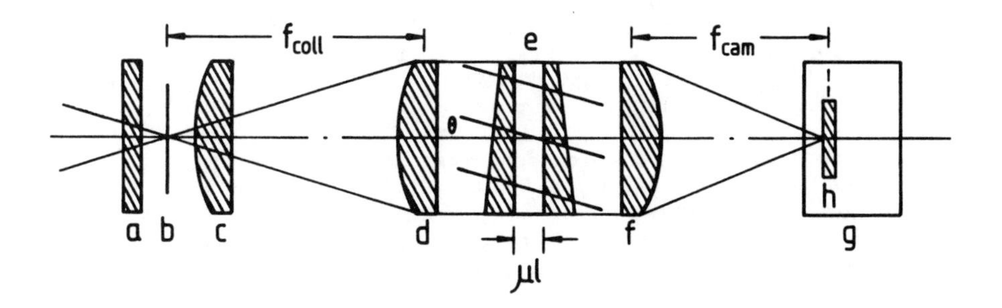

Fig. 1. Schematic drawing of an imaging Fabry-Perot interferometer.

2.1 Detector Choice

The relative merits of CCDs and PCDs are well known (Mackay 1986; Jenkins 1987). The particular choice of imaging detector for an IFPI system depends on the read-noise limit of the CCD and on the degree of atmospheric stability at the observing site. We demonstrate this by deriving the exposure time at a single etalon spacing for the signal-to-noise in a single pixel of the CCD to exceed that of the PCD. Let this time be τ_o. Therefore, we write

$$(\varepsilon_C \tau_o O)/\sqrt{(\varepsilon_C \tau_o S + \sigma_R^2)} > (\varepsilon_P \tau_o O)/\sqrt{(\varepsilon_P \tau_o S)}$$

where O and S are the detected counts from the object and sky per pixel. The read-out noise of the CCD is σ_R and the efficiencies of the CCD and PCD (telescope + instrument) are ε_C and ε_P. Rearranging the above leads to

$$\tau_o > \sigma_R^2/[S\, \varepsilon_C^2(1/\varepsilon_P - 1/\varepsilon_C)]. \quad (3)$$

We now substitute reasonable estimates for the above parameters, which are

$\sigma_R = 6.5$
$\varepsilon_C = 0.1$
$\varepsilon_P = 0.02$
$S = 0.9$

where S has been computed for a 5 meter telescope and 15μ pixels in an f/2 beam. This gives a changeover time of $\tau_o \approx 120$ sec. Typically, the CCD is the preferred detector for incremental exposure times, τ_Z, greater than about 2 minutes. Clearly, this relation has a strong dependence on detector read noise with only a linear dependence on the sky brightness. The above inequality does not take into account the read-out time, τ_R, of the CCD. The last few years has seen a push to larger detector formats with a corresponding increase in τ_R. The read time effectively sets a lower limit on τ_Z otherwise dead time will exceed the actual observing time. It is possible to bring down τ_R using on-chip binning if high spatial resolution is not required, for example, under conditions of poor seeing.

A more immediate problem concerns the stability of the atmosphere above the telescope. With a PCD, it is possible to scan rapidly in wavelength such that we observe at each gap spacing many times for, say, 3 sec $< \tau_Z <$ 6 sec. In this way, we average out moderate variations in atmospheric transparency. With a CCD, however, it is necessary to integrate long enough for the photon statistics to exceed σ_R. Consequently, we are restricted to slow scanning in wavelength such that only a few observations at each plate separation are possible. In a later section, we discuss practical ways to mitigate the effects of atmospheric fluctuations. With new developments in CCDs, the detector read noise may reach values of 1 e^- or less (Janesick 1988; Westphal 1988). Equation (3) would seem to imply that, eventually, it will be possible to scan rapidly with CCDs thereby undermining the use of PCDs all together. But, for reasons outlined above, the read-out _time_ of the CCD now restricts the scanning rate of the Fabry-Perot system.

2.2 Focal Reducer and Detector Characteristics

In equation (2), the instrumental angle, θ, is the physical angle subtended at the detector which, for a pixel dimension, $p\mu$, is proportional to $(p\mu/f_{cam})$. For a telescope focal length, f_{tel}, the angle subtended on the sky at the detector, ϕ, is proportional to $\theta(f_{coll}/f_{tel})$. For optimal throughput, we match the f/ratio of the collimator and telescope such that the angle subtended on the sky by a detector pixel is specified once we specify (f_{coll}/f_{cam}). It is provident to design around a 50mm beam cross-section as many optical components are routinely manufactured to this size and the cost of etalons increases in proportion to area for larger diameters. A further constraint on the pixel size is due to the angular dispersion over the detector area. As a change in off-axis angle, θ, induces a change in λ at the detector, we ask that the incremental change in wavelength over a single pixel element does not exceed the instrumental resolution, $\Lambda\lambda$. Differentiating equation (2) leads to

$$d\lambda/d\theta = -\lambda \tan\theta \ .$$

We define an element of the detector by a small change in angle, $\delta r \approx f_{cam}\delta\theta$. It follows that the magnitude of the spectral dispersion across a pixel is given by

$$\delta\lambda = \lambda\ (p_{\mu}/f_{cam})\tan\theta\ . \qquad (4)$$

This sets a lower limit on the instrumental resolution as Nyquist sampling over the full detector area requires that

$$\Lambda\lambda \geq 2.\delta\lambda\ . \qquad (5)$$

Conversely, for a specific etalon, equations (4) and (5) may be used to constrain the camera optics and detector arrangement.

2.3 Etalon Choice

By inspection of equation (1), it is seen that the Airy profile is a periodic function. The distance between successive orders is known as the free-spectral-range, $\Delta\lambda$. We normally make the wavelength free-spectral-range, $\Delta\lambda$, large enough to cover the intrinsic velocity structure of the envisaged source. The bandwidth of the blocking filter is selected to minimise inter-order confusion through wraparound with neighboring spectral lines, a discussion of which is found elsewhere (Atherton et al. 1982). We express the effective finesse, N_E, of the etalon as

$$N_E = \Delta\lambda/\Lambda\lambda\ .$$

The reflective finesse, N_R, is set by the etalon coating such that

$$N_R = \pi R^{1/2}/(1-R)$$

Larger finesses and, thus, higher resolving powers for a given free-spectral-range, are achieved using coatings with higher reflectivity. Queensgate Instruments Ltd. provide for quarter-wave coatings and dielectric multilayer coatings. The former provide a useful wavelength range of roughly 0.1λ (MacLeod 1969) whereas the latter may be used over the visible spectrum (Trauger 1976). The QI broadband coatings have either R = 0.90 or R = 0.95 such that only two values of N_R are possible. One of the advantages of quarter-wave coatings is that any value of the finesse is possible (Atherton et al. 1981). However, the efficiency of the etalon, ε, defined by

$$\varepsilon = N_E/N_R$$

decreases quite rapidly for N > 50. This is because distortions and irregularities within the etalon plates limit the finesse we can achieve in practice. In other words, a true finesse of 100 requires that the plate parallelism is better than λ/200, which includes both the gross and fine structure of the plates.

An important consideration arises from the shape of the Airy profile defined in equation (1). Roughly speaking, the Airy profile has a Gaussian core and Lorentzian wings. At low finesse, the wings of the profile predominate; at high finesse, the instrumental profile becomes more Gaussian. The distribution of power in the instrumental profile has important consequences for certain applications, for example, absorption line studies of continuum sources. The contrast, C, between the maximum and minimum transmission is defined by

$$C = (1+R)^2/(1-R)^2.$$

For large N, $R \approx 1 - \pi/N$ such that

$$C \propto N^2$$

which shows that a small change in finesse leads to a large increase in contrast (Roesler 1974).

2.4 Etalon Characteristics

The spectral characteristics of the etalon are essentially fixed once Δλ and the resolving power of the etalon are chosen. By expanding equation (2), and substituting for N_E, we find

$$\begin{aligned} \Lambda\lambda &= \lambda^2/(\,2\mu N_E l \cos\theta\,) \\ &= \lambda/(N_E n) \end{aligned}$$

which shows that the order of interference, n, is now determined. Thus, we can idealise the etalon characteristics in terms of the variables λ, Λλ and Δλ. In summary,

$$R_\lambda = \lambda/\Lambda\lambda$$
$$N_E = \Delta\lambda/\Lambda\lambda$$
$$n = \lambda/\Delta\lambda$$
$$l = \lambda^2/(2\mu.\Delta\lambda)\,.$$

A further constraint comes from sampling completeness over a free-spectral-range which requires that

$$\Delta\lambda/\delta\lambda \geq 2.N \tag{6}$$

where $\delta\lambda$ is now the sampling increment. A large finesse may result in an excessive number of CCD frames given that the exposure time per frame, τ_Z, is set by σ_R. As we scan in wavelength, we notice that the interference rings migrate across the detector area. Clearly, for full field coverage, it is necessary to scan a full free-spectral-range if, for a fixed etalon spacing, more than one order appears on the detector. If a fraction of an angular free-spectral-range, $\Delta\theta$, is observed, we need only sample the equivalent range in $\Delta\lambda$. In essence, this requirement constrains the order of interference at which we work. By expanding both sides of equation (2), we find

$$(n+1)\lambda = 2\mu l\cos(\theta+\Delta\theta)$$

which, for small θ, leads to

$$1/n \approx \theta^2(\Delta\theta/\theta).$$

For a fixed etalon spacing

$$M \approx n.\theta_{max}{}^2$$

where M is the observed number of rings at the detector for a fixed etalon spacing and θ_{max} is the physical angle subtended by the detector with respect to the on-axis pixel.

3. OPERATION AND DATA ACQUISITION

The Fabry-Perot interferometer is a straightforward device to operate under controlled conditions. For normal configurations, all elements of the Fabry-Perot optical train are kept rigid on the observing bench. In the early evening, the CCD dewar is cooled in the normal way. Before focussing the telescope on a bright star, the camera lens is focussed on the interference pattern of a diffuse, monochromatic source. After installing the etalon which will be used for the observation, we flush the air gap with a continuous supply of dry nitrogen. This stabilises the refractive index of the gap which is particularly important as otherwise thermal changes affect the wavelength response of the etalon. Before scanning, we stabilise the CS100 controller which drives and monitors the etalon to fix the zero point of the plate separation and parallelism.

During operation, a code is sent to the CS100 controller which identifies a unique etalon gap spacing. We require this linear scale to remain stable throughout the course of a night. There is essentially no hysteresis within the system such that a gap position is reached almost instantaneously while the controller maintains parallelism to better than $\lambda/100$. At the telescope, it is not practical to work with either l or λ. Instead, we define two variables, t and z, which are linearly related to λ. The variable t represents the code sent to the etalon controller. Initially, rough estimates to most quantities are measured on site using this parameter. The variable z is the CCD frame position in the final observed data structure and becomes important during the stages of data calibration and reduction. These variables are related to the gap through

$$l(z) = l(z_0) + c.z$$
$$t(z) = t(z_0) + c.z \qquad (7)$$

where c is an arbitrary constant and the subscript denotes an on-axis value, such that

$$n\lambda_0 = 2\mu l(z_0). \qquad (8)$$

With an IFPI system, our objective is to obtain spectra at many positions over an area of sky. We represent these as a <u>data cube</u>, $I(\alpha,\delta,\lambda)$, where α and δ are the spatial dimensions and I is the intensity of the datum. By scanning the etalon, what we actually measure is a series of narrowband CCD images $I'(x,y)$ taken over a fixed grid of etalon spacings. This grid is defined by the sampling inequality in equation (6). In practice, we sample randomly over the fixed grid in order to avoid systematic effects in our data from temporal variations in atmospheric transparency. The limiting read-out noise of the CCDs available in Hawaii necessitates 3-10 minute exposures allowing for only one or two observations at each etalon spacing.

In the next section, we show how to calibrate the observed data cube $I'(x,y,z)$ before any of the dimensions are used to measure physical quantities. The IFPI has a characteristic response which we need to remove from the data. When the etalon gap is scanned, each pixel of the detector maps the convolution of the Airy function and the filtered spectrum at that point. The filter is placed slightly out of the focal plane so that the filter structure is not imaged at the detector. But there will be differential transmission over the filter which we calibrate using a <u>whitelight cube</u>, $W(x,y,z)$. This is obtained by illuminating the optics with a current-stabilised, flat continuum source and scanning over the same spacings used for the data. In a similar manner, we procure a <u>calibration cube</u>, $C(x,y,z)$, by observing a diffuse, monochromatic source over a free-spectral-range. This is required to calibrate the wavelength response of the etalon.

An essential feature of Fabry-Perot observations is the ability to monitor system stability and performance. We achieve this through the calibration ring, R(x,y), which is an occasional observation throughout the night of the monochromatic source using the same etalon spacing. A major advantage of the Hawaii IFPI systems is the ability to perform reliable spectrophotometry. This is possible through the use of wide free-spectral-range (~100Å), high finesse (60) etalons and well-matched blocking filters. These allow us to separate out a narrow bandpass without pollution from neighboring orders and thereby to regain the photometric integrity of the original spectrum. Flux calibration is achieved by observing a standard star placed out of focus so as to distribute the counts over as many pixels as possible. We need scan only a few star frames, S(x,y), as the stellar continuum will be flat over our small bandwidth. If the astronomical source fills the field-of-view, it is important to obtain a long exposure of a blank patch of night sky. We normalize this frame, N(x,y), and subtract from each of the frames which make up the data set. We generally observe these calibration sets for each night of observation.

4. CALIBRATION

The observed data contain a number of instrumental artifacts. We now describe how these are calibrated and removed from the data. The observed data, I'(x,y,z), are essentially 4-dimensional with one photometric, one spectral and two spatial dimensions. In general, these are treated independently assuming no rotation has taken place over the field-of-view. We begin by describing the process of wavelength calibration, called phase correction. There follows a discussion of whitelight correction, flux calibration and procedures specific to CCD imaging. Finally, the CCD frames are registered spatially as with current scan rates, there is sufficient dead time between exposures for small drifts in the telescope tracking to occur.

4.1 Phase Correction

As an imaging device, the etalon is essentially neutral and simply determines which wavelengths are allowed through at particular etalon spacings depending on the order and angle subtended by the incoming ray to the optical axis. That is, if we illuminate the Fabry-Perot optics with a monochromatic source, the peaks in transmission of the line profile occur at different etalon spacings as a function of radial distance from the optical axis. To derive the constant wavelength Airy surface in terms of the cube dimensions, we write

$$l(z_0) = l(z_0+\delta z)\cos\theta .$$

We expand this using equation (7) to find

$$\delta z = n\lambda_0(\sec\theta-1)/2\mu c . \qquad (9)$$

From equation (8), it follows that

$$(n+1)\lambda_0 = 2\mu l(z+\Delta z_0)$$

where Δz_0 is the on-axis free-spectral-range. After expanding with equation (7), we subtract equation (8) to obtain

$$\Delta z_0 = \lambda_0/2\mu c\,. \tag{10}$$

We now substitute equation (10) into equation (9) to read

$$\delta z = n.\Delta z_0.(\sec\theta - 1) \equiv \Phi$$

where the phase surface, Φ, has the units of the free-spectral-range, Δ. For square pixels,

$$\theta = (p_\mu/f_{cam}).\sqrt{[(x-x_0)^2+(y-y_0)^2]}$$

where (x_0,y_0) is the center of the Airy surface at the detector. Therefore, for small θ,

$$\Phi_Z = (n/2).\Delta z_0.(p_\mu{}^2/f_{cam}{}^2).[(x-x_0)^2+(y-y_0)^2]\,.$$

The general form of the Airy surface is given by

$$\lambda = \lambda_0 + K_\lambda.r^2 \tag{11}$$

where

$$K_\lambda = (n/2).\Delta\lambda_0.(p_\mu{}^2/f_{cam}{}^2).$$

This surface has an important property: its curvature, K_λ, is wavelength invariant and depends on instrumental characteristics alone. At the telescope, it is straightforward to obtain a rough measure of the free-spectral-range simply by monitoring the Airy profile at a single detector element over two adjacent orders of interference. In this way, we find Δt and note that

$$\Delta t/\delta t = \Delta z/\delta z$$

where, by definition, $\delta z = 1$. It follows that

$$\Delta\lambda = \Delta z.\delta\lambda$$

where $\delta\lambda$ is found from observing two independent lines, λ_1 and λ_2, of a calibration source at the same order. The wavelength dispersion is

$$\begin{aligned}\delta\lambda &= \delta t.(\lambda_2-\lambda_1)/(t_2-t_1)\\ &= (\lambda_2-\lambda_1)/(z_2-z_1)\,.\end{aligned}$$

The offset in wavelength, λ_0, is known from a knowledge of the observed calibration spectrum. Since we have computed the scale factor, $\delta\lambda$, and thereby, $\Delta\lambda_0$, the wavelength calibration scale in equation (11) is now defined.

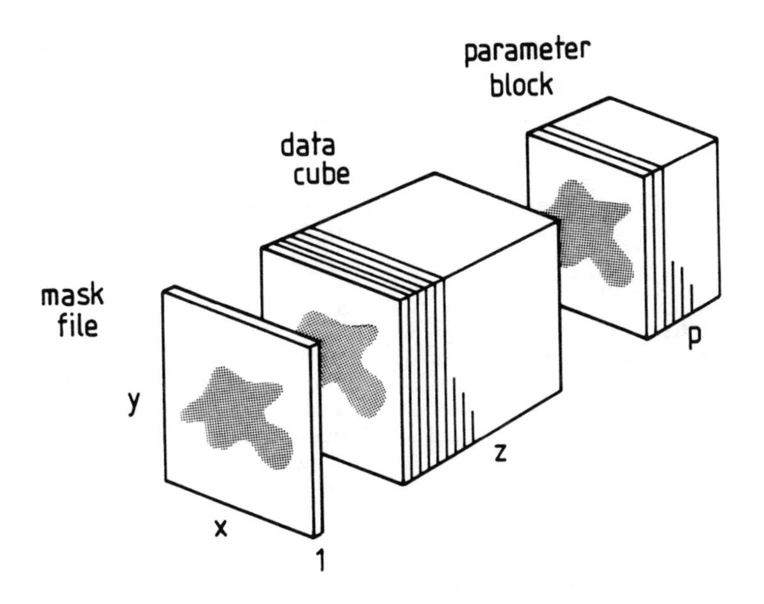

Fig. 2. Data structures associated with an imaging Fabry-Perot interferometer.

4.2 Monitoring Instrumental Stability

The calibration ring provides a powerful test of the rigidity of the IFPI system (Taylor and Atherton 1980). They are easy to observe and provide a 5-dimensional stability check on the optical system. Several times an hour, we observe a monochromatic source always using the same etalon gap spacing. We ensure that the ring is bright, with ~1000 counts in the peak, and covers as much of the detector area as possible. From each ring, we measure 5 parameters: x_0, y_0, r, e, $\Lambda\theta/\theta$. Throughout the night, a constant ring center, (x_0,y_0), indicates that no shifts have occurred within the optics. The parameter e monitors the geometric stability by measuring the pixel eccentricity; this is unity for most CCD detectors. The ring sharpness, $\Lambda\theta/\theta$, measures the instrumental resolution, where $\Lambda\theta$ is the half intensity points of the ring at an angle θ from the center. The most crucial parameter is the ring radius, r. A change in r signifies the magnitude and direction of any drifts in the wavelength response of the etalon through equation

(11). Generally, these drifts are due to changes in the optical gap, μl, arising from fluctuations in temperature, pressure or humidity.

4.3 CCD Frame Correction

Much has been written concerning methods to analyse CCD data (e.g. Young et al. 1979). For a CCD frame $I'(x,y)$, we subtract the bias, B, and divide by the normalized flatfield, $F(x,y)$, such that

$$I(x,y) = \{I'(x,y)-B\}/\{F(x,y)-B\} - (\tau_Z/\tau_D).\{D(x,y)-B\}/\{F(x,y)-B\}$$

where the dark frame, $D(x,y)$, taken in an exposure time, τ_D, is often used to monitor the constancy of bias over the chip. By integrating the whitelight cube over z, we obtain the CCD flatfield. Thus

$$F(x,y) = \sum^{z=n} W(x,y,z),$$

Typically, this frame has a high signal-to-noise ratio (~300) which is necessary to remove the pixel-to-pixel variations in sensitivity over the chip. Most solid state detectors suffer from cosmetic defects, for example, beaded and drained columns. We remove these using spline interpolation over the neighboring columns. Hot and cold pixels are treated with minimal median filtering, for example, a 5-point (up-down-left-right) filter, so as not to throw away information. Normally, a median filter is applied to specific regions, for example, in the vicinity of cosmic ray events.

4.4 Photometric Calibration

Variations in atmospheric transparency are moderated by sampling randomly over the fixed grid of etalon spacings. Stars or continuum sources which happen to fall in the field-of-view are particularly useful in quantifying these fluctuations. Stellar or point sources provide information on the stability of the atmospheric seeing. When observing at large angles from the zenith, it is important to correct for differential air mass. By co-adding spectra away from regions of strong emission, we obtain a spectrum which measures deviations from mean sky brightness. Even after subtracting a normalized night sky frame, $N(x,y)$, from each data frame, there may be residual sky brightness, in particular, when observing near astronomical twilight. Low frequency variations in mean sky brightness are due to the whitelight response of the etalon, treated in the next section. Finally, the standard star frame, $S(x,y)$, yields a scale factor which is, in essence, the efficiency of the entire optical system (telescope + instrument). We use this scale factor to convert the CCD counts to photon counts.

4.5 Whitelight Correction

In some respects, the whitelight cube, $W(x,y,z)$, provides a

3-dimensional flatfield. W(x,y,z) measures the change in response of the filter at each pixel position. The convolved whitelight spectrum has a characteristic modulation in intensity which we need to remove from the data. This is particularly important in regions of strong continuum. If the data have already been phase-corrected, then it is important to phase-correct the whitelight cube also. The modulation trough has a characteristic wavelength and so traces an Airy surface within the whitelight cube. Since the filter is placed out of focus, we convolve each x-y frame of the whitelight cube with a 2-dimensional Gaussian so as to improve the signal-to-noise ratio. The distance of the filter from the focal plane and the speed of the optical beam determine the size of the convolving gaussian. We then normalize and divide W(x,y,z) into the observed data cube.

Commonly, there are two anomalies to watch for. The whitelight division may produce spurious intensity gradients in the corrected data. This may indicate a phase shift has occurred between the two cubes owing to drifts in the etalon optical spacing. If so, the discrepancy should be evident in the calibration rings taken over the same period. A useful check is to perform spectral cross-correlations between the whitelight and the data in regions of weak line emission. Thereafter, we need simply phase-correct the whitelight cube again taking into account the anomalous phase difference before dividing into the data cube. This assumes that the rings taken before and after the whitelight cube show that the optical gap was stable during the observation. Another potential problem is deferred charge or the charge transfer efficiency of the CCD at low light levels. This may result in a mismatch between the faint emission from the source and the whitelight spectrum, with typically ~5000 counts per pixel in each frame.

4.6 Spatial Registration

As each CCD frame is taken independently, we now consider the spatial registration of the CCD frames. Of the basic geometric operations of rotation, translation and magnification, the rigidity of the optical system normally precludes the latter. Field rotation is easily checked using point sources, for example, H II regions, compact nuclei and so on. If there is a substantial delay between exposures, this may result in slight drifts in the telescope pointing. We monitor these drifts using point sources within the field. If these sources are not to be found, we apply a spatial cross-correlation between adjacent frames to determine relative shifts. Each frame is translated using spline interpolation to a common pixel position. To minimize loss of spatial information, we use the median values of the x and y offsets measured from the frames. Finally, every spectrum in the registered data cube is convolved with a minimal Hanning (1/4 - 1/2 - 1/4) function. This removes any Nyquist frequencies (spectral spikes) incurred by the alignment process.

5. REDUCTION AND DATA ANALYSIS

As the earlier sections demonstrate, the techniques of data acquisition and calibration for an IFPI system are well defined (Atherton et al. 1982). In the previous section, we described how the observed data, I'(x,y,z), are processed by interpolation to yield $I(\alpha,\delta,\lambda)$ where each variable is now linear and corresponds to a physical dimension. The term reduction is reserved for the process of reducing the information stored in the data cubes to a few maps measuring the distribution of various physical quantities. Extracting useful information from a large data cube is a lengthy process as astronomical sources possess a wide variety of spectral characteristics. The AUTOPILOT algorithm (Bland 1985) was an early attempt to generalize the problem and performs well in a large number of cases. Here, we parameterise the spectral properties into peak intensities (h), line dispersions (σ), radial velocities (v) and continuum intensity (b). These are stored in a parameter block, P(x, y, p), with spatial dimensions matched to the data cube as illustrated in Figure 2. The parameter block comprises a series of parameter maps: h(x, y), σ(x, y), v(x, y) and so on. For a given spectrum, the value of each parameter, once determined, is written away to the relevant plane at the pixel position corresponding to the data cube. The p-dimension is usually large enough to accommodate more than one spectral line. The mask array, M(x, y), shown in Fig. 2, is useful for delimiting particular regions of interest. A crucial stage in the reduction is the process of cleaning the parameter blocks. This involves recognizing least-squares fits which represent poorly their underlying spectra and removing the offending parameters from the parameter block. (For a discussion, see Bland et al. 1987). The reduction process is vindicated once large-scale, structural or physical symmetries are revealed within the observed data. Ultimately, such patterns lead to fresh insights into the phenomenon under study.

REFERENCES

Atherton, P. D., Reay, N. K., Ring, J. and Hicks, T. R. 1981 Opt. Eng. **20**, 806.

Atherton, P. D., Taylor, K., Pike, C. D., Harmer, C. F. W., Parker, N. M. and Hook, R. N. 1982 Monthly Notices. Roy. Astron. Soc. **201**, 661.

Bland, J. 1985 D. Phil Dissertation, University of Sussex and Royal Greenwich Observatory.

Bland, J., Taylor, K. and Atherton, P. D. 1987 Monthly Not. Roy. Astron. Soc. **228**, 595.

Boksenberg, A. and Burgess, D. E. 1973 Astronomical Observations with Television-type Sensors, J. W. Glaspey and J. A. H. Walker, eds., University of British Columbia, Vancouver.

Courtes, G. 1960 Ann. Astrophys. **23**, 115.

de Vaucouleurs, G. and Pence, W. D. 1980 Astrophys. J. **242**, 18.

Hicks, T. R., Reay, N. K. and Atherton, P. D. 1984 J. Phys. E.

17, 49.
Janesick, J. R. 1988 Bull. Amer. Astron. Soc. **19**, 1104.
Jenkins, C. R. 1987 Monthly Not. Roy. Astron. Soc. **226**, 341.
Jenkins, F. A. and White, H. E. 1976 Fundamentals of Optics, McGraw-Hill, New York.
Mackay, C. D. 1986 Ann. Rev. Astron. Astrophys. **24**, 255.
MacLeod, H. A. 1969 Thin Film Optical Filters., Adam Hilger Ltd., London.
Pence, W. D. 1981 Astrophys. J. **247**, 473.
Roesler, F. L. 1974 Methods of Experimental Physics, **12A**, N. Carleton, ed., Academic Press, New York, p. 531.
Taylor, K. and Atherton, P. D. 1980 Monthly Not. Roy. Astron. Soc. **191**, 675.
Trauger, J. T. 1976 Appl. Opt. **15**, 2998.
Tully, R. B. 1974 Astrophys. J. Suppl. Ser. **27**, 415.
Westphal, J. 1988, private communication.
Young, P. J., Sargent, W. L. W., Kristian, K., and Westphal, J. A. 1979 Astrophys. J. **234**, 76.

DISCUSSION

GENET: How much does this British device cost?

BLAND: In the present economic climate, $20,000 for the etalon controller (same for all etalons). The cost of etalons depends on their diameters, primarily: $8,000 for 28 mm, $13,000 for 50 mm, $20,000 for 68-75 mm and $30,000 for 110 mm. Beyond here, the price scales with area! There are a few hidden extras, e.g. $1,000 for outer surfaces polished to $\lambda/100$ and $1,000 for a broadband coating which allows the etalon to be used from 4000 Å - 7500 Å.

HAYES: What is the mechanism of scanning and maintaining paralleling?

BLAND: The Queensgate Instruments etalons employ capacitance micrometry and piezo-electric transducers. Between the etalon plates there are five capacitors along the perimeter: two to monitor changes in x, two in y and one in z. There are three piezo stacks which are spaced equally along the perimeter. A voltage is applied to the transducers which is determined from the capacitors.

HAYES: What range of wavelength can be scanned?

BLAND: As discussed in the paper, roughly speaking, all spectral parameters are fixed once you choose λ, $\Lambda\lambda$ and $\Delta\lambda$. The wavelength range which can be scanned over one order is $\Delta\lambda$. We normally choose the blocking filter bandwidth, $\Delta\lambda_F$, such that $\Delta\lambda_F < \Delta\lambda$, otherwise order confusion arises due to wraparound from neighboring orders. Note that $\Delta\lambda = \lambda/n = \lambda^2/2l$. Queensgate is able to maintain a high degree of paralleling for etalon gaps of a few microns up to 30 mm! For the visible spectrum, this implies 0.5 Å $< \Delta\lambda < 500$ Å.

HORNE: Could you compare the efficiency of observing with an Fabry-Perot compared with a long-slit spectrograph in which the slit is scanned across the source.

BLAND: In my experience, it is easier to recover the photometric integrity rather than the spatial integrity of the data, e.g. the Fabry-Perot does not suffer from light losses due to the seeing disk. Moreover, for a given spectral resolution, the Fabry-Perot has a much higher throughput than the long-slit spectrograph. Even though the latter provides greater wavelength coverage, the useful line information is restricted to a few columns of the detector.

SCHEMPP: How do you ensure that telescope drift is not a problem from frame to frame in strong emission line sources since the appearance of the image can change dramatically and self registration may not be possible?

BLAND: For typical fields of view, there are almost always a few compact sources for registration. Otherwise, spatial cross-correlation between frames in regions of faint, smoothly varying structure will do the trick.

THE MIRA MULTICHANNEL SIMULTANEOUS PHOTOMETER

Wm. Bruce Weaver

Monterey Institute for Research in Astronomy

ABSTRACT: A self-scanned photodiode array is a nearly optimal sensor for multichannel photometry. The diodes have a linear response over all intensities, have a higher quantum efficiency at all wavelengths compared to that of a photomultiplier, and show no residual images or fatigue. For the high signal-to-noise usually desired in photometry, the high quantum efficiency compensates for the significant readout noise.

When used to synthesize the Mitchell and Johnson 13 - color system, the MIRA MCSP should be 68 times faster than a noiseless S - 20 photomultiplier for an A star, 20 times faster for a flat spectrum, and two times faster for a K star. For typical stars, over an average of all spectral classes, the decrease in measurement times of the MCSP seems to be conservatively equal to about half the number of photometric bands for the Ga-As photocathode and two to three times better than that for an S - 20 cathode.

The MIRA MCSP uses a prism disperser to avoid the problems associated with gratings. The techniques planned for wavelength control will be discussed.

1. THE DETECTOR

A silicon-based detector is a good choice as a sensor for the optical region of the spectrum because of its high quantum efficiency and linear response. Budde (1983) has verified the linearity of photodiodes over eight decades (20 stellar magnitudes) within 0.1%. For Reticons in particular, Vogt et al. (1978) have confirmed its linearity of the full range of their system - 10 magnitudes. This compares favorably with photomultiplier tubes which have been shown to have gain instabilities which can affect pulse counting measurements at the 1% level (Rosen and Chromey 1984). The MIRA Reticon Spectrophotometer is used routinely over 15 magnitudes. In addition, photodiode arrays are insensitive to magnetic fields, are geometrically stable, and are light sensitive over their entire surface.

The Reticon's quantum efficiency curve, measured by the manufacturer, is shown in Fig. 1. The relatively high ultraviolet

A. G. Davis Philip, D. S. Hayes and S. J. Adelman, (eds.)
New Directions in Spectrophotometry 257 - 266

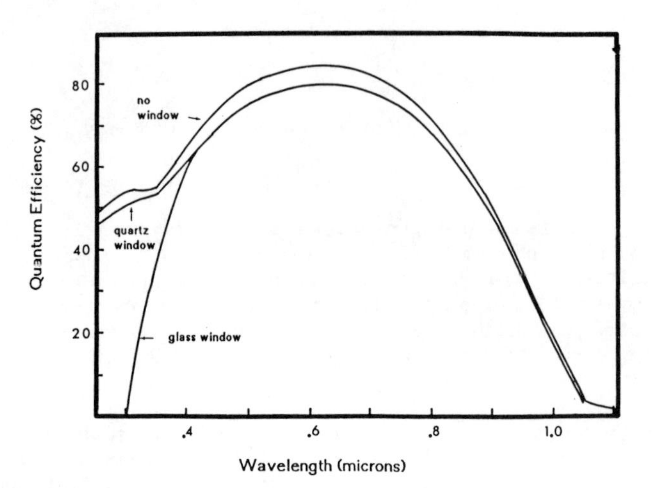

Fig. 1. The quantum efficiency of a Reticon self-scanned photodiode array as measured by the manufacturer (Adapted from Reticon Ap. Note No. 121)

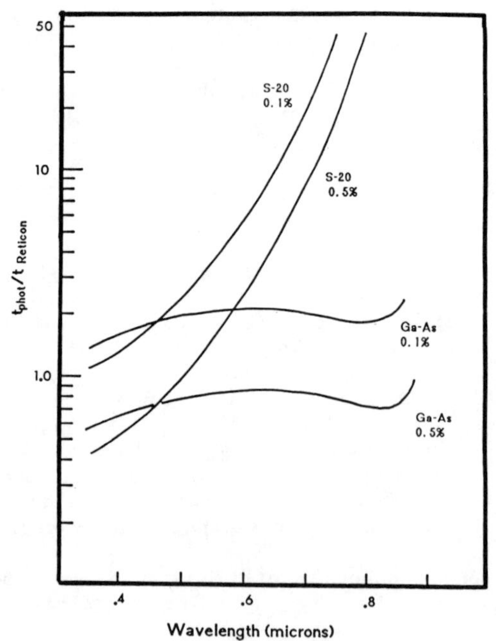

Fig. 2. The relative time required for the specified noiseless photomultiplier to complete a measurement to a specified precision compared to the time required for a <u>single</u> Reticon diode modeled with a 500 electron readout noise.

sensitivity is particularly noteworthy; it is above 50 per cent from the atmospheric cutoff to about 9000 Å. This is higher at all wavelengths than that of a typical gallium-arsenide photocathode and is much higher in the UV and blue than an unsensitized CCD. The main drawback of a silicon diode array is the relatively high readout noise. However, Vogt et al. (1978) have shown that, even with this relatively high readout noise, for high signal-to-noise observations a single photodiode, because of its high quantum efficiency, can be superior to a noise-free photomultiplier. Curves are shown in Fig. 2 for the noise characteristics measured for the MIRA Reticon detector compared to two photocathodes modeled with no instrumentation noise.

The noise of a single diode readout with the MIRA system is typically in the 500 electron range. This is a little lower than that reported by Walker et al. (1985) or Vogt (1981) because the 512-element array has half the line capacitance of these other devices and the bandwidth is limited by reducing the readout rate to one kilohertz. Walker (personal communication) has recently reduced his noise to the order of 300 electrons. Johnson (1985) reports a noise level of less than 400 electrons for a Reticon RL 1872F/20.

Although this noise is high compared to state - of the - art noise levels achieved with CCDs (about ten electrons), this difference is offset by a number of other factors favoring the Reticon. These include the above-mentioned high UV quantum efficiency and the large aspect ratio of the Reticon diodes. The 512-channel Reticon diodes are 1 mil wide along the dispersion but 17 mils long perpendicular to the dispersion, permitting all of the light at a given wavelength to fall on the same diode. Thus several diodes at the same wavelength need not be added, with their associated noise, to sum the contribution at that wavelength.

In the high signal-to-noise regime (0.1 per cent), where most photometry is performed and where Poisson noise dominates over readout noise, the relative importance of the readout noise alone is reduced to less than a half a stellar magnitude. However, other factors reduce this advantage. For example, multiple CCD pixels must be summed to obtain all the signal at one wavelength. The full-well capacity of most CCDs is inadequate to accumulate the necessary number of electrons to obtain the required signal-to-noise level just for shot noise considerations; the required multiple readouts accumulate noise as well as signal. The full-well of a Reticon is about 20 million electrons. When these additional factors are taken into account, the advantage of the CCD is reduced to about a quarter of a magnitude. Of course, at those wavelengths where the quantum efficiency of the Reticon is higher than that of a CCD, this difference is further reduced or even reversed.

Some practical aspects of self-scanned photodiode arrays are also attractive when compared to CCDs, especially for observatories with limited resources. They are readily available commercially; there is

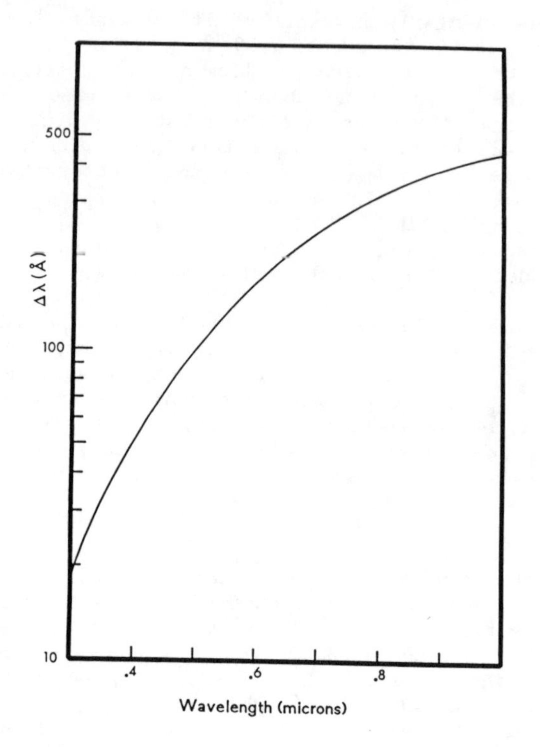

Fig. 3. The bandpasses subtended by a single diode as a function of wavelength for the MIRA MCSP with the 110 mm focal length UV-IR catadioptric camera. The 55 mm focal length camera bandpasses are twice those plotted.

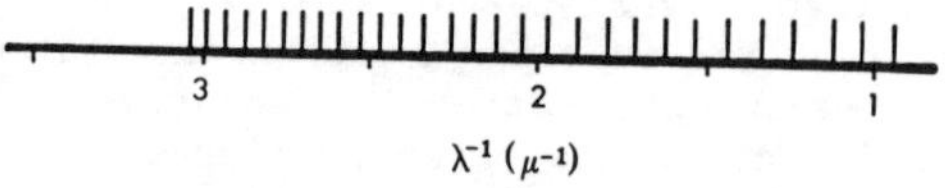

3 2 1

λ^{-1} (μ^{-1})

Fig. 4. Typical band centers for the MCSP with the 110 mm focal length UV-IR catadioptric camera. Band centers for other focal length cameras would appear at different frequencies but would have the same proportionate spacing.

no commercial source of thinned, back-illuminated CCDs. They are mechanically and electrically robust and require no temporary hypersensitization to achieve useful quantum efficiencies in the UV and blue regions of the spectrum. The relatively small number of pixels make data acquisition and reduction a practical matter with a personal computer.

2. THE PRISM

Use of a prism as a disperser has several advantages over a grating. In the first place, the prism is more efficient as it has no other orders in which to disperse the light. The quartz prism used at MIRA has a very well-behaved transmission as a function of wavelength, in contrast to gratings which have strong blaze functions, high frequency anomalous behavior such as Wood's anomaly, and strongly different behavior for differently polarized light. Although the dispersion of a prism varies rapidly in wavelength measure, as is shown in Fig. 3, the dispersion is nearly linear in inverse microns. Fig. 4 shows typical band centers for the 110 mm camera in inverse microns. Wave numbers (inverse wavelengths) are often preferred by photometrists as well as spectroscopists, possibly because black body curves are so well behaved in wavenumbers or because wavenumbers relate directly to energy units. The density of spectral features seems more uniform in wave numbers than in angstroms. Thus, it seems fair to characterize the MCSP from Fig. 4 as having relatively uniformly-placed bandpasses.

3. CURRENT MECHANICAL CONFIGURATION

Fig. 5 shows the completed optical and mechanical features of the MCSP. The short focal length side of the dual-ported spectrograph is used. The grating is replaced with a first-surface mirror which is mounted on an extra-thick grating cell to compensate for the deflection of the prism. The fused-silica 13-degree prism is mounted in front of the 55 mm focal length camera. A cylindrical lens acts both as the window to the evacuated cold box and as a pseudo-Fabry lens perpendicular to the dispersion. This version of the instrument has been completed and tested in the MIRA shop and at the telescope. An example of raw output for three stars of different spectral types, corrected only by quasicorrelated double sampling and for dark residual clock signal with the data reduction program RETINA and is shown in Fig. 6. This is compared to a spectrum shown in Fig. 7 that has been fully reduced by RETINA which was taken at the lowest resolution (30 Å) available with a grating on the Blue Spectrograph. The prism mount is designed to fit also with the 110 mm focal length catadioptric UV camera.

The entrance slit to the spectrophotometer can be opened bilaterally to apertures of 32 x 128 arcseconds. The "decker" direction has fixed choices by powers of two (1, 2, 4, ..., 128 arcseconds) while the slit width direction is continuous and is controlled by a micrometer. The largest aperture that can be used

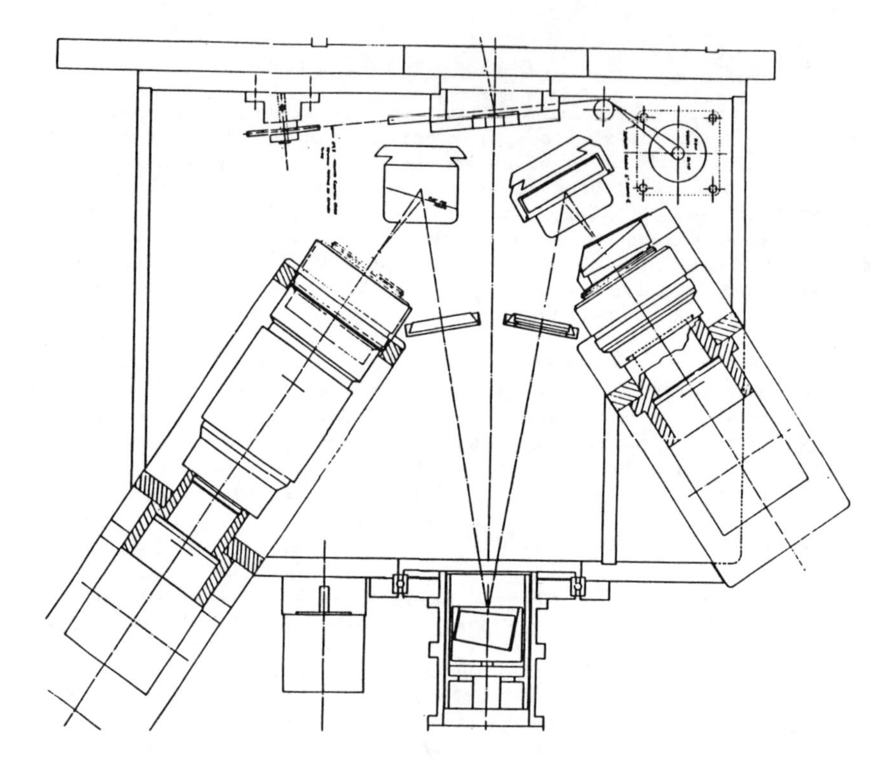

Fig. 5. Completed optical and mechanical features of the MIRA dual-ported Blue Spectrograph showing the MCSP mode. The 13-degree fused-silica prism is shown in front of the 55 mm focal length camera.

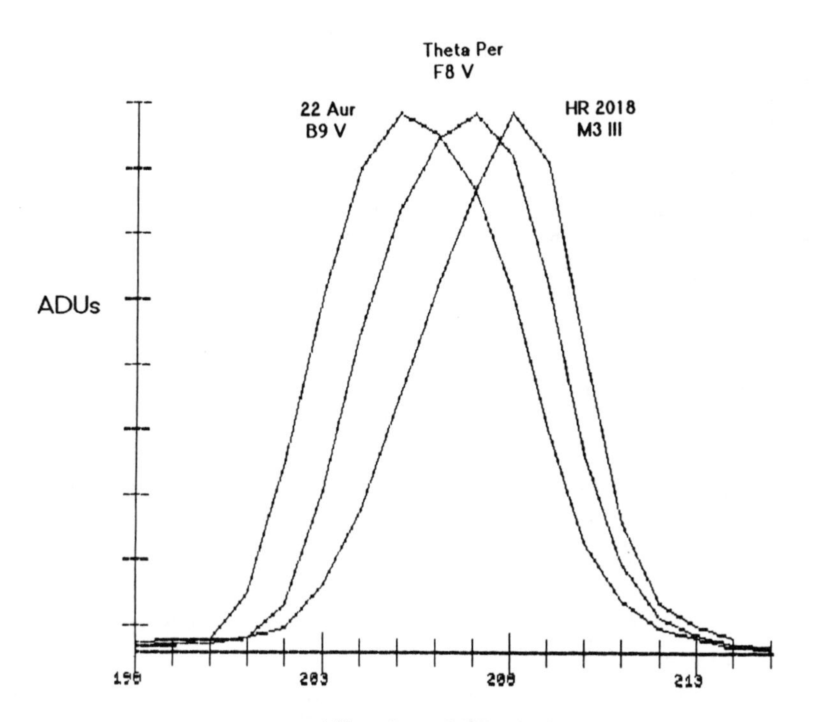

Fig. 6. Energy distributions for three stars observed with the MCSP with the 55 mm focal length camera. Only quasicorrelated double sampling and dark residual clock pattern corrections have been applied to the data.

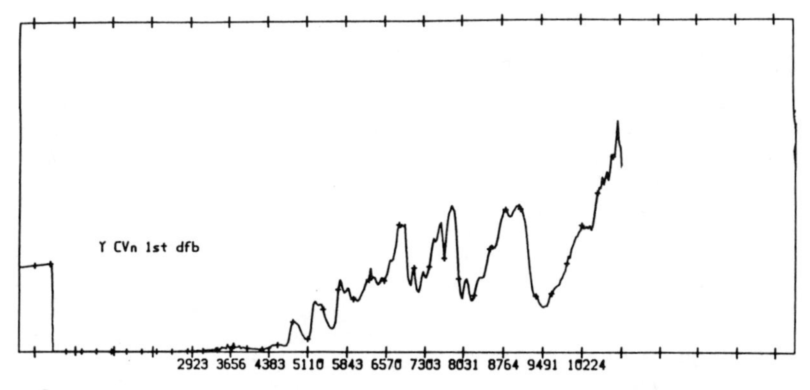

FIg. 7. A RETINA-reduced spectrum observed at the lowest grating dispersion possible with the Blue Spectrograph (30 Å/diode).

without overflowing the diodes is 16 x 28 arcseconds. This is roughly equivalent to a 24 arcseconds diaphragm. From diffraction image considerations, this should exclude less than 2 per cent of the light (Young 1974).

4. SPEED AND ACCURACY

The multiplexing advantage is the most important gain provided by the Reticon system. The entire wavelength range can be measured simultaneously and, in addition to the native photometry system, a large number of medium and large-bandpass photometry systems can be synthesized from this single observation. Because of the large dynamic range of the system (40,000), most stars can be measured in a single scan. When the current 55 mm focal length camera is used with the prism, the light between 0.32 and 1 micron is dispersed over 25 diodes. The 110 mm focal length catadioptric camera will increase that to 50 diodes. The bandwidth sizes for the catadioptric camera are shown in Fig. 4. From these data, a wide variety of photometric systems can by synthesized. The gain in speed actually realized (compared to a single channel photometer) will depend on the number of colors being synthesized, the precision of the measure, the photocathode to which the Reticon is compared, and the spectrum of the underlying star.

For the purposes of such a comparison, the total relative speed of a Poisson noise-only photomultiplier was calculated at each band using published quantum efficiencies. Three spectra were examined, those of an A star, a mid-K star, and a flat spectrum. No time was allocated for filter changes and only the initial precision of the measurement was considered. The sum of the exposure times for this ideal photometer was then compared to the time required for the MCSP to reach the required precision in its slowest photometric band. Twelve of the 13 colors in the Mitchell and Johnson (1969) system can be measured with the MCSP, eleven with a Ga - As photocathode, and ten with an S - 20.

At 0.1 per cent precision with the 110 mm focal length camera, the MCSP is 8 times as fast as a Ga-As while measuring an A0 star, 12 times as fast for a flat spectrum but only twice as fast for the K star. For the S-20, the corresponding values are 68 (15 excluding the 0.8 micron band), 20, and 2. The relatively modest speed gain for late type stars is due to the domination of the measurement time by the extreme ultraviolet bands. For typical stars over an average of all spectral classes, the gain seems to be conservatively equal to about half the number of bands for the Ga - As photocathode and two to three times better than that for an S - 20 cathode. As a very rough estimate, the speed gain is numerically equal to about half the number of bands. Naturally, several different photometric systems can be synthesized from the same MCSP observation, permitting a large number of medium- and wide-band indices to be examined. Here, of course, the speed gains are cumulative.

Synthetic photometry is becoming increasingly common with the use of flux-calibrated spectrophotometry. For example, Greenstein (1984) used the multichannel spectrophotometer on the Hale 5 m reflector to derive six colors for white dwarfs. Since broad-band photometric systems and many intermediate band systems are well sampled in the sampling theorem context by the MCSP (contrary to filter photometry) more accurate data reduction may be possible (Young 1984). In particular, the local spectral slope will be available. For fully sampled spectral systems, the same linear transformations would apply to all stars, regardless of the intrinsic or atmospheric reddening. The fact that all the colors are measured at the same airmass under identical seeing will also aid in the data reduction.

5. WAVELENGTH REGISTRATION

The primary difficulty expected in using the MCSP to emulate filter photometry is accurate registration of the spectrum on the Reticon. This is the remaining effort to be completed before the MCSP can be used. Errors of a few Å can seriously affect attempts to obtain high precision photometry. Since the Reticon system will continue to be used as a grating spectrophotometer, the first surface mirror, which replaces the grating when the instrument is being converted to the MCSP mode, will be frequently in and out of the spectrograph, and must be rotated very accurately to the correct angle when inserted.

This problem is not unique to the MIRA MCSP; any system not using filters must maintain correct wavelength registration to about an Ångstrom (about 0.01 diodes in the MCSP) to permit high accuracy photometry. We expect to use a microstepping system donated by Compumotor Corporation connected to a worm drive to properly position the mirror.

A coaxially-mounted encoder will be used to set the mirror rotation. Calibration and confirmation will be made with light from a laser which will be fed into the spectrograph through the usual f/10 calibration feed optics. The mirror will be tilted under computer control until the signal is exactly divided between the same two reference diodes. This relative measurement should be repeatable to better than 0.1% error in the signal level from the diodes. This should place the mirror so that monochromatic star light near the same wavelength as the laser will fall on the correct diode to a photometric measurement precision of also better than 0.1 per cent.

ACKNOWLEDGMENTS

Development of the Reticon Spectrophotometer has been made possible through the support of the Reticon Corporation, The Research Corporation, The Margaret and Herbert Hoover Jr. Foundation, the Flora and William Hewlett Foundation, Compumotor Corporation, and the Friends of MIRA. The development of the Reticon readout electronics would not have been possible without the substantial efforts of Gary Love.

REFERENCES

Budde, W. 1983 Appl. Optics **22**, 1780.
Greenstein, J. 1984 Astrophys. J. **276**, 602.
Johnson, R. 1985 Bull. Amer. Astron. Soc. **17**, 45.
Mitchell, R. I. and Johnson, H. L. 1969 Com. Lunar Planet. Lab. **132**, 1.
Rosen, W. and Chromey, F. 1984 in Proceedings, Workshop on Improvements to Photometry, W. Borucki and A. Young, eds., NASA Ames, p. 125.
Vogt, S. 1981 Proc. S.P.I.E. **290**, 70.
Vogt, S., Tull, R. and Kelson, P. 1978 Appl. Optics **17**, 574.
Walker, G., Johnson, R. and Yang, S. 1985 Adv. Electron. Electron Phys. **64a**, 213.
Young, A. 1972 in Methods of Experimental Physics, Vol. **12** part A, N. Carleton, ed., Academic Press, New York, p. 123.
Young, A. 1984 in Proceedings, Workshop on Improvements to Photometry, W. Borucki and A. Young, eds., NASA Ames, p. 217.

DISCUSSION

ETZEL: Do you cool your Reticon to lower the noise? Any signs of fatigue?

WEAVER: We cool to liquid nitrogen temperatures. We can do a dark scan after full saturation with no measurable effects due to the saturated state.

ETZEL: Do you (or will you) have sufficient resolution to calibrate wavelengths with the Balmer lines of A stars?

WEAVER: Not with the lowest resolution shown in the figure. However we may be able to resolve them with one of the longer focal length cameras.

CRAWFORD: I think what I push is the Tao of spectrometers or photometers (rather than motorcycles). One should be in tune with one's scientific problem, one's spectrometer, and one's budget. One can do excellent science without maximizing the efficiency of each subset. It's the 80/20 (or 90/10) law. We strive to get most of the performance of the ultimate, with care, at costs we can afford.

WEAVER: Sure.

SPECTROSCOPY USING A 0.6 METER REFLECTOR AND A TI 800 X 800 CHARGED-COUPLE DEVICE

Kelly McDonald

University of Colorado at Boulder

ABSTRACT: In this report I present a description of the spectroscopic system at the Sommers-Bausch Observatory on the campus of the University of Colorado at Boulder. The charge coupled device was donated to the Observatory by the National Science Foundation. I also present results of some selected observations, including S/N ratios, spectral resolution and some estimates of the efficiency of the entire system. I conclude with some comments on some current and future applications of our charged coupled device.

A. G. Davis Philip, D. S. Hayes and S. J. Adelman, (eds.)
New Directions in Spectrophotometry 267 - 268

DISCUSSION

KURUCZ: This is just what I need. You can measure Balmer profiles for the whole Bright Star Catalog with a S/N of 1000. You could also measure the field horizontal-branch stars that Philip and Hayes work on.

MCDONALD: That would be a good project, but there are limitations on our observing time. Maybe that would be a good thesis topic.

DESIGN AND PROGRAM PLANNING FOR AUTOMATED SPECTROPHOTOMETRY OF GALACTIC NUCLEI

William C. Keel

University of Alabama

1. INTRODUCTION

Optical variability of quasars and Seyfert nuclei has been known for over 20 years, and it has been recognized for just as long that these variations can provide unique clues to the structure and size of these objects. The few objects so far observed in spectroscopic time series of any length have borne out this expectation handsomely. The strong broad-emission-line variability of Arakelian 120, a type 1 Seyfert, indicates that these lines arise in a region only a light-week in size, much smaller than ionization models had suggested and in fact posing some problems for such models (Kollatschny et al. 1981; Schulz and Rafanelli 1981; Peterson et al. 1983, 1985). UV and optical spectra of the nearby Seyfert NGC 4151 by Ulrich et al. (1984a,b) and by Antonucci and Cohen (1983) have shown an apparent sequence of varying spectral components, perhaps causally connected: first variation in a UV "bump" is seen (fuelling rate?), then the power-law continuum dominant at lower energies, followed by the broad emission lines.

The basic scheme in all these cases is that of mapping an ionization echo of flares in the UV (ionizing) continuum from a small central source (see Capriotti and Foltz 1982). The continuum sources certainly exhibit occasional rapid flares approaching the idealized case of a light pulse whose effects could be traced outward through the surrounding emission-line regions, but the irregular nature of most objects' continuum variations means that time-series, cross-correlation or Fourier techniques must be used to derive structural information from the observations. Results to date have been fruitful, but limited by the frequency of observations and total number of observations available. Since quality spectra of these objects have frequently required instruments found on 1-3m telescopes, time is rarely allotted in blocks sufficient to trace variations over more than a few days, much less for longer-term studies.

It now appears feasible to produce a system for monitoring many active galactic nuclei (AGNs) automatically, with the concomitant advantages of time coverage and long-term project security. I will

A. G. Davis Philip, D. S. Hayes and S. J. Adelman, (eds.)
New Directions in Spectrophotometry 269 - 274

discuss briefly the scientific goals, required instrumental performance, and sketch an instrument and operating scheme suitable for this program.

2. SCIENTIFIC GOALS AND INSTRUMENTAL NEEDS

Some discussion of the aims of this program will illuminate what is required of the telescope, instrument, and software for useful results. Consider the basic goals of measuring continuum changes (at various wavelengths), changes in broad-line fluxes, and checking the (more or less) narrow lines for constancy. We then require adequate signal-to-noise in the continuum to reach the accuracy limit set by calibration uncertainties, and preferably similar limits for emission lines of interest. Having large enough samples of QSOs and Seyfert nuclei measurable can be assured by reaching a limiting magnitude V=16 (see Table I of Keel 1986), which yields several hundred objects accessible from mid-northern latitudes.

A photomultiplier-grating scanner still delivers the ultimate absolute spectrophotometric accuracy, but a device of this type is inappropriate here on two grounds. First, it lacks the enormous multiplex advantage of array detectors, thus driving up the observing time needed even for a small, carefully chosen set of passbands. Second, and most damaging, the active nuclei to be observed have a large redshift range - z=0 to 1.5 - and hence a very large number of grating settings, and associated calibration sequences, would be needed during each night.

The relative faintness of these objects requires very accurate sky subtraction, and limits the acceptable detector sensitivity and readout noise. A CCD system is favored for these reasons alone, as most Reticons have a readout noise much too high unless used with an image intesifier (which are generally taboo for unattended operation). A long-slit or multiaperture mode offers better sky subtraction in less time than a sky-chopping scheme, offsetting the larger data-reduction chore it produces.

The dispersion and wavelength range to be used are set be a compromise between enough resolution to split lines of interest (such as Hα and the red nitrogen lines in low-redshift objects) and begin to resolve broad-line profiles, and the desire to have a single instrumental setup applicable over a wide redshift range (as wide as the detector will cover, until atmospheric absorption bands become troublesome near 1 micron). These constraints suggest a pixel size near 10 Ångstroms and a wavelength coverage of about 4000-8500 Ångstroms (the blue end depending on the specific detector to be used).

The fact that many of the Seyfert nuclei to be observed are in rather bright galaxies adds another constraint - a known and consistent amount of light from the surrounding galaxy must always be included to avoid spurious variations due to seeing changes. This renders accurate spectrophotometry of these objects trickier than for quasars, where such a galaxy is, by definition, faint or unseen. This adds a further allowance in aperture size or slit width to those given by typical seeing allowance and tracking error, as the only feasible approach seems to be trying to include all the host galaxy's light (except for the very nearest systems), which could be removed after once performing multiaperture measures in good seeing at another telescope.

An instrument meeting these criteria, on a dedicated telescope, could do much to answer the following key questions: What is the nature of continuum variations in a "typical" active nucleus? Why do the broad-line regions of Seyferts appear much smaller than derived from straightforward ionization-level arguments? Is the situation similar in QSOs? Do any of the "narrow" lines come from regions of a few light-years' size? Do any purely narrow-line objects (Seyfert 2s) vary? (This might require someday introducing a polarimetric capability). Indeed, are there any AGNs that do not vary when watched over several years?

3. INSTRUMENT DESIGN

The scientific considerations just described indicate that one needs a high-efficiency spectrometer with pixels 10 Ångstroms or a bit smaller, covering as much of the 0.4-1.0 micron range as feasible. Sky subtraction will be critical, and should be done from simultaneous (multiaperture or long-slit) observations. Apertures or slit widths of 15 - 20 arcseconds are needed to allow for pointing and tracking jitter, seeing, and always including a constant contribution from the host galaxies when these are detected; the spectral resolution of 20 Ångstroms (2 pixels) should be reached even for such large entrance apertures.

Choice of a detector is influenced by expected object and sky count rates for "reasonable" exposure times (10-30 minutes uninterrupted, with several per object). Some kind of CCD is preferred to a Reticon both because of the latter's high readout noise, and to allow sky subtraction from a larger area of sky (gaining a factor 1.41 in precision). Use of liquid nitrogen at a remote site would impose a considerable burden on site support, so a system not requiring infusions of cryogens would be a boon. Performance figures for a Thomson chip from Photometrics Ltd. indicate that a thermoelectrically cooled CCD (at 220 K) has a dark current comparable to the expected sky level within a 20" slit and 5" spatial increment. This would add significantly to the sky noise, but this is tolerable if the dark current is stable enough (at the 1% level). Also, a thermoelectrically cooled system is physically much smaller than a dewar.

A very efficient spectrometer is required, for maximum photon grasp in attainable 1-m class telescopes and to offset the high dark count of a device without cryogens. Gratifying results have been obtained with transmission grism spectrometers on large telescopes (KPNO Cryogenic Camera, ESO's EFOSC), and a miniature design of this type should serve well. Lenses act as collimator and camera, and a grating-prism (grism) in a temporarily parallel beam is the disperser. Compactness is a prime consideration, since APT designs favor small spaces behind the primary for structural and cost reasons. For example, a device for use on one of the Mt. Hopkins 0.75-m telescopes would have a beam width at the grism of no more than 25mm without folding, which would add reflection losses. For the moderate dispersion needed, this should pose no optical problem, though alternate multiaperture designs using optical fibers are certainly worth pursuing. In principle, multiple grisms (in a wheel) could be added for various spectral ranges.

Acquisition and guiding are important in observing such relatively faint objects - "peaking up" with a coaligned photomultiplier would take unreasonable long for most target objects. The procedure used at most general-purpose telescopes - recognition of the object by pattern-matching on a TV display and pointing via a reflection off the jaws of the spectrograph slit-would require very sophisticated software and otherwise superfluous moving parts within the instrument, to operate in an automatic mode. Simplified approaches are possible if the telescope can point within a few arcseconds, for example upon offsetting by a degree from a bright star. Pointing within 3-4 arcseconds would put the target within the central part of the slit, obviating any direct need to verify acquisition. Typical SAO or AGK3 stars are bright enough that a single photomultiplier can be used to peak up on them to an arcsecond in short times. For this application, a "fuzzy" mask might be used for this instead of a sharp aperture, so the star could be found further from the nominal aim point. This mask might well take the form of a sharp mask placed out of focus for the photomultiplier's Fabry lens. Elaborations are, of course, possible if acquisition cameras and simple recognition software can be added - the reference star could then be skipped. Note that open-loop tracking must be good over 30-minute periods to avoid the risk of losing an object in mid-exposure, though east-west slit alignment could increase the tolerance in this direction.

Lacking an observer to judge the quality of a night, this system would measure standard stars more frequently than is usual for many spectrophotometric programs (especially extragalactic ones). Wavelength calibration should be needed only infrequently, with night-sky lines available as checks (especially at the coarse wavelength scale of the basic configuration).

4. OPERATIONS AND DATA HANDLING

A system of weights based on several factors - expected brightness, zenith distance, time since last observation, and history of variability - will need to be adopted to control the observing sequence during each night. The brighter objects could be followed throughout the lunar month, but mean sky brightness will usually be a factor in target selection.

The raw-data volume produced by a 2-D array dwarfs that from single-channel photometry, and goes beyond what could be sent over phone lines. However, since we need only one-dimensional spectra, that of each object after sky subtraction and optimal extraction could be sent in this way (for 20 or so targets per night). This requires application of well-developed techniques for finding the object, sky subtraction, and bad-pixel rejection, preferably to be carried out during the night. Sufficient statistics would be carried along to tell whether any problems were encountered - questionable data are easy to reject if one hasn't taken them personally.

Measurement of line and continuum fluxes from marginally undersampled data will be aided by some knowledge of the spectral components involved, particularly for lines. For greatest accuracy, known components of the relevant spectral shapes can be fit where appropriate. Maintenance of a long-term archive of such measurements is an important goal, which could do much to show us what the systematic properties of variations in AGNs really are, and tell much about the properties of the nuclei themselves.

REFERENCES

Antonucci, R. R. J. and Cohen, R. D. 1983 Astrophys. J. **271**, 564.
Capriotti, E. R. and Foltz, C. B. 1982 Astrophys. J. **255**, 48.
Keel, W. C. 1986 in Automatic Photoelectric Telescopes, D. S. Hall, R. M. Genet and B.L. Thurston, eds., Fairborn Press, Mesa, p. 176.
Kollatschny, W., Fricke, K. J., Schleicher, H. and Yorke, H. W. 1981 Astron. Astrophys. **102**, L 23.
Peterson, B. M., Foltz, C. B., Miller, H. R., Wagner, R. M., Crenshaw, D., Meyers, K. A., and Byard, P. L. 1983 Astron. J. **88**, 926.
Peterson, B. M., Meyers, K. A., Capriotti, E. R., Foltz, C. B., Wilkes, B. J. and Miller, H. R. 1985 Astrophys. J. **292**, 164.
Schulz, H. and Rafanelli, P. 1981 Astron. Astrophys. **103**, 216.
Ulrich, M.-H., Boksenberg, A., Bromage, G. E., Clavel, J. Elvius, A., Penston, M. V., Perola, G. C., Pettini, M. Snijders, M. A. J., Tanzi, E. G. and Terenghi, M. 1984a, Monthly Notices Roy. Astron. Soc. **206**, 221.
Ulrich, M.-H., Boksenberg, A., Bromage, G. E., Clavel, J. Elvius, A., Penston, M. V., Perola, G. C., Pettini, M. Snijders, M. A. J., Tanzi, E. G. and Terenghi, M. 1984b, Monthly Notices Roy. Astron. Soc. **209**, 479.

DISCUSSION

WEAVER: Reticon chips are available with two rows of diodes which permit simultaneous sky and galaxy measures. What S/N do you require? On which telescopes were the spectra taken?

KEEL: The S/N needed is 20 per pixel or better. One can get better results for specific features once the spectral shape of a component is known; it may be measured with weighted contributions from multiple pixels (many in the continuum).

I showed data taken with 1.5 to 4 m telescopes. The most extensive studies were done with the Lick 1 m and IDS and with IUE, so there is no question that 1 m instruments with high-efficiency array detectors would be suitable.

MILLISECOND SPECTROPHOTOMETRY OF LUNAR OCCULTATIONS

Nathaniel M. White

Lowell Observatory

A new direction in spectrophotometry is the application of high time resolution spectrophotometry to the measurement of stellar angular diameters of cool stars by lunar occultation. The two questions to be addressed are is it worth doing and can it be done? The answers are both yes. However, first let us review the history of spectroscopy of lunar occultations.

On January 4, 1865 Sir William Huggins (1865) observed the immersion of Epsilon Piscium at occultation with a spectroscope. He was attempting to determine if the Moon had a measurable atmosphere and from the observations inferred that it did not. This was probably the first spectroscopic observation of an occultation, but unfortunately, it had to rely on the observer's volatile memory.

Approximately 100 hundred years later, the problem of volatile memory was solved but the only sufficiently efficient, fast, low noise detectors were photomultipliers. Used in parallel, two, three and sometimes four different colors could be recorded simultaneously. Several telescopes used together could also increase the wavelength coverage and gain some redundancy in the observation of these transient events. The efforts were fruitful in that the color dependence of cool star diameters began to be documented. In 1981, White suggested the use of the developing CCD array to detect and record the spectrally resolved and time resolved intensity fluctuations of a star at occultation. Schmidtke (1986) successfully applied this technique to the occultation of Antares.

To answer the question is it worth doing, we should first consider the theoretical suggestions and then the observational support. There are three causes for the color dependence of angular diameters of cool stars in the optical region of the spectrum. The most subtle is the classic limb darkening effect observed over a large wavelength range as measured in several hundred Ångstrom band passes. Less subtle and less well documented is the rapid increase in angular diameter in the blue and ultraviolet region that may occur for those stars with

A. G. Davis Philip, D. S. Hayes and S. J. Adelman, (eds.)
New Directions in Spectrophotometry 275 - 282

circumstellar dust shells. Models by Rowan-Robinson and Harris (1982) predict a spatial redistribution of blue light will occur due to scattering in the dust shell giving the star an apparent greater angular extent at bluer wavelengths. Finally, dramatic differences in a cool star's angular diameter should be measured in and out of absorption lines and bands. Although opacities in cool stars are still not well known, Scholz's (1985) models provide some relative information suggesting a 50% difference in diameters measured in and out of spectral features.

Using these theoretical suggestions, we developed a schematic representation of the angular diameter as a function of color for a cool star with a dust shell. A representative plot of the relative flux of an M supergiant is shown in Fig. 1. The bold line labeled 1.0 is a hand drawn attempt at placing a continuum and the dashed line labeled 1.5 is a similar attempt at defining a locus of absorption maxima. According to the previously mentioned models, one might also interpret the scale 1.0 to 1.5 as representing the relative color-dependent angular diameter of the cool star! With color dependent angular diameter measurements, cool star atmospheric models could be constrained not only by the observed flux spectrum but by an angular diameter spectrum as well.

There is some observational evidence for an angular diameter spectrum. An occultation of the M supergiant, 119 Tauri was recorded in several 200 Å band passes and a 3 Å band pass centered on Hα (White, Kriedl and Goldberg 1982). The angular diameter was nearly 50% larger in Hα light than in the "continuum".

The results of other interferometric techniques (speckle, amplitude and Michelson interferometry) have given some indication of an angular diameter spectrum for Alpha Orionis (White 1980). The data show a large amount of scatter in Fig. 2 which may be due to the fact that the observations were made with a variety of new and developing techniques over a time interval of years. A weighted inverse wavelength square law is fitted to the data for reference. Statistically this relation fits better than a simple inverse dependence, however, we are not inferring a physical meaning yet. Alpha Orionis is not occultable by the Moon as seen from the Earth, so we must wait for the detailed application of multi-aperture interferometry to confirm its color angular dependency in the optical spectrum.

In addition to showing the aforementioned extended Hα diameter, occultation measurements in broad bands of the supergiant, 119 Tauri, show a similar trend of increasing angular diameter with decreasing wavelength (White 1980). The same color diameter dependence is seen in the occultation results for Alpha Scorpii (White 1986). What is particularly unique about the data shown in Fig. 3 is that all but one of the points blueward of 0.55 microns were recorded simultaneously removing any time dependent effects. The inverse wavelength square dependency persists.

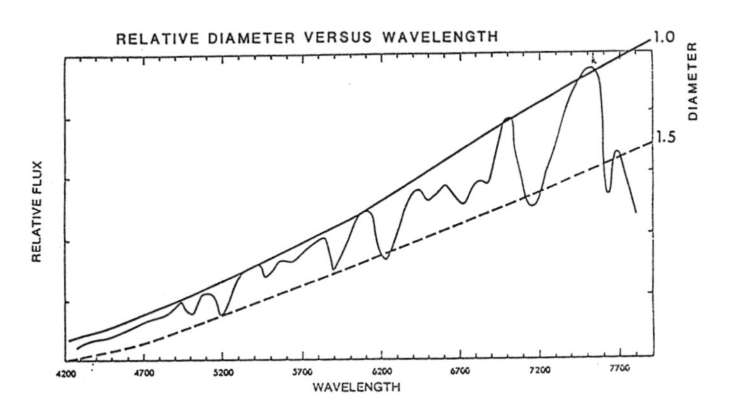

Fig. 1. A schematic representation of the angular diameter as a function of wavelength for a cool star with a dust shell. The squiggly curve represents relative flux, the left ordinate, and relative angular diameter, the right ordinate.

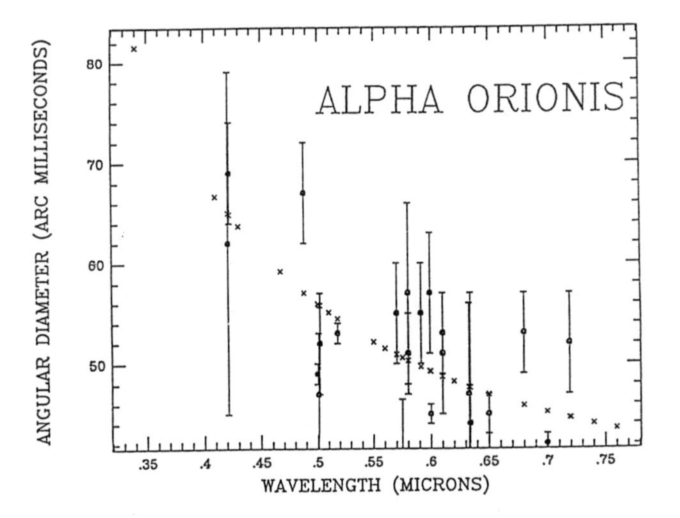

Fig. 2. The angular diameter of Alpha Orionis measured by a variety of interferometric techniques. The "X"'s indicate an inverse wavelength suqare law, weighted fit.

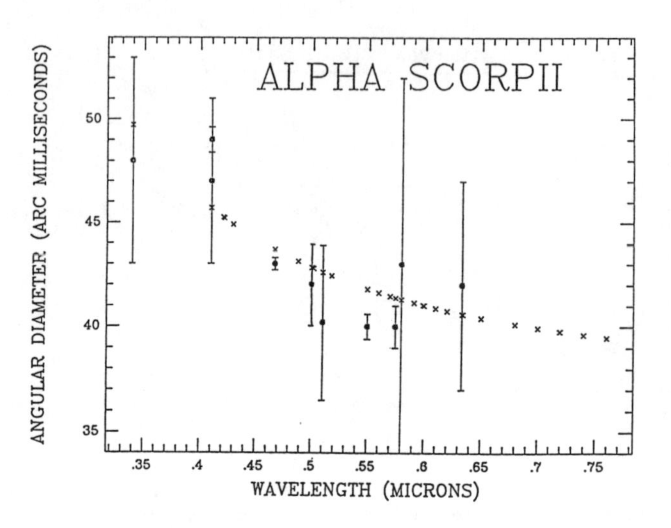

Fig. 3. The angular diameter of Alpha Scorpii. All but one measurement blueward of 0.55 microns was measured simultaneously at occultation. The meaning of "X"'s is the same as in Fig. 2.

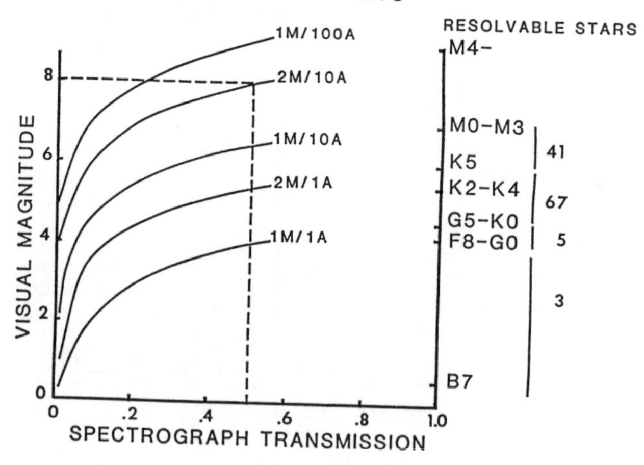

Fig. 4. Resolvable angular diameters as a function of visual magnitude, spectral type, telescope aperture, bandpass per pixel and spectrograph transmission.

Finally, there exists some dramatic evidence of stellar diameter changes as recorded within spectral lines. Schmidtke and colleagues (1986), using the Kitt Peak 4 m telescope, recorded at 1 Å resolution the spectral occultation of Alpha Scorpii centered on the infrared calcium triplet. The results show a continuous increase in the angular diameter from the continuum value to nearly a factor of two larger in the core of the lines. The angular diameter verses wavelength plot of these results give the appearance of an emission line spectrum as the diameter values increase within the absorption features.

There are ample preliminary indications that observable angular diameter-color dependencies exist for cool stars. What is needed to make these observations effectively? Clearly the multiplex advantage of a spectrograph and CCD array detector over the color filter and photomultiplier is necessary. Although the stars of interest are brighter than 8 mag, the sampling rate is 500 times per second. In addition to the multiplex advantage, high through-put and quantum efficiency are necessary to attain a sufficient signal to noise ratio.

The instrumental requirements have been summarized in Fig. 4. The magnitude limit for a given telescope aperture, spectrograph transmission, and resolution in Ångstroms per pixel are shown. Based on experience with high speed filter/photomultiplier photometers in the photon counting mode, the star signal per integration must be about 50 counts and the total random background signal must be less than or equal to the star counts to determine an angular diameter to about ±0.1 milliseconds of arc. This was assumed in deriving the curves in Fig. 4 as well as nominal values of efficiency for reflection and quantum detection.

The top curve in the figure indicates that a telescope of 1 meter aperture, and a spectrograph of 20 percent efficiency producing 100 Ångstroms per pixel resolution would be capable of measuring the angular diameter of M 4 stars brighter than 8 mag. The righthand side of the diagram indicates the number of stars occultable for a given magnitude limit and spectral type with a predicted angular diameter greater than 2 milliseconds of arc. Thus all of the occultable stars from spectral type M 3 to B 7 which are bright enough to have a measurable diameter could be observed with a 1 meter telescope at 100 Å per pixel. A 1 m telescope with a 10 Å per pixel, 50 percent efficient spectrograph would be limited to stars brighter than 6.5 mag.

We have described a practical, new direction in spectrophotometry. Time and spectrally resolved observations of lunar occultations can be recorded and new observational constraints can be determined for cool star atmosphere models.

REFERENCES

Huggins, W. 1865 _Monthly Not. Roy. Astron. Soc._ **25**, 60.

Schmidtke, P. C. 1986, personal communication.
Scholz, M. 1985 Astron. Astrophys. **145**, 251.
Rowan-Robinson, M. and Harris, S. 1982 Mon. Not. Roy. Astron. Soc. **200**, 197.
White, N. M. 1980 Astrophys. J. **242**, 646.
White, N. M. 1981 In Current Techniques in Double and Multiple Star Research, R. S. Harrington and O. G. Franz, eds., Lowell Obs. Bull., No. 167, p 60.
White, N. M. 1986 Bull. Am. Astron. Soc. **18**, 916.
White, N. M., Kreidl, T. J. and Goldberg, L. 1982 Astrophys. J., 245, 670.

DISCUSSION

AKE: One of the uses of angular diameters is to determine effective temperatures. What does T_{eff} mean if different values are obtained at different wavelengths?

WHITE: One has to measure the "continuum" angular diameter to derive a meaningful effective temperature. In a cool star where the photosphere is distended, the chances are that an angular diameter larger than the "continuum" diameter will be measured in the visual region. The effective temperatures thus derived will be too cool. This would be less of a problem for the K-type and warmer stars.

KURUCZ: You have not mentioned the error limits you assume in your plot. What does measurable mean?

WHITE: Empirically, a σ of ±0.1 to ±0.2 milliseconds of arc is considered "usual" under normal occultation conditions. I used a value of $\sigma = \pm 0.2$. To achieve that requires the star signal to be at least equal to the background signal.

A measurable angular diameter by lunar occultation is one > 2 milliseconds of arc for the purpose of this discussion. The smallest angular diameters measured by the technique have been near 1.3 millisecond of arc. The limits, of course, increase with magnitude.

BOYD: Is the CCD read repeatedly during the occultation or is the CCD continuously clocked perpendicular to the spectrum?

WHITE: We have tried two ways. One uses the CCD as a ring buffer where the spectrum is displayed across the top row and clocked down at the integration rate and discarded continuously until the occultation is complete. The rapid clocking is instantly stopped and then the several hundred spectra remaining on the CCD are read out normally. The other uses the "snap shot"system described by E. Dunham et al. (Dunham, E. W., Baron, E. L., Elliot, J. L., Vallerga, J. V., Doty, J. P. and Richer, G. R. PASP **97**, 1196, 1985) which is based on a programmable clock controller and is read out continuously.

JOHNSON: For several stars you chose a $1/\lambda^2$ line as the best fit to the wavelength- dependent angular diameters. What is the significance of this relation?

WHITE: Simply as a point for comparison, I fitted by weighted least squares the functions $1/\lambda$ and $1/\lambda^2$. In each case $1/\lambda^2$ gave the best correlation for the three M-supergiant data sets.

ETZEL: In effect, isn't the dependence of angular diameter on wavelength really a measure of limb-darkening differences as a function of wavelength, particularly for the Ca II triplet region? This is exciting stuff for us who work on eclipsing binaries.

WHITE: There is a limb darkening effect, but the depth of cool star photospheres seems to be a factor of five to ten times larger.

DEVELOPMENT OF A FOUR-CHANNEL FIBEROPTIC SPECTROPHOTOMETER

K. H. Mantel, H. Barwig and S. Kiesewetter

University Observatory, Munich

ABSTRACT: For high precision photometry of rapid variable objects such as cataclysmic variables, pulsars or burst sources, a four-channel fiberoptic photometer has been designed with high time resolution. Four input channels allow simultaneous observation of object, two nearby comparison stars and the sky background to compensate for atmospheric extinction effects and the changing sky contribution. The input channels are realized by four optical fibers which can be positioned in the focal plane of the telescope by means of computer control. A Fabry lens at the entrance of each fiber transforms the 2 mm entrance pupil to the 400 μm fiber diameter. The four fibers feed the slit of a single spectrograph. To reach high efficiency the total wavelength region (3400 - 9000 Å) is split by means of a dicroic filter into two beams which are dispersed by separate prisms. A spectral resolution of about 30 is reached allowing one to synthesize different photometric systems. The resulting spectra of the four input channels are registered by a two-dimensional photon counting resistive annode detector (mepsicron). The detector has a resolution of 400 x 400 pixels and a dark current of about 50 counts/s at an operating temperature of -30° C. The signals of the resistor anode are processed by a position computing system resulting in an overall time resolution of better than 1 ms. Online recording and reduction of data on hard disk, optical disk or tape is done by a VME-bus computer system.

A. G. Davis Philip, D. S. Hayes and S. J. Adelman, (eds.)
New Directions in Spectrophotometry 283 - 284

DISCUSSION

WEAVER: What is the fiberoptic throughput? Is there a fiber packing problem? How much does the bundle cost?

MANTEL: The fiberoptic throughput will be about 0.8. We hope to get a packed fiber bundle where the cladding of the individual monofibers is removed at both ends. Such a bundle costs about $500.

WARREN: What is the total light loss of the instrument? For example, what would an approximate limiting magnitude be with a 1 m telescope?

MANTEL: The instrument is still under construction, so I can only make a rough estimate. The transmission of the fiber bundle will be about 0.8 and that of the spectrograph 0.8. The detector has a quantum efficiency of 15 - 20% at 5000 Å. The limiting magnitude for a time resolution of 1 - 2 sec. will be about 15th to 16th mag with a 1 m telescope.

LODEN: Do you think it will be difficult to find a suitable comparison star? Could this be an obstacle to successful performance?

MANTEL: Normally, it's no problem finding one or two comparison stars in the neighborhood of the object. It might be difficult to find one with a flux distribution similar to the object, but this effects only the correction of second-order extinction effects.

ADELMAN: Do you have any problem with the UV transmission of the fibers?

MANTEL: The fibers are made of quartz and have a high transmission in the UV.

OPTIMAL EXTRACTION AND OTHER CCD SPECTRUM REDUCTION TECHNIQUES

Keith Horne

Space Telescope Science Institute

ABSTRACT: Optimal techniques for reducing and calibrating stellar spectroscopy with a CCD can substantially improve the quality of the results. The methods described here are considered optimal because they consistently employ unbiased minimum-variance statistics and avoid resampling the raw data. Topics of particular interest include optimal spectrum extraction, robust rejection of cosmic ray hits on the spectrum, and elimination of the telluric absorption bands in near-infrared spectra.

1. INTRODUCTION

This talk is intended to be a brief survey of data reduction techniques for stellar spectroscopy with a CCD. I will touch briefly on aspects that are in general use and focus on several techniques that I have found useful but which are not yet or have only recently been implemented in general spectrum reduction packages such as Keith Shortridge's FIGARO or Frank Valdes's APEXTRACT package in IRAF. These include the optimal extraction algorithm (Section 2, Horne 1986, Robertson 1986), the double-loop sigma-clip algorithm for robust rejection of cosmic ray hits (Section 3), and removal of telluric water line absorption in the near infrared (Section 4, Wade and Horne 1988).

The main advantages of a CCD detector for stellar spectroscopy are its two-dimensional format, high quantum efficiency, and linearity. Its flaws include charge-transfer inefficiencies at low exposure levels, sensitivity to cosmic ray particle hits, cosmetic defects, and readout noise. The algorithms described here assume exact linearity, ignore charge-transfer problems, and handle cosmic rays and other blemishes under the premise that it is better to detect and ignore the bad pixels than to attempt a repair by interpolation.

The raw CCD image $R(x,\lambda)$ is corrected for electronic bias, pixel-to-pixel sensitivity differences, and sky background by subtracting a bias image $B(x,\lambda)$, dividing by a flat-field image $F(x,\lambda)$, and subtracting a sky background image $S(x,\lambda)$:

A. G. Davis Philip, D. S. Hayes and S. J. Adelman, (eds.)
New Directions in Spectrophotometry 285 - 296

$$D(x,\lambda) = \frac{R(x,\lambda) - B(x,\lambda)}{F(x,\lambda)} - S(x,\lambda) \ .$$

Here the indices x and λ refer to the rows and columns of CCD pixels, which are approximately aligned with the spatial and wavelength dimensions respectively. The bias and flat-field images combine many individual exposures (typically 40) so that the statistical noise introduced by subtraction of B and division by F are much less than the noise in R. Cosmic ray hits affecting the individual bias and flat-field images can be effectively eliminated by using the median rather than the mean to combine the individual exposures. The flat-field F can be renormalized by polynomial division to give values around 1.

The true spatial and wavelength dimensions of the data are never precisely aligned with the rows and columns of pixels, due to geometric distortions, differential refraction, etc. In long-slit spectroscopy of extended objects, e.g. rotation curves of galaxies, the preferred treatment is complete rectification of the 2-dimensional image so that rows and columns coincide precisely with constant spatial position and wavelength. However, the required resampling of the raw data introduces statistical correlations between adjacent pixels.

The algorithms discussed here for stellar spectroscopy take small geometric distortions into account without resampling the data. For example, the sky background in row λ may vary slowly with x as a nearby night sky line shifts into this row of pixels. We, therefore, represent the sky background at each λ by a low-order polynomial in x fitted to the bias-subtracted and flat-fielded data in selected blank sky regions between the spectra of objects on the slit. The method used here and throughout to fit functions to data is inverse-variance weighted least-squares, with pixel variance estimates from a noise model (Section 2.2) and a sigma-clip iteration to detect and reject inconsistent data (Section 3).

2. OPTIMAL SPECTRUM EXTRACTION

The spectrum extraction step produces a 1-dimensional spectrum $f(\lambda)$ which effectively integrates over the spatial x dimension of the data. The most straightforward extraction method is to sum contributions from a range of spatial pixels X_1 through X_2 containing the object spectrum:

$$f_{standard}(\lambda) = \sum_{x=X_1}^{X_2} D(x,\lambda) \quad ; \quad \mathrm{var}\left[f_{standard}(\lambda)\right] = \sum_{x=X_1}^{X_2} \sigma^2(x,\lambda) \ .$$

The expression here for the variance spectrum assumes that the individual pixel variances $\sigma^2(x,\lambda) \equiv \mathrm{var}[D(x,\lambda)]$ are known and statistically independent. Section 2.2 discusses methods for determining $\sigma^2(x,\lambda)$.

But now we face the following dilemma: we are tempted to sum over a wide range of pixels to be sure that we count all of the starlight, but the sum then extends far into the wings of the starlight profile, where the signal is weak compared with the background noise. Thus wide object limits deliver a noisy spectrum that is photometrically accurate, while narrow limits offer a less noisy spectrum that omits part of the light. The astronomer choosing the object limits must consider the relative importance of low statistical noise (narrow limits) versus low systematic error (wide limits), bearing in mind that this trade-off will differ with the signal-to-noise ratio at each wavelength.

Can we avoid this dilemma? Yes, but to do so we must know the spatial profile of the starlight at each wavelength pixel, $P(x,\lambda)$. Section 2.1 will discuss methods for determining $P(x,\lambda)$. As a probability density in x at each λ, $P(x,\lambda)$ satisfies the normalization condition

$$\sum_x P(x,\lambda) = 1 \ .$$

We can use $P(x,\lambda)$ to construct unbiased and statistically independent estimators of the spectrum $f(x,\lambda)$ for each pixel x:

$$f(x,\lambda) = \frac{D(x,\lambda)}{P(x,\lambda)} \quad ; \quad \mathrm{var}[f(x,\lambda)] = \frac{\sigma^2(x,\lambda)}{P(x,\lambda)^2} \ .$$

The estimators $f(x,\lambda)$ are statistically independent because they depend on data from different pixels. They are linear in the data. They are also unbiased since the division by P corrects for the fact that each spatial pixel receives only a fraction P of the starlight. The expression for the variance of $f(x,\lambda)$ assumes that $\sigma^2(x,\lambda)$ and $P(x,\lambda)$ are exactly known. Note that for pixels far out in the wings, the variance of $f(x,\lambda)$ approaches infinity as P approaches zero. This division by P spoils the mathematical rigor of our derivation, but P will not appear in the denominators of the final formulae.

Since we now have independent and unbiased spectrum estimators from each data pixel, we can construct the most general linear unbiased estimate by simply taking a weighted average:

$$f_{linear}(\lambda) = \frac{\sum_x w(x,\lambda) f(x,\lambda)}{\sum_x w(x,\lambda)} \quad ; \quad \mathrm{var}[f_{linear}(\lambda)] = \frac{\sum_x w(x,\lambda)^2 \mathrm{var}[f(x,\lambda)]}{\left[\sum_x w(x,\lambda)\right]^2} \ .$$

Furthermore, we can minimize the variance of this weighted average by choosing weights that are inversely proportional to the variances:

$$w(x,\lambda) = \frac{1}{var[f(x,\lambda)]} = \frac{P(x,\lambda)^2}{\sigma^2(x,\lambda)} \ .$$

By substituting the expressions for $w(x,\lambda)$ and $f(x,\lambda)$ into the weighted average, we arrive at the optimal (minimum variance unbiased) spectrum estimator:

$$f_{optimal}(\lambda) = \frac{\sum_x \frac{D(x,\lambda)P(x,\lambda)}{\sigma^2(x,\lambda)}}{\sum_x \frac{P(x,\lambda)^2}{\sigma^2(x,\lambda)}} \quad ; \quad \mathrm{var}\left[f_{optimal}(\lambda)\right] = \frac{1}{\sum_x \frac{P(x,\lambda)^2}{\sigma^2(x,\lambda)}} \quad .$$

This estimate is unbiased and minimum variance by construction. Wide object limits can be used to include all of the starlight, and the signal-to-noise ratio will always be less than in a standard extraction because lower weights are given to the noisy pixels far in the wings of the starlight profile. The largest gain in signal-to-noise ratio occurs when the noise is dominated by background rather than by the star's photon-counting statistics. An additional gain is realized in practice because the availability of a spatial profile at each wavelength makes it possible to reliably reject bad data pixels caused e.g. by cosmic ray hits on the spectrum (Section 3).

Optimal extraction is equivalent to performing an inverse-variance weighted least-squares fit at each λ of a known spatial profile $P(x,\lambda)$ to the data points $D(x,\lambda)$ with known uncertainties $\sigma^2(x,\lambda)$. The results should, therefore, be very similar to methods which fit an assumed spatial profile, such as a Gaussian, provided the true spatial profile is well described by the adopted function and appropriate care is taken to use correct weights in fitting.

Our derivation of the optimal extraction formula assumed that the starlight profile $P(x,\lambda)$ and pixel uncertainties $\sigma(x,\lambda)$ are known. In practice they need not be known exactly, but should be estimated with an accuracy that is small compared with the statistical fluctuations in the data. Section 2.1 deals with the problem of estimating P, and Section 2.2 section considers estimates for σ^2.

2.1 Estimating the Starlight Profile $P(x,\lambda)$

There are two basic approaches to modeling the starlight profile. One is to adopt a parameterized analytic function, for example a Gaussian, to represent the x profile at each λ. The parameters of the chosen function, for example the width and spatial centroid of the Gaussian, are then determined by fitting to the data as functions of wavelength. The model parameters can be allowed to vary only smoothly with wavelength.

Here we describe a more empirical approach to starlight modeling, which assumes no particular functional form for the spatial profile. We begin with a simple estimate

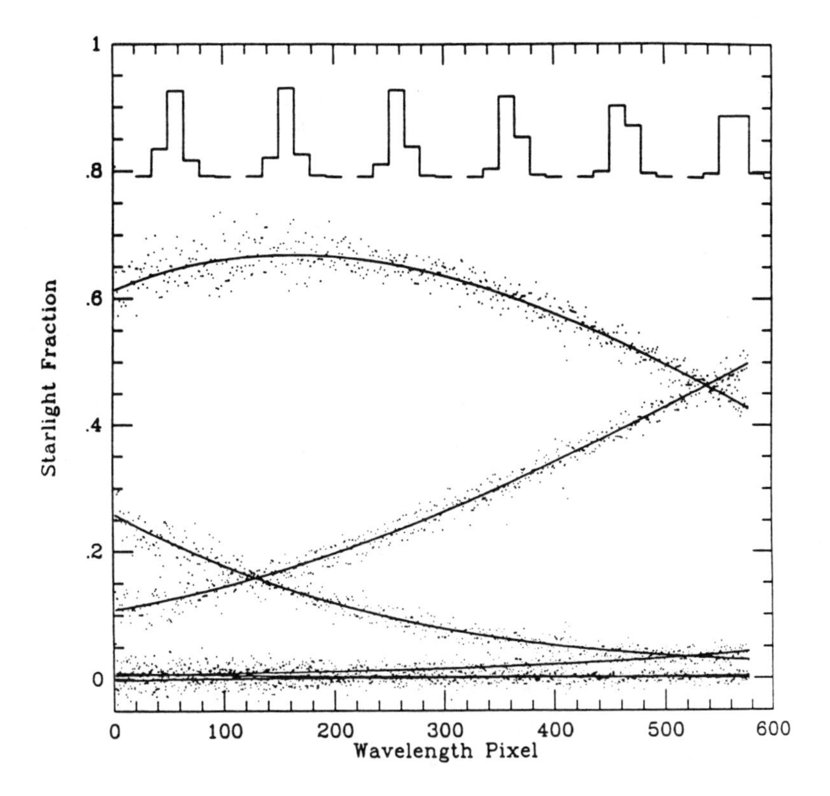

Fig. 1. The spatial profile of starlight detected by a CCD spectrograph is shown at the top for six different wavelengths, and the fraction of the starlight collected by individual CCD pixels is shown below. The smooth curves are weighted polynomial fits. A conventional spectrum extraction would in this case require a sum over four pixels at each wavelength. The two outer pixels contribute only 10 - 15% of the starlight, but nearly double the statistical variance. Optimal extraction uses the polynomial model of the starlight profile to assign optimal pixel weights that minimize the statistical noise while preserving the photometric accuracy of the spectrum.

$$P(x,\lambda) = \frac{D(x,\lambda)}{\sum_x D(x,\lambda)} \quad .$$

This has the correct normalization, $\Sigma_x\, P(x,\lambda) = 1$, but is too noisy, with statistical errors in $P(x,\lambda)$ that are highly correlated with those in $D(x,\lambda)$. Fortunately, we have profiles at numerous wavelengths which can be averaged together,

$$P(x,\lambda) = \left\langle \frac{D(x,\lambda)}{\sum_x D(x,\lambda)} \right\rangle_\lambda \quad .$$

An average over all λ won't work because P is in general not constant with λ due to changes in focus, seeing, differential refraction, etc. We allow for a *slow* variation of P with λ by fitting at each x a polynomial in λ (or some other appropriate function) to the initial estimates of $P(x,\lambda)$.

Fig. 1 shows an example of an empirical starlight profile constructed from a CCD spectrum with moderate signal-to-noise ratio. The starlight shifts continuously across the rows of pixels because the spectrum is not precisely aligned with the pixel grid. Thus moderate geometric distortions are automatically accounted for, obviating the need to resample the image prior to extraction.

The only assumption about the spatial profile is that it varies smoothly with wavelength, so that a polynomial is a good approximation. The advantage of this empirical approach is its ability to treat a great variety of spatial profiles, which may arise for example from differences in focus, accidental tracking errors, or deliberate trailing of the spectrum within the entrance aperture during the exposure. The empirical method has some problems with very weak exposures, in which there is insufficient information to estimate both the spectrum and the spatial profile. In such cases a starlight profile from a previous exposure on a bright star may be used to extract the spectrum of the fainter star. This practice may also be used to advantage with fixed-format spectrographs, for example when fiber optics are used to guide photons from the telescope focal plane to fixed positions on the spectrograph slit.

2.2 A Noise Model for CCD Pixel Variances $\sigma^2(x,\lambda)$

To apply the optimal extraction algorithm, it is essential to have a well-calibrated model describing the noise characteristics of the 2-dimensional data. The pixel variance estimates from this model are also used to calculate the uncertainty spectrum. My experience with CCD spectra suggests that the noise is adequately modeled by

$$\sigma^2(x,\lambda) = \frac{\sigma_0^2 + \left[\frac{S(x,\lambda) + f(\lambda)P(x,\lambda)}{Q}\right]}{F^2(x,\lambda)}$$

The term σ_0 is the root-mean-square (RMS) noise in data numbers introduced by the CCD readout electronics. The second term represents Poisson noise due to photon-counting statistics. The Poisson noise variance is proportional to the expected signal, which includes contributions from both the background sky $S(x,\lambda)$ and from the stellar spectrum $f(\lambda)$ $P(x,\lambda)$. The parameter Q is the number of photons required to produce 1 data number. The division by F^2 accounts for the effect of flat-field corrections on the noise (a detail which may be omitted if the flat field image was renormalized to values around unity).

Fig. 2 illustrates a graphical method of finding the two noise parameters σ_0 and Q from the fluctuations present in the individual bias and flat-field exposures. These bias and flat-field images are first de-biased and flat-fielded to eliminate the fixed background pattern and pixel-to-pixel sensitivity differences, following exactly the same procedure that is applied to the spectrum images. The bias and flat-field images are then partitioned into boxes. The mean and the RMS of the data in each of the boxes are then calculated and plotted on a log-log diagram such as Fig. 2 in order to see how fluctuations in the data depend on the signal level. At high signal levels the fluctuation amplitude grows with a slope of 1/2, reflecting Poisson statistics. At low signal levels the fluctuation amplitude is limited to a finite value by the readout noise. The smooth curves represent the readout + Poisson noise model for two values of σ_0 and Q.

A subtle but serious mistake that is often committed in data reduction is the use of uncertainty estimates calculated from noisy data values, rather than from the model that is fitted to the data. For example, one might be tempted to assume

$$\sigma^2(x,\lambda) = \sigma_0^2 + \frac{D(x,\lambda)}{Q} .$$

This is similar to the expression as above except that the noisy observed data $D(x,\lambda)$ is used in the Poisson term instead of the predicted data $S(x(,\lambda)$ + $f(\lambda)$ $P(x,\lambda)$. This is very bad because the uncertainty estimates are then strongly correlated with fluctuations in the data $D(x,\lambda)$. If a particular data value is low because of a statistical fluctuation, that point is assigned a smaller uncertainty, and is therefore given more weight than it deserves. The result is a systematic bias toward low data values. The bias may not be apparent in a single spectrum, since the bias is generally only a fraction of the statistical noise, but it will become apparent when many spectra

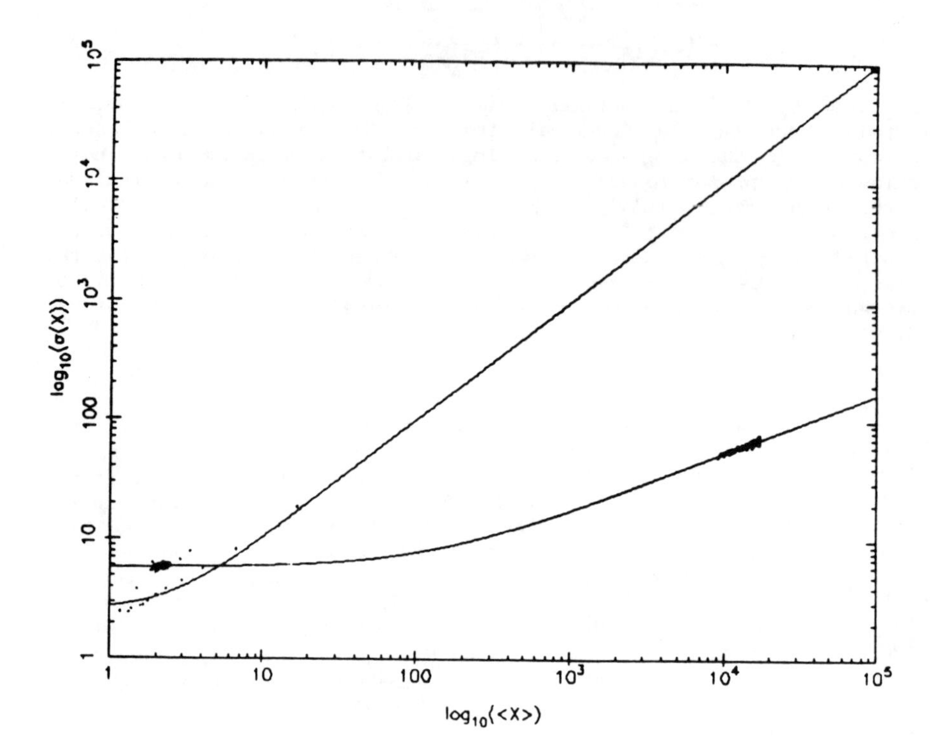

Fig. 2. The noise characteristics of a CCD detector are analyzed graphically by making a log-log plot of noise against signal. The noise is represented by the root-mean-square (RMS) variation from pixel-to-pixel within a small rectangular region of a CCD image, after applying bias and flat-field corrections. Data points are from a flat-field image with data numbers around 10,000 and RMS fluctuations around 100, and from a bias image with data numbers around 2 and RMS fluctuations around 6. The transition from slope 0 (readout noise) to slope 1/2 (Poisson) occurs near 100 data numbers. The noise parameters σ_0 = 5.8 data numbers and Q = 3.5 photons per data number are determined by fitting the continuous curve through the data.

are averaged together. The problem is easily avoided by always calculating uncertainty estimates from the fitted model, never from the noisy data.

3. ROBUST REJECTION OF COSMIC RAY HITS

There are many occasions in data reduction when a small number of bad data pixels can spoil the results. In CCD data the main culprit is the cosmic ray, which deposits a spurious signal in one or more adjacent pixels as it passes through the CCD. When fitting polynomials in x to the sky background, polynomials in λ to the starlight profile, and when fitting the starlight profiles to the data at each λ, it is necessary to identify and give zero weight to bad pixels. What is required ideally is a robust, hands-off algorithm that is efficient at detecting bad pixels without being too cavalier about losing good data. The double-loop sigma-clipping algorithm described below has proven to be effective in a wide variety of model fitting problems. The first loop is as follows:

1. Fit the model to the data by least-squares with equal weights.
2. Compute σ = RMS residual relative to the fit.
3. Reject the *one* pixel with *largest* residual if $> K_1\sigma$.
4. Repeat steps 1, 2 and 3 until no further pixels are rejected.

This first loop uses uniformly-weighted fits and rejects only 1 pixel per iteration. The noise model is not used to weight the fits at this stage because the pixels affected by cosmic ray hits do not conform to that model. Only a single pixel is rejected in each iteration to avoid the loss of good data in the early iterations when the fit may still be affected by bad pixels. Inverse-variance weights are used in the second loop:

5. Compute pixel variances from the noise model using the current fit.
6. Fit model to data using inverse-variance weights.
7. Compute χ^2 relative to the variance estimates. The pixel variance estimates may optionally be scaled by a factor $(1-p)+p\chi^2/\nu$, where ν is the number of data points minus the number of fitted parameters, and p is the probability that a χ^2 random variable with ν degrees of freedom differs from the expected value ν by more than the observed value.
8. Reject *all* pixels with residuals if $> K_2\sigma$, where σ is calculated from the noise model using the current fit.
9. Repeat steps 5 thru 8 until no further pixels are rejected.

Step 7 allows for the possibility that the noise model used for the error bar estimates is wrong by a scale factor. The double-loop is controlled through the two sigma-clipping thresholds K_1 and K_2. Values of 2.5 to 6 are typical, with higher values for cases when the omission of good data is more serious that the inclusion of bad data.

CCD's excell in the near-infrared spectral region where their quantum efficiency is near its peak. In the near infrared there are numerous zones of telluric absorption (Fig. 3). An atlas of the atmospheric absorption spectrum (Curcio, Drummeter and Knestrick 1964) covering 5400 Å to 8520 Å at a resolution of 0.2 Å shows that the absorption bands comprise numerous highly-saturated lines mainly from water vapor and molecular oxygen.

Attempts to remove the telluric absorption bands from CCD spectra usually employ observations of hot stars, typically of spectral class B, whose spectra have few intrinsic lines in this region. Each such B-star spectrum is divided by a fitted continuum to calibrate the depth of the atmospheric absorption at the airmass of the B-star observation. Other spectra are then divided by the absorption spectrum, after correcting for the difference in airmass, to remove the telluric lines.

The main point I wish to make here is that unlike the continuous component of atmospheric extinction, the depths of the telluric absorption bands do not increase linearly with airmass. At moderate resolution each spectral pixel spans a few Ångstroms and includes clear continuum regions and one or more highly-saturated telluric lines. The depth of the absorption averaged over the pixel, therefore, grows approximately as the square-root of the airmass. CCD observations of B-stars at different airmasses indicates that the empirical curve of growth for the telluric lines is close to $(\text{airmass})^{0.6}$. The telluric bands can be removed very effectively from CCD spectra if this airmass dependence is taken into account.

REFERENCES

Curcio, J. A., Drummeter, L. F. and Knestrick, G. L. 1964 Applied Optics **3**, 1401.

Horne, K. 1986 Publ. Astron. Soc. Pacific **98**, 609.

Robertson, J. G. 1986 Publ. Astron. Soc. Pacific **98**, 1220.

Wade, R. and Horne, K. 1988 Astrophys. J. **324**, 411.

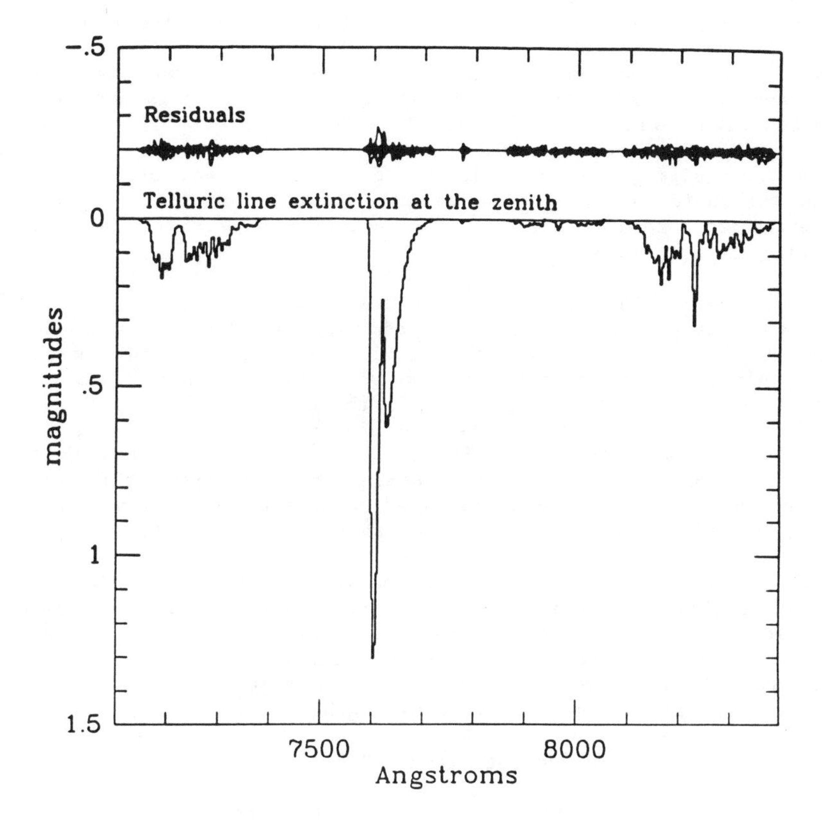

Fig. 3. Telluric absorption bands in near-infrared spectra as calibrated from CCD observations of B-stars at several airmasses. The A band at 7600 Å is due to molecular oxygen, the other bands are due to water vapor. At this resolution the band depths scale as $(\text{airmass})^{0.6}$ because each CCD pixel spans several windows of transmission between narrow and highly-saturated absorption lines.

DISCUSSION

BOHLIN: How do you choose k_1 and k_2? In the second iteration, do you use the error estimate from the previous fit and the noise model?

HORNE: The rejection threshold settings depend on the relative importance of avoiding type 1 and type 2 errors. If k is set too high, some bad pixels will not be rejected, a type 1 error. If k is set too low, some good pixels will be rejected, a type 2 error. I generally use 4 or 5 sigma thresholds in the extraction step, and 2.5 to 3.5 sigma thresholds when fitting to the starlight profile. The error estimates are calculated from the noise model using the spectrum estimate from the previous iteration.

AKE: I have a few comments. At the IUE RDAF, there are similar profile fitting routines, as well as some to allow one to manually reject the cosmic ray hits. The profile fitting is a quick method to reject hits, but it doesn't work well on weak exposures due to the uncertainty of the fitting. I also think you are fooling yourself somewhat in saying the method removes the noise since the technique is essentially a smoothing one where you are forcing the data to follow an analytical function. To retain wavelength resolution, the crosscuts must be very close to being parallel with the absorption lines, but that is a problem with any extraction routines.

HORNE: I am not very familiar with the IUE RDAF program with which you are having these problems. I have had good results with the cosmic ray rejection algorithm I've described, which uses a double iteration. The profile fitting method will certainly reduce statistical noise. It does not smear out the resolution of the extracted spectrum. Only the starlight profile is smoothed in wavelength.

KURUCZ: What is the effect of bleeding?

HORNE: Bleeding is a problem. It violates the assumption that the starlight profile is a smooth function of wavelength. To some extent the cosmic ray detection algorithm will also find and reject pixels affected by bleeding, but this is not perfect. I think the best thing to do with a saturated CCD spectrum is to throw it away.

BELL: Does your method take more people time and computer time to use?

HORNE: Optimal spectrum extraction takes much more CPU time than a normal extraction, due to all the least-squares polynomial fitting. But the extraction step is only a small part of the total processing, and so the extra CPU time is well worth it. I think the human time is about the same for optimal and normal extraction. Both have control parameters (reject thresholds, object limits) that generally need to be set by a human.

H-ALPHA SPATIAL COVERAGE OF HUBBLE'S NEBULA

A. B. Schultz

University of Nevada, Las Vegas

James W. Christy

Hughes Aircraft Company

W. K. Wells, U. Fink and M. DiSanti

LPL, University of Arizona

ABSTRACT: Narrow band filter CCD images of R Mon/NGC 2261 (Hubble's Variable Nebula) were obtained during the 1985/1986 and 1986/1987 observing season. The H-alpha filter has a halfwidth of 38 Å and centered at 6563 Å. We used a filter centered at 6775 Å and a halfwidth of 1034 Å to sample the continuum. Images constructed by scaling and subtracting the continuum images from selected H-alpha images show that the majority of the H-alpha emission is centered on R Mon and not in the nebula.

1. INTRODUCTION

R Mon/NGC 2261 (Hubble's Variable Nebula) has been observed by many observers since Hubble (1916) discovered that the nebula varied both in brightness and in apparent form. NGC 2261 is a comet shaped reflection nebula which extends about 3' to the north of R Mon, while R Mon is considered a young stellar object (YSO) going through a phase of evolution on its way to the main sequence. Variations in the morphology of the nebula since Hubble have resulted in observers postulating a dark body or bodies orbiting R Mon and thereby casting a shadow pattern on the nebula. Johnson (1966) as well as Kazaryan and Kachikyan (1972) conclude from the available observations that there is no correlation between the variations of the shadow patterns.

Radio observations (CO) of NGC 2261 (Canto et al 1981) suggest that the nebula is the northern lobe of a bipolar source. The northern lobe is the interaction between a strong stellar wind and the interstellar medium. We only observe this northern flow by reflected light, while the southern lobe is being obscured by a molecular disk surrounding R Mon. This model suggests that R Mon/NGC 2261 is a more

A. G. Davis Philip, D. S. Hayes and S. J. Adelman, (eds.)
New Directions in Spectrophotometry 297 - 306

complex system than earlier supposed, and any decoupling of the morphological changes to verify dark bodies orbiting R Mon will require a concerted effort in both spectroscopy and direct imaging.

We have obtained spectroscopic and direct imaging observations of R Mon/NGC 2261 during the periods 1969 - 71 and 1985 - 88. In this paper, we will discuss one sample of our H-alpha narrow band imaging and the procedure used to flux calibrate our images. The other observations will be discussed in a later paper.

2. OBSERVATIONS

The observations were taken at the UAO 61" telescope, Mt. Bigelow, Arizona using an off-axis reducing camera system, which reduces the image scale from 0.15 arc seconds/pixel to 0.94 arc seconds/pixel. The detector employed was a CCD camera with a Texas Instrument 800 x 800 three phase chip having 15 micron pixels. The CCD camera, data acquisition system, and off-axis reducing optics were built by U. Fink at LPL, Univ. of Arizona (Fink et al 1986).

The H-alpha images span two observing seasons 1985/1986 and 1986/1987. The data discussed here consists of three H-alpha and four continuum images obtained during January 1986. The two filters used were: H-alpha 6568/38 Å and continuum filter 6775/1034 Å. A summary of our H-alpha observations is given in Table I.

Table I

H-alpha Observations

Date	Exposure time (sec)	Date	Exposure time (sec)
1986 Jan 13	30	Dec 02	15
	120		30
	300		60
			300

3. TWO-DIMENSIONAL SPECTROPHOTOMETRY

Images constructed by scaling and subtracting the continuum images from the H-alpha images give spatial information about the nebula and, therefore, give information complementary to spectroscopic studies of NGC 2261. Contour maps scaled to a relative intensity can be used in determining the geometric structure of the nebula, but we must remember that we are dealing with a two dimensional projection of a three dimensional structure. Therefore, care must be taken when contour maps are used to test different types of structure models for the nebula. We also must mention that in some cases, and for NGC 2261, contour plots give a better representation of bow shocks than a photograph.

Fig. 1 (see next page) is a contour plot of the H-alpha distribution of R Mon/NGC 2261 in relative intensity to show structure.

A one-dimensional profile created from a two dimensional image is equivalent to a H-alpha spectral line profile from a spectrometer with the slit positioned across the object identically as the extraction aperture for the one dimensional profile. Thus, we have the flexibility of creating a spectral line profile from any direction and position we chose. We have created one dimensional profiles of R Mon/NGC 2261 by taking apertures or strips across the data frame. Figs. 2 and 3 display a series of these strips taken from the H-alpha frame and overlayed on the same plot.

A one-dimensional profile created from strips has greater signal-to-noise over a single slice and will both mimic a spectroscopic observations and aperture photometry across the nebula. The advantages of this technique are complete spatial coverage, increased spatial resolution with the CCD camera and selecting the aperture at a later time. However, the disadvantages include a greater effort in reduction techniques, limited spectral coverage, and difficulties in using narrow band filters. A few of these difficulties are discussed in the following paragraphs.

4. IMAGE PROCESSING

The following analysis is based on the two-dimensional spectrophotometry technique developed by astronomers studying planetary nebulae: Kupferman (1983) and Jacoby (1987). We have employed the image reduction and analysis facility (IRAF) computer code developed by NOAO, Tucson for the data reductions.

All the image frames were flat fielded using a standard method of first subtracting an average dark frame from all frames followed by ratioing with the appropriate flat field frame. Dark frames are images taken during the observing run in which no light reaches the CCD chip, and flat-field frames are images taken through the corresponding filter of an illuminated target, where the illumination is provided by a quartz lamp. Several flat field images are taken and averaged together to create one flat-field frame.

Flat fielding of the data frames is followed by registering all the frames to a common coordinate system. One H-alpha frame is used as a control and all the other image frames including the continuum images are shifted, rotated, and corrected for any plate scale difference. As many stars as possible that surround R Mon/NGC 2261 and are not saturated in each frame were used to derive the translation. Registration is necessary when a continuum frame is subtracted from an

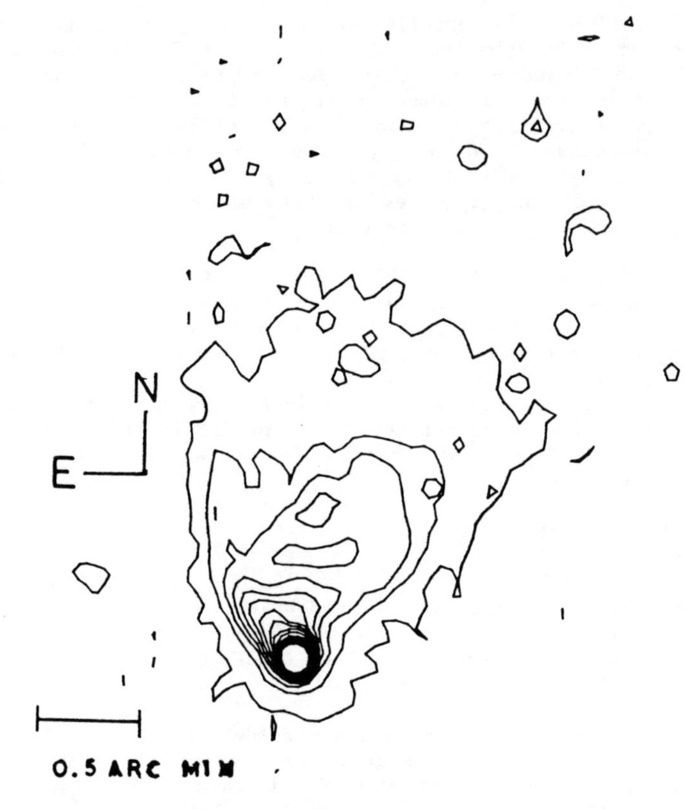

Fig. 1. Contour diagram for R Mon/NGC 2261. H-alpha spatial distribution after subtracting a scaled continuum image from an H-alpha image. Contour levels are evenly spaced in units of relative intensity.

H-alpha frame.

4.1 Flux Calibration

Stars chosen from the KPNO IRS Star Manual (Barnes and Hayes 1982) were imaged through the same filters as the nebula. Basic aperture photometry was performed on these images to obtain counts/sec. The counts/sec value is related to the published flux distribution for the given star through the total energy passed by the filter. The idea here is to use the standard star to flux calibrate the individual data frames taking into account the flux of the standard star, the transmission curve of the filter, the exposure time of the data frame, and the atmospheric extinction. In the following paragraphs, a detailed description of this technique is given for our observations of R Mon/NGC 2261.

The total energy passed by a filter for a standard star is found by convoluting the filter transmission curve with the flux distribution given for a standard star in the IRS Star Manual. See equation 1. In practice, some approximation can be made for the transmission curve of the filters and the flux distribution. For the narrow band filter, we approximated the transmission curve with a Gaussian curve and assumed the flux through this filter was of constant value. For the continuum filter, which has a much larger band pass, we used a constant value for the transmission curve (a flat response) and a straight line fit for the flux distribution of the standard star.

$$E(\mathrm{erg/s/cm^2}) = \int F(\lambda) * T(\lambda)\, d\lambda \qquad (1)$$

The total energy passed by each filter is divided by the respective counts/sec found for the standard star to obtain a value that has units of $\mathrm{erg/counts/cm^2}$. Each data frame is divided by its exposure time to obtain counts/sec and then multiplied by the respective value found for $\mathrm{erg/counts/cm^2}$. The resultant image will have units of $\mathrm{erg/sec/cm^2}$.

The bandpass of a narrow band filter may shift due to telescope characteristics and temperature. One must take correcting steps in the calculations for any changes in the filter response. The integral of the filter transmission curve is divided into the data frame to obtain the flux calibrated image. Finally, each image is corrected for atmospheric extinction taking into account the difference in air mass between the data frame and the corresponding standard star frame.

The steps described above for flux calibrating a CCD image are represented by equation 2:

$$\left[\frac{\mathrm{image}}{\mathrm{counts}}\right] \cdot \left[\frac{1}{\mathrm{exp}}\right] \cdot \left[\frac{\mathrm{erg}}{\mathrm{counts/cm^2}}\right] \cdot \left[\frac{1}{\int T(\lambda)}\right] \cdot \left[10^{.4\, k_{\nu}(AM_{obj} - AM_{std})}\right] \qquad (2)$$

where exp is the exposure in seconds, $T(\lambda)$ is the integral of the transmission curve, and k is the atmospheric extinction in magnitudes per air mass. We used a seasonal atmospheric extinction for our calculations.

R Mon has a radial velocity of about 10 km/sec and, therefore, no correction was necessary for any shift in the H-alpha peak relative to the envelope of the filter profile. No correction for interstellar reddening was applied to our images.

4.2 Continuum Subtraction

The last step in the data reduction process is to remove the sky level and nebula continuum from the H-alpha frames. NGC 2261 does not fill our field of view, which allows us to measure the sky level in each frame. We found several areas around the nebula that were free of background stars and nebula continuum. These areas were sampled and averaged in each frame. The resulting averaged value of sky in each frame was subtracted from the entire respective frame.

Ideally, one would chose a continuum filter that samples the continuum near the emission line of interest without the bandpasses of the two filters overlapping. This is not the case for our observations. Therefore, before we could subtract the continuum frame from the H-alpha frame, a correction was needed to take into account the flux detected in the overlap area. We multiplied the continuum frame by the ratio of the transmission integrals: narrow band filter to the continuum filter.

5. SPATIAL PROFILES

Spatial profiles were created by summing rows or columns from the corrected H-alpha images. The most difficult aspect of this analysis is deciding on the size of the aperture to use. We decided to sample the spatial profile of stars in the field that were not saturated. In this way, our aperture would correspond to the size of the seeing disk for the observations, and thereby simulate aperture photometry. The profiles represent the spatial extent of a measurement of radian intensity of a small section of the nebula.

For the purpose of the profile scaling, we define the radiant flux per unit solid angle times the full width half maximum (FWHM) of the narrow band filter as the Line Intensity. The radiant flux per unit solid angle is called the radiant intensity (I), and multiplying by the FWMH value is essentially an integration of the radiant flux over a small wavelength interval. The line intensity has units of erg/sec/cm^2/sr and can be considered the effective radiant flux per solid angle. The final result is now detector independent and directly comparable to other observations.

The derived profiles of R Mon/NGC 2261 are presented in Figs. 2

and 3. Each profile represents either the sum of 11 rows or 11 columns in the image. This corresponds to an aperture of 9.6 square arc seconds. Fig. 2 (see next page) displays profiles across the nebula arranged to achieve a three dimensional effect. The first profile is 10.3 arc seconds north of the peak photo center in the nebula, which represents the position of R Mon. Each succeeding profile is 10.3 arc seconds north from the previous one. An arbitrary bias has been added to each profile for positioning purposes. Fig. 3 displays profiles that are centered on R Mon, but are strips that run North-South and East-West. The scale has been expanded to show detail in the nebula close to R Mon. The profiles in Fig. 2 have been smoothed with a box car of five, while the profiles in Fig. 3 have not been smoothed. Fig. 3 shows a noisy background left after flat fielding and continuum subtraction, but an acceptable SNR for determining the Line Intensity close to R Mon.

6. DISCUSSION

Imaging R Mon/NGC 2261 with a narrow band filter centered at H-alpha (6563 Å) has allowed us to isolate this important emission line, and provided us with a tool to probe the entire nebula. Theoretically, it would be possible to integrate the line flux over the entire nebula to obtain a total line flux; however, care must be taken when averaging fluxes from different areas in the nebula that are at different ion temperatures and density (Jacoby 1987).

The contour map of R Mon/NGC 2261 does show that the majority of the H-alpha emission is centered on R Mon and not in the nebula. The contour levels are evenly spaced in units of intensity. We interpret the geometric spacing between the levels as normally attenuated light reflected from the nebula with the closely spaced contours as an indication of bow shocks caused by a stellar wind from R Mon. A strong jet is indicated on the eastern edge of the nebula. The overall impression is that the interstellar medium is not uniform in the neighborhood of R Mon.

Spatial profiles of the nebula depict a nebula that is a complex and dynamic structure and possibly evolving with time. The jet on the eastern edge of the nebula is clearly defined in the profile map, as well as other more prominent spatial features. The profiles were created from one 30 second exposure; the longer exposure images should yield a better SNR.

The second set of images taken in December 1986 combined with the results obtained in this analysis may yield the desired clues to correlating changes in the morphology of the nebula to physical processes. In the future, observations with a H-beta filter along with the H-alpha observations are needed to map out the temperature gradients of the nebula.

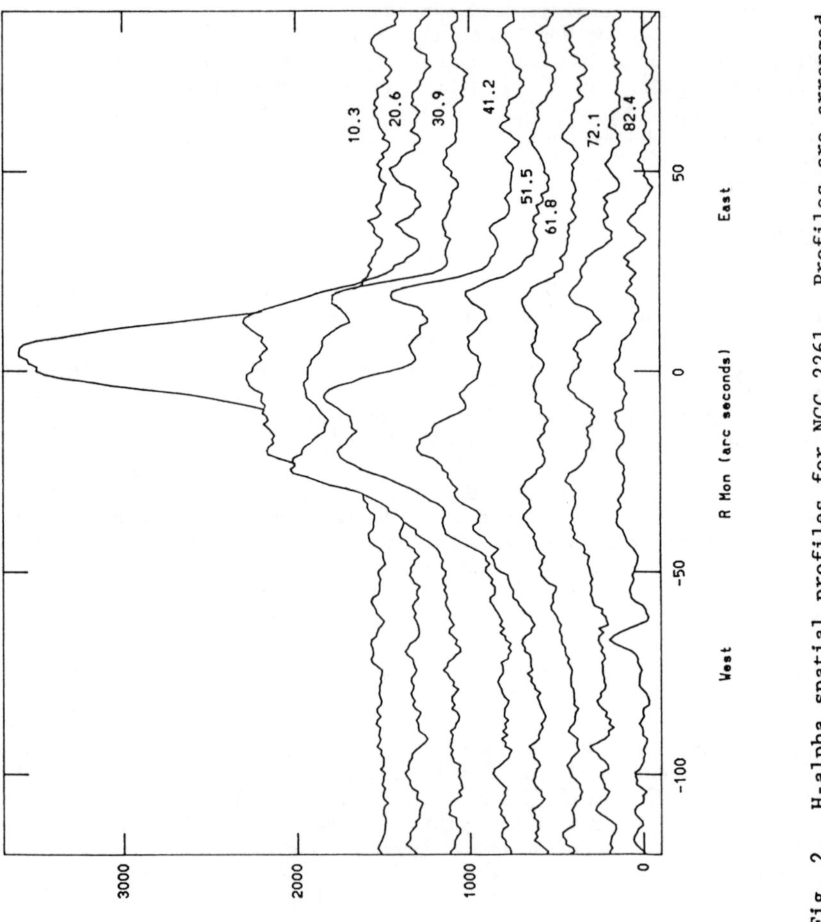

Fig. 2. H-alpha spatial profiles for NGC 2261. Profiles are arranged to achieve a three dimensional effect. Numbers indicate arc seconds north from R Mon.

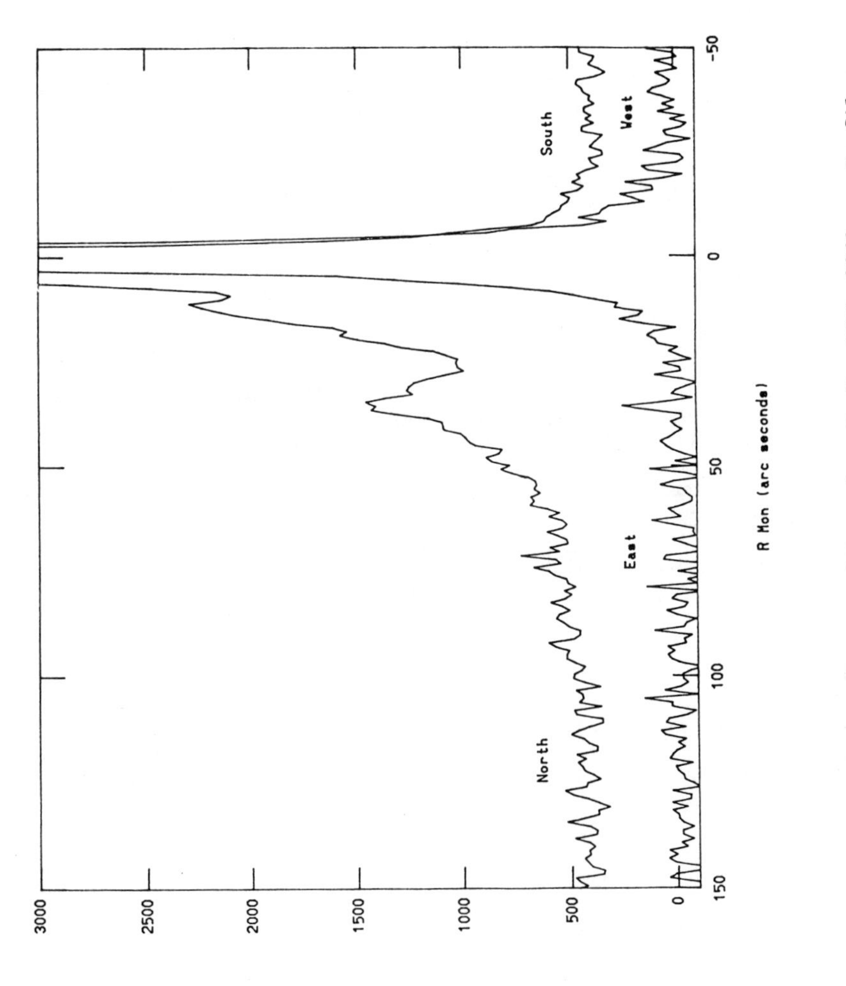

Fig. 3. H-alpha spatial profiles for R Mon/NGC 2261. Profiles centered on R Mon.

ACKNOWLEDGEMENTS

We wish to thank Dr. Sidney Wolff, Director, NOAO for the loan of the H-alpha filter and H-continuum filter and on occasion access to the NOAO computer facility. Special thanks goes to Jeannette Barnes and Ed Anderson, NOAO, Tucson for their valuable assistance in the instruction in the art of IRAF operation. A debt of gratitude goes to George Jacoby for the pre-print of his paper and the many conversations on narrow band imaging.

The CCD camera is operated under NASA grant NSG 7070.

REFERENCES

Barnes, J. V. and Hayes, D. S. 1982 IRS Standard Manual, Kitt Peak National Observatory.
Canto, J., Rodriguez, L. F. and Barral, P. 1981 Astrophys. J. **244**, 102.
Fink, U., DiSanti, M. and Schultz, A. 1986 in 20th ESLAB Symposium on the Exploration of Halley's Comet, Vol. 2, ESA, Paris, p. 485
Hubble, E. P. 1916 Astrophys. J. **44**, 190. (see also 1986 Proc. Acad. Sci. **2**, 230)
Jacoby, G. H., Quigley, R. J. and Africano, J. L. 1987 Publ. Astron. Soc. Pacific **99**, 672.
Johnson, H. M. 1966 Astron. J. **71**, 224.
Kazaryan, M. A. and Khachikyan, E. E. 1972 Astrofizika **8**, 17. (English translation in Astrophysics **8**, 8.)
Kupferman, P. N. 1983 Astrophys. J. **266**. 689.

DISCUSSION

GRAYZECK: What is the ejection speed of the filament?

SCHULTZ: The filament velocity has a profile which goes up to 180 km/s and then drops slowly with distance from R Mon.

AUTOMATIC SMALL TELESCOPE OPERATIONS

Russell M. Genet, Donald S. Hayes and Louis J. Boyd

Fairborn Observatory

The era of small ground-based automatic optical telescopes began in the late 1960's with an 8-inch system at the University of Wisconsin. The Wisconsin system has been described by McNall et al. (1968). Its operation consisted of observations of a fixed sequence of bright nonvariable extinction stars. The star list was on punched paper tape as there was only 4K words of memory in the computer and it was all needed to operate the telescope. The observatory was automatic in its operation, with weather sensors, roof control, etc. Output data were punched on paper tape for later analysis. This excellent pioneering effort gathered useful extinction data, sparing the larger telescopes from this task. However, the main reason for the project was to help Wisconsin into the age of space telescopes, and in this it was an unqualified success. We have it on good authority that operations were finally suspended because of frequent false roof closures. Students picnicking on the observatory grounds had discovered that if they poured beer on the rain sensor the roof slammed shut!

Memory restrictions were less severe when Skillman (1981) began operations with a semi-automatic 12-inch system with an Apple computer. While manual startup and shutdown was required, and the system was essentially limited to a single variable/comparison/sky combination per night, Skillman put it to very effective use in observing eclipsing binary stars. He placed the system on a binary going into eclipse in the eastern sky in the early evening, went to bed, and in the morning would shut the observatory down. Data collection and analysis were computerized.

A fully automatic system described by Boyd et al. (1984) began operation in 1983. This system had two features of operational importance. First, it was able to observe stars anywhere within its "observing window." Second, it had a more or less permanent list of possible objects to observe, and used algorithms to decide which, of all possible, it would observe at any given time. These two features allowed long-term observing programs with many program stars to be undertaken - each one being observed about once per night. Hall, at Vanderbilt University, has used this capability to good effect on the RS CVn binaries. Genet brought a similar system into operation in

A. G. Davis Philip, D. S. Hayes and S. J. Adelman, (eds.)
New Directions in Spectrophotometry 307 - 310

Fairborn, Ohio, in 1984.

The next improvement in operations was the grouping together of a number of automatic telescopes at a remote mountaintop observatory to serve a number of astronomers. This was done when the Fairborn Observatory and the Smithsonian Institution formed the Automatic Photoelectric Telescope Service in 1985. Currently the APT Service has three automatic telescopes in regular operation, with some 20 institutions in the US, Canada, and Europe receiving observations. The remote location of the APT Service necessitated fully automatic operation of both the telescopes and the observatory itself. A phone call is automatically made each morning by the observatory to the Phoenix area, some 160 miles away to report on equipment status, observations made, quality of the observations, and remaining storage space for the data. The APT Service has been described by Genet et al. (1987).

While the current APT Service operations have been successful in making routine, long-term programs requiring once a night differential photometric observations, we thought that the usefulness of such automatic telescopes could be improved if their instructions were adaptable to a wider range of observing tasks, if the requests could be made more directly, and if the results could be quickly retrieved. To these ends we have devised an automatic telescope instruction set (ATIS) that places very few limitations on what can be done with the systems, and a remote access capability that will allow astronomers to make a request via modem one afternoon (with just a few minutes of telephone time) that will result in observations being automatically made that night. They can then retrieve the results the following morning or anytime after that at their convenience. Once remote access automatic telescopes are located at several sites, it would not be difficult to coordinate their observations such that stars could be kept under constant surveillance or observations could be made of almost any portion of the sky at any time - a useful capability in support of space telescopes or larger ground-based telescopes.

In the past APT's have been 0.4 m or smaller in size, but larger APT's are now under construction. Shown in Fig. 1 is the first of four 0.75 m telescopes planned for the APT Service. Recently the Danes have brought a 0.5 m telescope at La Silla under automatic operation. It seems that even larger automatic telescopes are likely in the future.

ACKNOWLEDGEMENTS

The authors are grateful for the support of the National Science Foundation and the Smithsonian Institution.

REFERENCES

Boyd, L. J., Genet, R. M. and Hall, D. S. 1984 IAPPP Comm., 15, 20.

Genet, R. M., Boyd, L. J., Kissell, K. E., Crawford, D. L., Hall, D. S., Hayes, D. S. and Baliunas, S. L. 1987 Pub. Astron. Soc. Pacific, **99**, 660.
McNall, J. F., Miedaner, T. L. and Code, A. D. 1968 Astron. J. **73**, 756.
Skillman, D. R. 1981 Sky and Telescope, **61**, 71.

Fig. 1. One of the four 0.75 m APT telescopes under construction. When pointed to the zenith the height of the telscope is less than 2 meters. Photo by R. Genet on Mt. Hopkins.

DISCUSSION

JONER: Are you currently observing with the APTs exclusively in the differential mode? Are there potential problems involved in switching it to a true all-sky mode?

GENET: Primarily, the observations are in the differential mode. Quasi-all-sky observations are made of standard stars. We do not foresee any serious problems switching to a true all-sky mode. The algorithms for automatic all-sky photometry have been developed.

KURUCZ: Are you over subscribed?

GENET: No. There is some time available on the 10-inch UBV system.

WEAVER: As the telescope sizes increase, the limiting magnitude of the program stars will get fainter. How will you handle the increasingly difficult process of finding and centering?

GENET: There are two approaches to improving the acquisition and centering process. First, one can center up initially on a nearby brighter star, and then move with increased certainty to the star in the crowded field. Second, we plan to develop a CCD acquisition and centering system that will operate on fainter stars.

HORNE: What is the cost of bringing up a 30-inch APT?

GENET: Not having finished bringing up the first one yet, this is difficult to say for sure.

PETERS: How much do you charge to carry through one observation of a single object? (i. e., how much does it cost per star per night?)

GENET: The current charge for observing three stars and sky in three colors (33 total observations) on one night is $2.00. The observational charges for a star for a year can be less than the page charges to publish the results.

PRELIMINARY SPECIFICATIONS FOR AN AUTOMATIC SPECTROPHOTOMETRIC TELESCOPE

Donald S. Hayes

Fairborn Observatory
Institute for Space Observations

Saul J. Adelman

The Citadel

Russell M. Genet

Fairborn Observatory

ABSTRACT: We present a preliminary design study and specifications for an automatic spectrophotometric telescope (ASPT). The system would be fully automatic, in that it would operate at a remote mountain site without any observers being present. Observing programs would be loaded in by communications from an observer's computer over phone lines via modems at intervals of weeks or months; the data would be downloaded over the phone lines at intervals which could be as short as a day and as long as two or three weeks. The spectrophotometer would make observations covering the wavelength range for 3200 to 11,000 Å with a resolution of 8 Å in the second-order blue and 16 Å in the first-order red.

1. INTRODUCTION

The advent and success of the automatic photoelectric telescope (APT) (Genet 1986; Genet, Boyd, and Baliunas 1986; Genet, Boyd and Hayes 1988 in this Symposium) suggests that the time has come for the application of automatic telescopes and of automatic instrumentation to a wider field of astronomical observations from the Earth's surface. The major advantages of an automatic telescope-instrument combination compared with conventional operations are a lower cost of operation, more efficient data taking, and very high degree of consistency in how the data are obtained. The extension of APT-type operations to spectrophotometry to produce an automatic spectrophotometric telescope (ASPT) could well produce an instrument capable of having a great scientific impact in the next decade.

A. G. Davis Philip, D. S. Hayes and S. J. Adelman, (eds.)
New Directions in Spectrophotometry 311 - 327

2. AN AUTOMATIC SPECTROPHOTOMETRIC TELESCOPE

The idea of using a fully-automated telescope for spectrophotometry involves extending telescope and instrument technology beyond its current state. The extension is modest, however, since the most critical part of the ASPT concept is the capability of the system to do fully-automatic observations of any type. Fully-automatic operation of a photometric telescope doing UBV differential photometry of variable stars was achieved by the Fairborn Observatory five years ago; currently three small telescopes are operating in the fully-automatic mode on Mt. Hopkins (Genet, Boyd and Hayes 1988). A key aspect of fully-automatic photometry is finding and centering the stars. The current systems use the output of the photometer during the acquisition phase; after the telescope has moved to the new position it begins a spiral search. When a signal is detected which matches the expected brightness of the target star, the system goes into a "centering" mode (Genet, Boyd and Baliunas 1986).

Spectrophotometry requires observations which are very similar to filter photometry in the operational sense: the star is acquired and centered in a diaphragm of from 15 arcsec to 60 arcsec diameter and the telescope must keep it within a few arcsec of the center for times from 10 sec to about five minutes while the integration proceeds. Since the detectors which are proposed to be used in the spectrophotometer do not give a real-time signal which can be monitored, the current method of acquisition and centering cannot be used. A simple uncooled, unintensified integrating CCD camera can be used instead, however. A description of the method of using a CCD is given by Boyd and Genet (1987); newly-introduced products allow using an unintensified, integrating camera instead of the intensified, TV-frame-rate camera described in that paper.

Since spectrophotometry involves a higher resolution of the spectrum, the observations cannot be made as faint as by filter photometry for a given size telescope. This means that acquiring and centering, the star will not be as difficult as in the case of filter photometry. In either case, however, if an object is to be observed which is too faint for the acquisition and centering process to handle, a nearby brighter "navigation" star can be acquired and centered and a blind offset made to the position of the target. Spectrophotometry differs from the current mode of automatic photometry in that it involves all-sky photometry, rather than differential photometry. Three new 0.75 m APTs are being built for operation on Mt. Hopkins, and the more general control system being designed for them will include the capability to do all-sky photometry; the new systems are termed programmable automatic telescopes (Hayes, Genet and Boyd 1988). CCD acquisition and centering is being developed for these systems. Thus, by the time construction would be started on an ASPT, the key new capabilities of all-sky photometry and CCD acquisition and centering will have been proven for filter photometry.

Beyond the capabilities discussed above, the extensions to current technology which must be developed include the cooling of an array detector and the collection, storage, transmission and reduction of the data, which will be produced in larger quantities than by the current filter-photometers. These capabilities will be discussed in later sections.

It appears to be wise to limit the extensions to the technology required by the ASPT, so this proposal will be based upon the 0.75 m telescope design which is being constructed now, and upon a spectrophotometer following current practice in optics and detectors. Eventually, it may become appropriate to build a larger telescope and use a different optical layout for the spectrophotometer.

2.1 The Telescope

Fig. 1. An APT-Service 0.75 m automatic telescope

The telescope design has been described by Genet, Boyd and Genet (1987) and by Genet, Boyd and Hayes (1988) in this Symposium. We will review only enough of the design to give completeness to our system design and to describe the optical and physical environment of the spectrophotometer. The telescope has a rather unusual configuration, and is very compact for its size. In Fig. 1 is shown a photograph of one of these telescopes. The telescope stands only about 6 feet tall as a result of the compact and inaccessable instrument bay and the fast primary (f/2). The former characteristic results from the decision to limit the telescope to fully-automatic operation; there is no provision for access to the focal plane by an observer using an eyepiece, for example.

The fast primary results in a very short structure, in spite of an effective f-ratio of f/8. The short structure is important because it results in a lower moment of inertia which allows for quick motions during automatic operation. The structure is open, which with its short length lowers the wind loading. Since the optical demands of photometry and spectrophotometry are not high, a lightweight primary of slumped 1 5/8-inch pyrex is used, lowering the weight of the entire telescope, which is about 500 lbs, lowering the thermal inertia, and lowering the cost. A classical Cassegrain optical layout is used. Friction drive is utilized, so that backlash is essentially zero; this again is another characteristic vital to automatic operation.

The Cassegrain focus occurs in an instrument bay which is about 18 inches square by about 9 inches deep. This is quite adequate for a photometer, but it limits the size of a spectrophotometer. Thus, one of the prime characteristics of the spectrophotometer to be presented below is its small size. The scale of the image at the Cassegrain focus is 35 arcsec/mm. We estimate that the telescope with a CCD acquisition system will be able to acquire stars to about 13th mag, and center to 1-2 arcsec. The tracking should be good to 1-2 arcsec for exposures of up to about five minutes.

2.2 The Spectrometer

The spectrophotometer to be described here is a variation of a design discussed by Hayes (1987). In that paper, a design intended for a telescope of 1.0 m aperture with an f-ratio of f/3.75 was presented. The particular design adopted there required a very fast camera which would be expensive to build. Since the present design is intended for an f/8.0-telescope of 0.75 m aperture, it has been possible to produce a configuration with a considerably slower camera.

As in the spectrophotometer discussed by Hayes (1987), the design is constrained by the need for a compact instrument which will fit in the instrument bay of the telescope, and for an instrument which is simple to design, build and operate under fully-automatic conditions. Further, it should not be expensive to build.

A number of conflicting constraints determine the design of a spectrophotometer, and the resulting parameters must necessarily represent a compromise. Among the important constraints are: 1) the resolution and wavelength coverage required by the scientific objectives to be met: for the present design we wanted a resolution of about 10 Å and a wavelength coverage of from 3200 to 11,000 Å. 2) the characteristics of the detector to be used, including particularly the size of a resolution element and the total number of resolution elements in each direction. In this case, we will assume a generic array detector with square pixels which are 30 microns on a side; the array is assumed to have at least two rows 1000 pixels long. The detector will be discussed further in the next section. 3) the requirement that image motions and blowup due to seeing, and image wander due to imperfect tracking and mount alignment will not have significant photometric effects. The parameters of the spectrophotometer which reflect these constraints are listed in Table I.

TABLE I

Parameters of the spectrophotometer

Wavelength Range	Order	Dispersion		Resolution	Filter
2500- 6500 Å	2	4 Å/pixel	133 Å/mm	8 Å	$CuSO_4$
5000-13000 Å	1	8 Å/pixel	267 Å/mm	16 Å	lowpass at 5500 Å

Let us expand on 3). In the case of a conventional photoelectric photometer, a Fabry lens is placed ahead of the photomultiplier so that image motions do not interact with irregularities in the photocathode sensitivity to produce low-frequency noise in the signal. In an array detector, a Fabry lens is impractical. The simplest strategy is to project the image onto the detector such that a pixel is several arcsec across. With the present telescope, the scale corresponds to 1 arcsec per 29 microns. That is, if the camera and collimator focal lengths were the same, the stellar image would project to a scale of about 1 arcsec/pixel on the detector. This is not acceptable, and the design presented here has the collimator focal length 3 times the camera focal length, so that the scale projects to 3 arcsec/pixel. It would be preferable to have an even larger ratio of focal lengths, but to do so results in making the camera too fast for reasonably easy and inexpensive construction.

The final design is shown in Fig. 2. As also noted there, the collimator has a clear aperture of 35 mm and a focal length of 266 mm. The light is reflected off a folding flat onto the grating, which has 400 lines/mm and a clear aperture of 35 mm. The camera has a clear aperture of 41 mm, a focal length of 89 mm, and a resulting f-ratio of f/2.16. Note that the longest dimension of the spectrometer is about 11 inches, so the instrument is very compact.

The spectrophotometer is designed to achieve full wavelength coverage from 3200-11,000 Å by making two exposures, one in the first-order red and one in the second-order blue. The orders will fall on the detector along the same row, so they must be separated by filters. In the first-order red, a lowpass filter cutting on at about 5500 Å is used to eliminate the second-order blue; the first order will be contaminated longward of about 1.1 micron, but it is unlikely that the detector will have good response beyond this wavelength. In the second-order blue, a copper sulfate filter is used to pass light from the short-wavelength limit to about 5500 Å. The choice of wavelength limits given in Table I is largely determined by the atmospheric limit on the short-wavelength end, red limit of the copper-sulfate filter, and the long-wavelength limit of detector sensitivity; the apparent waste of the short-wavelength end of the detector in the second order and the long-wavelength end of the detector in the first order is the inevitable result of these limits and the use of overlapping first and second orders.

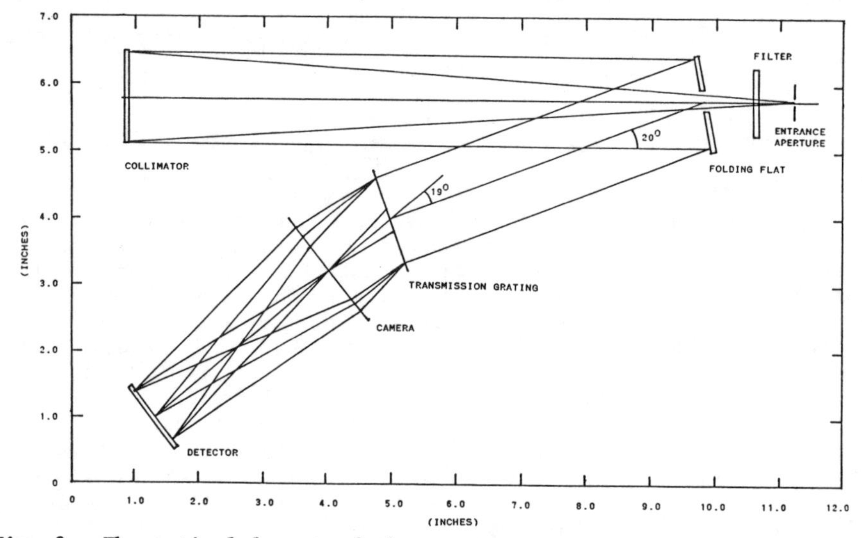

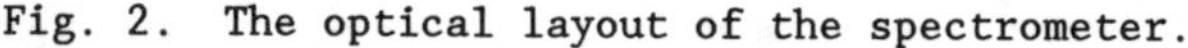
Fig. 2. The optical layout of the spectrometer.

Two exposures are thus to be made to achieve full wavelength coverage. The filters will be mounted in a simple slide and moved under computer control. In general, the exposure in the second order will be much longer than the exposure in the first order, since most stars have less light in the blue and UV, the atmospheric transmission is less toward short wavelengths, and the detector has a lower response to the shorter wavelengths, even with a coating to enhance the response in this region. Thus, except for very blue stars, the penalty of having to make two exposures is the time spent reading out the chip a second time, and of having to store and process the data. An alternative would be to use cross-dispersion to separate the orders such that they could be measured simultaneously. This route would be somewhat more efficient with exposure time, but not enough to compensate for the added complication in the spectrometer and in the reduction of the data because of the resulting curved orders. In general, an automatic telescope is better suited to survey observations on brighter stars, and the observation of very faint objects with the utmost in efficiency in the use of exposure time is not well suited to an automatic system.

2.3 The Detector

There are only two reasonable choices for the detector, a self-scanned photodiode array, such as those from Reticon (we will follow common usage in astronomy and call these devices by the generic term "reticon," recognizing that Reticon is a trademark company name, that Reticon also makes CCDs, and that self-scanned photodiode arrays are available from other sources), or a CCD. A reticon has the following advantages: 1) it is available in a suitable format for spectrophotometry, with 2x1000 pixel arrays being easily obtained with rectangular pixels with their long dimension perpendicular to the long dimension of the array itself, 2) it has good quantum efficiency over the desired wavelength range, particularly in the UV, 3) it can be operated under astronomical conditions with thermoelectric cooling, which is better suited to automatic, unintended operation than cooling with cryogens, and 4) it is relatively inexpensive and requires only modest supporting electronics and computers. A reticon has one extremely serious disadvantage, its high read noise of about 500 electrons. To achieve precisions around 1%, the integration times are about seven times longer than they would be with no read noise for stars fainter than about 3rd mag. With programs made up mostly of stars brighter than 5th mag, a reticon would be an attractive choice. It is very likely, however, that this instrument would be used much of the time on programs with stars fainter than this, so a reticon does not appear to be practical.

CCDs, on the other hand, exhibit the advantages of low read noise plus the generic ones of array detectors. Although a CCD would be more expensive and would require more complex supporting electronics and computers, they are becoming well understood and the supporting system well developed as a result of their increasingly common use in astronomy for spectroscopy and imaging. CCDs require special treatment to achieve good UV sensitivity, but this process is now reasonably well developed, and the resulting sensitivity quite satisfactory for spectrophotometry. CCDs are less likely to achieve optimum performance with thermoelectric cooling, but it is undoubtedly possible to operate a CCD with thermoelectric cooling for astronomical observations. The most difficult problem is to obtain one with the most useful format, a long narrow array with about 1000 pixels along its length. Over the next year, cr so, it will be necessary to determine if a suitable CCD can be obtained for this instrument.

A CCD would have an advantage in multiple-integration sequences. For example, it is most effective to break the integration up into four shorter integrations. This results from the following considerations. It is most efficient to measure the star and sky simultaneously, and an array detector makes this possible, if two diaphragms are used to project the stellar spectrum along one row of the chip and the other to project the sky spectrum along another row of the chip. But since the sensitivities will not be the same along the two rows, one must cancel out the sensitivity ratio by "beamswitching" - moving the telescope to place the stellar spectrum along the former sky row, and vice versa. If the first case has the star in diaphragm "A", and the sky in diaphragm "B", then one measures an A-B sequence on the star. One gains some immunity to changes in sky transparency if a symmetrical ABBA sequence is used; this requires four integrations as specified above. The CCD has the advantage that, instead of using the time, and incurring the read noise, to read out the chip after each of the integrations, one can simply shift the rows along the chip for storage until the complete set is done. After this the chip is read out, and the sum of the sky rows is subtracted from the sum of the star rows.

Since an array detector produces a large quantity of data with each observation, it will be necessary to do the summing and sky subtraction after each integration before the data are written to disk. Even so, each observation of a star will produce two spectra (first and second order exposures) containing about 1000 numbers, each of which is two bytes long. Thus, each observation contains a minimum of 4 K bytes of data; adding the "header" information gives roughly 5 K bytes. If 200 such observations are made each night, about 1 M bytes of data are produced. As a consequence, a hard disk of modest capacity is needed to store the data for a couple of weeks between times that it is downloaded. The data storage (and later data transmission time) can be reduced by a factor of 5 by using a data compression routine. We are designing the new programmable automatic telescopes, which will do filter photometry, so that the users will download the data over the phone lines via modems. This will still be possible with the ASPT, in spite of the larger quantities of data. Downloading one nights worth of spectrophotometric data will take about 80 sec, assuming data compression of a factor of 5 and a 19.2 kilobaud modem (Although these fast modems are still new on the market, the Fairborn Observatory has a pair and they are in use). This is quite manageable, even with two or three weeks of data.

It is worth noting that the amount of data produced by the spectrophotometer, as calculated above, is only about 5 times the amount of data produced by current single-channel filter-photometer APTs. Thus, the extension of the technology and mode of operations to go from the current systems to an ASPT is not great, and certainly is not as great as the step to an imaging CCD would be (because of more critical tracking, the much greater quantity of data, and the more extensive data processing needed for de-biassing and flat-fielding). Inevitably, automatic telescopes will be used for imaging, but an ASPT will be a suitable intermediate step.

2.4 System Performance

The system performance has been estimated by adjusting numbers given by Hayes (1987), which were based upon the performance of spectrophotometers at Kitt Peak National Observatory and elsewhere. The basic values are that an A-type star with a monochromatic magnitude per unit frequency interval of 10.0 mag. will give count-rates of 6.5 and 22 counts /sec/pixel at 3500 and 4000 Å, respectively. Thus, to achieve a photon-statistics precision of 1% at 4000 Å would require an integration time of 455 sec., or 7.6 min. An A-star is roughly 1.5 mag. fainter at 3500 Å, so to achieve 1% precision at this wavelength for a 10th mag star will require 4 times longer, or about 30 min. Thus, for 8-hour nights during which 70% of the time is spent on program stars (with the remainder being spent on standards and extinction stars), about ten 10th mag. stars per night could be observed, allowing for some overhead for slewing, acquisition and centering. In the case of 5th mag. stars, this figure rises to about 130 stars per night. In this case, an overhead of about 2 min. dominates, and brighter stars would not be observed much more rapidly.

The performance figures given above only take into account the sensitivity of the detector, and not the contribution of read noise. As noted in the preceding section, a reticon has a read noise of about 500 electrons; and the effect of this contribution to the noise will be to lengthen the exposure times given above by a factor of about seven. Clearly, a reticon would only be useful for very bright stars, if such a high precision were to be expected.

3. ADDITIONAL CONSIDERATIONS

3.1 Wavelength Coverage

To connect the optical region observations to the ultraviolet and to the infrared, it is desirable to observe as short and as long wavelengths, respectively, as possible. Shortward of 3300 Å, atmospheric ozone absorption makes it quite difficult to do good spectrophotometry. With a more adequate study of the atmospheric extinction it might be possible to go shortward to near 3200 Å. Longward of about 10,000 Å, the quantum efficiency of many of the optical region detectors drop. But for bright stars it should be possible to observe longward to near 11,000 Å. So we have a spectral region of order 7800 Å to study. Some wavelengths in the red and near infrared are seriously affected by telluric bands. Again more adequate study of the atmospheric extinction should aid the calibration. If initially it is not possible to properly model the atmospheric extinction in certain wavelength regions, then such problems can be made a subject for future research. Some interesting meteorological data may be hidden in the extinction measurements.

3.2 Modes Of Operation

With an ASPT, three modes of observations would be possible. First one could perform absolute all-sky photometry, which is generally what has been done with rotating grating scanners. But with the significantly faster ASPT one could also perform differential spectrophotometry of variable stars. This mode would resemble differential broad- and intermediate-band photometry in which one mixes observations of comparison, check and the sky with the program variable star. For variable stars this mode would probably be the most common way of performing observations. One can make the necessary observations with skies not as pristine as for all sky photometry. Finally one could operate in a spectroscopic mode when the sky conditions were worse than could be tolerated for differential spectrophotometry and yet were not so bad as to exclude observations. This mode would be useful for measuring relative changes in nearby spectral features. It might be useful for example in monitoring stars with large changes in emission or absorption features. As a large fraction of number of nights at any site are spectroscopic, but not photometric, a substantial amount of data could be obtained in this mode.

A major failing of rotating-grating scanner spectrophotometer reductions is the use of mean extinction coefficients. If one determined the extinction according to good photometric practice, it would have taken at least 75% of the observing time. At best with the HCO scanner on the KPNO No. 1 0.9 m telescope, we could obtain of order 40 scans per night of which typically 10 scans per night were of standards. If with the ASPT, 40% of the time were devoted to measuring the extinction and obtaining standard star measurements, then 48 scans per night would be used for these purposes. In an 8 hour night, this is six scans per hour which is the lower bound for good all sky photometry. On those nights the APTs at the same site are performing all sky photometry, the APT and ASPT observations could be combined to improve the extinction solution.

3.3 Amount Of Data

The amount of data that an ASPT could generate is quite large. To slew to and to observe a 5th magnitude A star will take at most 4 minutes. Thus at least 15 such observations are possible in a hour or about 120 per night. If there are 150 photometric nights per year, then of order 18,000 such measurements per year are anticipated. This is many times the good spectrophotometric data in the literature. In fact, all accessible stars in the Yale Bright Star Catalog and its supplement could be observed several times in one year. In addition it should be possible to obtain relative energy distributions on perhaps the equivalent of another 50 nights a year and obtain some 6,000 additional measurements. If the observing selection algorithm takes into account both the sky and stellar brightness, it should be possible to extend slightly the usable observing time by making measurements of the brightest accessible stars in twilight at the beginning and end of each night.

3.4 Operations

An ASPT could be developed and operated by either a single institution or by a consortium of institutions. As most astronomy departments and observatories consist of individuals with a wide variety of interests, we believe that a consortium would be an appropriate way of operation. Such a telescope could probably supply the data needs of a large fraction of the world's spectrophotometrists. By dividing the costs of building and operating such a facility among many institutions, a share could become affordable to even some of the smaller institutions. With the scheduling techniques being developed for APTs, all of the partners would not have to have equal shares.

In some senses an ASPT consortium would have to behave as a single institution. It is extremely important that the data be reduced using a uniform reduction procedure and that it be archived. The potential volume of data suggests that we have one astronomer supervising the data reduction with the procedures agreed upon by all the consortium astronomers.

We forsee two classes of projects, major projects in which the entire consortium participates and individual projects. Two major projects should be done during the first two years of operation: revision of the secondary standards and production of a catalog of representative stellar fluxes. We would anticipate that these projects would use about 50% of the observing time. After the completion of the these major projects part of time that was initially allocated to major projects might be used for collaborations with non-consortium astronomers.

The catalog of representative stellar energy distributions should include typical Population I and Population II stars covering the observed range of metallicity. Stars with good elemental abundance analyses and standards of all major photometric systems should be included in this sample. This will be invaluable in population synthesis studies which utilize ASPT observations of globular clusters, galactic clusters and associations, and galaxies. A major auxiliary project should be to synthesize magnitudes in all major photometric systems to link the photometry and spectrophotometry together and to understand the physical content of photometric systems. Another major auxiliary project should be to use this data in connection with other astrophysical data to determine the fundamental parameters (such as effective temperature, surface gravity, and metallicity) of as many stars as possible, especially those with known parallaxes. The rights to use the major project data in other ways would be divided among the consortium astronomers. We believe that it will be in the interest of the consortium to undertake additional major projects after the completion of these two although the share of major project time might well be reduced.

Individual projects would be done by one to a few consortium astronomers using their share of the observing time. The areas of the latter projects will have to be negotiated among the consortium members. Many such projects have been suggested at this meeting.

Four types of individuals will be involved in the running the ASPT: employees and contractors, participants, associates, and guest participants. The employees and the contractors could include individuals who run the telescope, who would help bring into operation, and who would reduce the data for us. The participants would be astronomers from institutions which belonged to the consortium. Associates would be astronomers from non-participating institutions who were invited by the consortium astronomers to collaborate on a major project. Guest participants would be astronomers who were collaborating with one or more consortium astronomers on some individual project.

We have been involved with the Four-College Automatic Photoelectric Telescope Consortium, one as a consultant, one as a co-investigator, and one as a subcontractor building the telescope and photometer. In this consortium, the title to the telescope belongs to one of the four institutions. This is one way to run a consortium, but at times it has lead to feelings of inequality among the partners. An alternative way to run a consortium is to set-up an non-profit corporation in which the various institutions are partners. This has the advantage that the telescope belongs to all, but at the same time to none of the institutions.

4. ASTROPHYSICAL CONSIDERATIONS

At New Directions in Spectrophotometry, many possible scientific projects which would be possible with an ASPT were presented. We would like to suggest a few others.

The next generation of optical region spectrophotometric instruments should be able to provide better time-resolved observations. With rotating-grating scanners and a reasonable number of wavelengths, it was hard to get five scans an hour with the HCO Scanner. But with an ASPT, it might well be possible to observe a bright star at least once every other minute. This is important for objects with physically important short time scales such as RR Lyrae stars, β Cepheids, δ Scuti stars, flare stars, dwarf novae, catacylsmic variables, and some Ap stars. For some of these types we would want to see the effects of stellar pulsations. Hydrodynamic effects should also be present in the energy distributions of Cepheids of Populations I and II, but in these stars although their periods are of order days what one is probably looking for occurs only in a very limited part of their period.

For unresolved binary systems, ASPT observations could be used to place constraints on the components. Multicolor light curves could be generated and solved. Some energy distributions will indicate the presence of hitherto unknown companions. For RS CVn systems, evidence of the dark wave should be found (see, e.g. Shore and Adelman 1984).

If the ASPT is programmed to find and subsequently track a slowly moving object, then spectrophotometric studies of solar system objects would become possible. For asteroids, one could study their energy distributions as a function of aspect in an attempt to deduce information on inhomogeneous surface compositions and their shapes. Spectrophotometric measurements of cometary comae and tails through various entrance apertures might be useful in understanding the chemistry of these objects. One could also obtain observations of the planets and perhaps the brightest moons, but this would probably require using the CCD camera images and appropriate offsets in a real time mode.

Much of the ASPT science with a resolution of order 10 Å will involve studies of continuum or quasi-continuum features. Only the equivalent widths of the strongest atomic lines can be deduced if the continua are sufficiently smooth. This will be important for the Ca II K line and for the strongest isolated He I lines, such as λ4922 and perhaps λ5875 (provided its non-LTE effects can be calculated). Objects with variable and/or strong emission lines such as in Hα can also be studied.

The long-term consistency of the data obtained with an ASPT is important for the study of stellar variability with periods of order a year or more. The development of dust in Mira atmospheres, the variability of solar type stars, and the striking changes of FK Comae stars should be monitored. Such stability is also important in being able to synthesize good photometric colors from spectrophotometric data.

Many spectrophotometric studies involve the determination of stellar effective temperatures and surface gravities via comparison with the predictions of model atmospheres. This type of study will continue to be important. Especially if the errors in the calibration of the primary and secondary standards can be reduced, the confrontation between observations and model atmospheric fluxes will provide critical tests of stellar model atmospheres and their input physics. The existence of non-LTE effects in the continua of hot and/or metal-poor stars can be studied as well as effects due to the non-plane-parallel nature of extended stellar atmospheres. Observations of stars with solar-type and non-solar-type compositions will indicate how the abundance anomalies affect the energy distributions. For hot stars the ability to link optical and ultraviolet observations is needed while for cool stars it is the ability to link optical and infrared observations. Stellar flux observations covering very extensive wavelength ranges are also quite useful for discovering unknown binary companions and for population studies via spectral synthesis. For hot stars spectrophotometric measurements longward of 7500 Å are few. An ASPT could well open this region to study.

Studies of all types of stellar variability can be performed with an ASPT. For Be stars one could look for the start of an emission episode and then trace its development. It would be necessary to develop a quick-look algorithm which would compare the data as observed with some canonical values and if significant differences were found the ASPT would begin a preplanned intensive monitoring program. For magnetic Ap stars, ASPT observations would be important in defining the changes in the continua as well as in the broad, continuum features. Continuous wavelength coverage is desirable to define the extent of such features (see Adelman and Pyper 1985 and references therein).

There are also going to be many nights which the ASPT will not be able to obtain even relative spectrophotometry measurements in an all sky mode, but still will be able to obtain useful data. As the stellar fluxes will be obtained simultaneously over a considerable wavelength range, programs which are concerned with the relative strengths of features could be performed. So on nights with thin clouds we can anticipate monitoring programs such as for example of the activity of Be and WR stars.

ACKNOWLEDGEMENTS

The authors are indebted to a number of their colleagues for valuable discussions during the design of these concepts. In particular, they would like to thank Lou Boyd, Dave Latham, Bruce Weaver and Nat White. SJA's work was supported in part by grants from The Citadel Development Foundtion.

REFERENCES

Adelman, S. J. 1987 in New Generation Small Telescopes, D. S. Hayes, R. M. Genet and D. R. Genet, eds. Fairborn Press, Mesa, p. 157.

Adelman, S. J. and Pyper, D. M. 1985 Astron. Astrophys. Suppl. **62**, 279.

Boyd, L. J. and Genet, R. M. 1987 in New Generation Small Telescopes, D. S. Hayes, R. M. Genet and D. R. Genet, eds., Fairborn Press, Mesa, p. 35.

Breger, M. 1976 Astrophys. J. Suppl. **32**, 1.

Genet, R. M. 1986 in Automatic Photoelectric Telescopes, D. S. Hall, R. M. Genet and B. L. Thurston, eds., Fairborn Press, Mesa, p. 1.

Genet, R. M., Boyd, L. J. and Baliunas, S. L. 1986 in Automatic Photoelectric Telescopes, D. S. Hall, R. M. Genet and B. L. Thurston, eds., Fairborn Press, Mesa, p. 15.

Genet, R. M.. Boyd, L. J. and Genet, D. R. 1987 in New Generation

Small Telescopes, D. S. Hayes, R. M. Genet and D. R. Genet, eds., Fairborn Press, Mesa, p. 28.

Genet, R. M., Boyd, L. J. and Hayes, D. S. 1988 in New Directions in Spectrophotometry, A. G. D. Philip, D. S. Hayes, and S. J. Adelman, eds., L. Davis Press, Schenectady, p. 311.

Hayes, D. S. 1987 in New Generation Small Telescopes, D. S. Hayes, R. M.Genet and D. R. Genet, eds. Fairborn Press, Mesa, p. 186.

Hayes, D. S., Genet, R. M. and Boyd, L. J. 1988, in preparation.

Shore, S. N., and Adelman, S. J. 1984, Astrophys. J. Suppl. **54**, 151.

DISCUSSION

WEAVER: Very few members of the interesting variable star classes you mentioned are as bright as 5th magnitude. I think that this is the reason that very few spectrophotometrists take six standards per hour.

ADELMAN: I agree in part. With the HCO scanner and/or the Cassegrain scanner at Palomar and a fair number of wavelengths to be observed it would have taken an hour to observe six standards.

TEAYS: Since you mentioned variable stars, I wonder if you could comment on your experience with, and expectations for the future, of dealing with time-critical observations such as getting a specific phase for a variable.

HAYES: We are designing the new systems to be able to handle this. As you recall from Genet's paper, the new Automatic Telescope Instruction Set has a variable called PRIORITY. Each "group" (a differential group or star plus sky, etc.) is assigned a PRIORITY, and the PRIORITY is used by the telescope/instrument controller in the process of selecting the next group to observe. You can simply specify a high PRIORITY and a time to observe the group, and it will be observed in preference to all other groups which could be observed at that time.

TAYLOR: Why use a transmission grating in your design? Can you use one of these without throwing away too many photons, especially in the UV?

HAYES: I chose a transmission grating for its favorable geometry, to get a compact instrument. I didn't look at the efficiency relative to a reflection grating.

TAYLOR: Would a reflection grating cost you seriously in terms of the design layout?

HAYES: Well, in the initial version of this spectrometer, which I described in a paper in New Generation Small Telescopes (Hayes, Genet and Genet, eds, Fairborn Press, Mesa, 1987), the transmission grating was very advantageous. That was for an f/3.75 telescope, whereas this one is for f/8. I didn't re-examine this issue when modifying the design, but I expect that a reflection grating could be used. Incidentally, the reflection grating has an advantage in giving a favorable geometry for guiding off of the zero-order image. This is an idea of Lou Boyd's, and I think it would be a valuable approach for an automatic spectrophotometric telescope.

JONER: For detectors that are thermoelectrically rather than cryogenically cooled, is the noise increase problem more serious for the CCD than the reticon?

HAYES: I think thermoelectric cooling is more disadvantageous for the CCD than for a reticon, but we will be observing in a range of brightnesses which give high signals relative to the dark counts, even with thermoelectric cooling, with either detector.

ETZEL: Are we going to see the Fairborn Observatory prototype a design and initiate construction, or are you interested in forming a consortium to support the development?

HAYES: The Fairborn Observatory has no active plans to prototype a design, and our intention is to form a consortium, if enough people with common instrumental needs can be found.

ADELMAN: The Fairborn Observatory is fully occupied with the construction of the three 0.75-m filter-photometric systems for the next year or so. A consortium would be necessary to begin the project in the near future.

LODEN: When do you consider it realistic to see this project in routine operation?

HAYES: In a few years.

ADELMAN: In about 1992.

KURUCZ: Couldn't this be run as a service? I could pay for a few stars.

HAYES: Well, yes. Fairborn Observatory runs such a service, which is called "rent-a-star," on one of the small telescopes doing differential photometry of variable stars. Certainly, there would be many more potential users than just the consortium members, particularly users with smaller programs. The Fairborn Observatory can't fund the construction of the system itself, but a consortium could run a "rent-a-star" program using some of the time.

ADELMAN: We need an "angel" to fund the construction of the telescope and instrument. Then you could fund the operation from the charges for observations.

A SUMMARY OF THE MEETING

R. A. Bell

University of Maryland

When I agreed to do this, I'm not sure that I realized what I was getting into. A quotation from Groucho Marx comes to mind. However, it does give me a chance to say how pleasant it is to see so many people, including a significant number of Europeans, at the meeting.

Some extraneous points come to mind. Firstly, this is really a pleasant place to have a meeting like this and whoever negotiated the room rate should go far. They should be buying telescopes. Secondly, we did learn a few new buzzwords, "A scanner is a failed spectrograph", "High dispersion Imperialism", "Aperture Fever" (and the even-more significant "Dark Time Aperture Fever") "Small Telescopes can be Fun". It's interesting that no one made corresponding comments about the supercomputer users, two (at least) of whom are chewing up goodly amounts of time.

I do want to comment on two of the supercomputer-based papers. Firstly, there was the paper by Puls, reporting on some of the Munich work which is led by Kudritzki. This paper pointed out the importance of studying mass loss by hot stars. The early work by Lucy and Solomon identified the mechanism of mass loss - a wind is driven by the pressure of radiation absorbed by spectral lines - and this work was developed by Castor, Abbott and Klein and by Abbott. The latter author used data for 250,000 spectral lines, computed for 30 elements in six stages of ionization. Puls identified the five requirements - mass and momentum conservation, the rate equations, the transfer equation and the energy equation - which must be satisfied by a self-consistent theory. He described the improvements which the Munich group have made - considering the angular distribution of the photons, improvements in the statistical equilibrium calculations, allowing for fact that lines of different elements can have the same wavelengths and allowing for the atoms in the wind radiating on the star. Puls also discussed the observational consequences of the work, pointing out cases where better agreement was obtained between theory and observation.

Kurucz discussed some of the new developments in his calculations of spectral lines. In addition to pointing out that some of the calculations give very good agreement with some of the very precise

A. G. Davis Philip, D. S. Hayes and S. J. Adelman, (eds.)
New Directions in Spectrophotometry 329 - 332

measurements of oscillator strengths made by Blackwell, it is clear that inclusion of the new lines gives much better agreement between computed and observed solar spectra. When you think that every neutral iron line which has been seen in laboratory spectra is seen in the solar spectrum (or is blended), it's obvious that we need Kurucz. It's also obvious from some of the other papers that Kurucz has had a profound influence on his field. For example, Dr. Malagnini gave us a detailed description of an application of Kurcuz's work.

Several people - Labhardt, Crawford, Horne - spoke of using catalogs of spectrophotometric observations to see if, or rather how well, colors computed from these data using the appropriate sensitivity functions match the colors of stars. This is clearly a very interesting approach because this is what the computational astronomers have been doing and it would allow us to assess the reliability of the sensitivity functions which are used in the color calculations. One basic problem which concerned Horne was, of course, the question of changes in the HST sensitivity functions with time. I do hope there will be an opportunity to measure the filter transmission profiles once more before launch.

Any use of the existing catalogs of the kind I've just described does rely on their being of very high quality. Some members of the audience reminded us that much of these data were obtained for other purposes and may not be ideal for what we have in mind. There is also the vexing problem of the lack of an observational calibration in some wavelength regions - caused by observational problems in studying Vega in the regions where there are either numerous Balmer or Paschen lines.

The question of the HST and ultra-violet calibration has clearly been addressed very carefully. Bohlin told us about his calibration work for IUE and the careful monitoring of the variation of sensitivity with time. Oke described his work in observing calibration stars for HST, supplying raw data for Horne to use. Oke's observations will obviously be of great value in other studies.

We heard more about IUE from other people - Johnson described his studies of the UV fluxes of a number of late-M giants. Fortunately, he did manage to get some ground-based support from a rather unexpected source, in one case. Carol Grady gave us a very thoughtful discussion about managing the large IUE data archive (I believe that its size has now been established as 30 Gigabytes) with its implications for managing data archives from automated telescopes. As one of those who occasionally writes proposals for using IUE, I am pleased that the catalog has been made machine-readable.

Warren spoke to us about the work of the Astronomical Data Center, located at Goddard. All of us on this side of the Atlantic owe Wayne a very great debt of gratitude for his work. I remember very well how he supplied the Gunn - Stryker catalog to me and helped with a problem I had with it.

Ake and Etzel discussed problems of binary stars - with implications for decomposing the spectra of composite objects. This led into Cacciari's talk, for if theory can't reproduce the observations of two or three (or even one!) star, it's going to be a weak crutch in studies of globular clusters of galaxies. Cacciari gave us a very nice description of the alternative methods of finding the population content of clusters and galaxies. This problem is complicated by our lack of knowledge of some aspects of stellar evolution - the structure of the horizontal branch and the way stars have of altering their surface-CNO abundance as they evolve. In addition, I personally believe that there are inconsistencies in the high oxygen content seen in subdwarfs and used to explain the turnoff and that needed to explain CO, CN and CH bands. Measurement of a library of spectra for cluster and galaxy synthesis is clearly a project for which an automated spectrophotometric telescope would be very valuable.

There are obviously several other people who would welcome an automated telescope - Peters for her work on Be stars and Pyper and Adelman on peculiar A-stars. I presume that some of Wing's work on infra-red spectrophotometry could be automated, as could the work of Little on $^{12}C/^{13}C$ abundances from near-IR CN bands.

Querci gave us an eloquent description of the need of spectrophotometric data to study the variability of late-type stars. Philip and Hayes told us about their work on the horizontal-branch stars using Strömgren photometry - with appropriate calibration this could be very valuable for obtaining globular cluster distances.

Among the several papers on instrumentation, Taylor enlarged our vocabularies and discussed the general problem of spectrophotometry. Bland described a very nice Fabry-Perot interferometer, Weaver described the virtues of solid-state detectors and Schempp discussed a new design for a holographic Fourier-Transform Spectrometer. McDonald described his spectrographic system and made the rather alarming comment that oxygen soaking only hypersensitized half of his TI CCD chip. The suggestion that a telescope of the size (24") he used could be devoted to line profile studies is a valuable one.

Keel explained the criteria which an automated spectrophotometric telescope would have to satisfy for a project of studying active galactic nuclei. White described how the angular diameter of some cool stars varied with wavelength, being dependent on the presence of spectral features. This gives a valuable constraint on model atmospheres, although it might be an optical depth effect rather than limb darkening.

Mantel described a spectrometer which seems to have a great deal of promise for studies of cataclysmic variable stars, owing to the high time resolution of 1 millisecond. These data might be used to synthesize broad- and narrow-band photometry.

Horne described some very useful improvements in the reduction of CCD spectroscopic data - improvements that should be widely used. Schultz described the analysis of some very interesting images of R Mon.

Finally, we had the session on automated telescopes. Genet reviewed the history of these instruments and described the present situation on Mount Hopkins. The available space is becoming crowded with quite sizeable telescopes. Hayes described a preliminary design for a spectrophotometric scanner for a 30" automated telescope.

I congratulate Boyd and Genet on their achievements. One critical contribution is that of Boyd who realized that it was not necessary to make an automated telescope point very precisely but that a good search procedure can compensate for approximate pointing. A tongue-in-cheek comment is that I'm glad the proposed spectrophotometric telescope will have a CCD camera - this may convince referees that the right stars have been observed.

The automated telescopes seem to me to be <u>the</u> way for us to establish a large photometric and spectrophotometric database which will allow us to make great progress in astronomy.

POSTER PAPERS

P CYGNI: AN EXCELLENT OBJECT FOR AN AUTOMATIC SPECTROPHOTOMETRIC TELESCOPE

Mart de Groot

Armagh Observatory

P Cygni is the type star of the P-Cygni-type (PCT) stars whose characteristics have recently been redefined by Lamers (1986). P Cygni is a hypergiant of spectral type B1 Ia^+ (Lamers et al. 1983). The star was discovered in the year 1600 when it had an outburst and reached third magnitude. After rather drastic brightness variations in the 17th century, the star settled down to $V \approx 4.9$ mag., showing only small variations with an amplitude less than 0.2 mag. in V (Schneller 1957; de Groot 1969). The PCT stars are located in the upper part of the HR diagram close to the empirical Humphreys-Davidson upper luminosity limit. PCT stars are believed to represent a relatively short-lived phase in the evolution of massive stars whose close investigation will contribute greatly to our understanding of the physical processes which play a crucial role in this part of the HR diagram. It appears that all PCT stars show high mass-loss rates resulting in photometric and spectroscopic variations. The relation, if any, between these two types of variation is not at all clear because very few suitable observations of these stars have been made so far.

The best-studied case is P Cygni, itself. Older photometric observations have been used to derive periodicities between 0.5 days and 18 years. Recent observations, obtained mostly with modest equipment by amateur astronomers (Percy et al. 1988) show that the star varies irregularly with a characteristic time scale between 40 and 80 days and with a typical amplitude of 0.2 mag. in V. Spectroscopic observations over the last decade, both at visual and UV wavelengths show structure in the shortward-displaced absorption components indicating the ejection of shells on a fairly regular time scale. However, whether this time scale is 60 days (Van Gent and Lamers 1986) or 200 days (Markova 1986) has not been established with certainty. Although polarization measurements of P Cygni are still very scarce, those by D. P. Hayes in 1985 allowed Percy et al. (1988) to show that an increase in polarization coincided with a decrease of the star's brightness after a sharp maximum in the light curve. This can be understood as the result of an asymmetric shell ejection. Further polarimetric observations should provide important clues about the nature of the shell ejection mechanism.

A. G. Davis Philip, D. S. Hayes and S. J. Adelman, (eds.)
New Directions in Spectrophotometry 335 - 336

The relation between the photometric and spectroscopic variations is not clear because simultaneous observations are so rare. The only report of such observations is by Baliunas et al. (1987) who obtained 19 Hα profiles during a 200-day interval in 1986. From their data one sees some indication of an increase in the visual brightness with a decrease in Hα emission intensity. A decrease in Hα intensity can be related to a decreased mass-loss rate, resulting in a thinner, more transparent shell which allows one to see deeper, hotter layers of the star's atmosphere. While this suggests that P Cygni's brightness variations are due to variations in the mass-loss rate, and while this seems to confirm the picture derived from the polarimetric observations, more simultaneous observations of these different kinds are needed to better specify the mass-loss mechanisms.

Since P Cygni is bright, high-resolution profiles of its Hα line can be obtained in a reasonable time. Baliunas et al. (1987) used the 1.5-m telescope at the Smithsonian's Oak Ridge Observatory to obtain a resolution of 0.015 Å. From the profile in their Fig. 7, one sees that a resolution of 0.05 Å would be sufficient for determining the existence of a correlation between Balmer-line intensities and brightness. Also, the variation of the radial velocity can be determined, which will lead to a more accurate value of the ejection time scale. Observations with an automatic spectrophotometric telescope would therefore be very useful, if this resolution could be obtained.

Photometry is being done by a number of observers, but more would be important, and polarimetric observations are very rare. This is a wide-open field for enthusiastic observers.

REFERENCES

Baliunas, S. L., Donahue, R. A., Loeser, J. G., Guinan, E. F. Genet, R. F. and Boyd, L. J. 1987 in _New Generation Small Telescopes_, D. S. Hayes, R. M. Genet and D. R. Genet, eds., Fairborn Press, Mesa, p. 97.

de Groot, M. 1969 _Bull. Astron. Inst. Netherlands_, **20**, 225.

Lamers, H. J. G. L. M. 1986 in _IAU Symposium No. 116, Luminous Stars and Associations in Galaxies_, C. W. H. de Loore, A. J. Willis and P. Laskarides, eds., Reidel, Dordrecht, p. 157.

Lamers, H. J. G. L. M., de Groot, M. and Casatella, A. 1983 _Astron. and Astrophys._ **128**, 299.

Markova, N. 1986 _Astron. and Astrophys._ **162**, L 3.

Percy, J. R., Napke, A. E., Richer, M. G., Harmanec, P., Horn, J., Koubsky, P., Kriz, S., Bozic, H., Clark, W. E., Landis, H. J., Milton, R., Reisenweber, R. C., Zsoldos, E. and Fisher, D. A. 1988 _Astron. and Astrophys._ **191**, 248.

Schneller, H. 1957 _Geschichte und Literatur des Lichtwechsels der Veranderlichen Sternen_, Akademie-Verlag, Berlin.

Van Gent, R. H. and Lamers, H. J. G. L. M. 1986 _Astron. and Astrophys._ **158**, 335.

JOINT SPECTROPHOTOMETRIC CATALOG OF THE STERNBERG STATE ASTRONOMICAL INSTITUTE AND FESENKOV ASTROPHYSICAL INSTITUTE: THE ANALYSIS OF COMMON STARS

I. N. Glushneva*, A. V. Kharitonov**, I. B. Voloshina*,
A. I. Zakharov*, L. N. Knyazeva**, and V. M. Tereshchenko**

*Sternberg State Astronomical Institute,
University of Moscow

**Fesenkov Astrophysical Institute of the
Kazakh SSR Academy of Sciences, Alma-Ata

The spectrophotometric catalog of the Sternberg State Astronomical Institute contains 875 stars; 8 of which are standards. A catalog of the Fesenkov Astrophysical Institute consists of 1123 stars. The data on the spectral energy distributions for about a half of these stars are published. A comparison of the energy distribution data for 473 common stars in the range 3200-7600 Å was made. The calibration of α Lyr from Hayes (1985) was used for all the stars of both catalogs.

Compared with previous publications the comparison includes many more stars and the energy distribution data were corrected where necessary. The agreement between the corrected data is better than the case of the earlier comparison. Reduction factors due to instrumental effects were taken into account for 275 stars from the Sternberg Institute catalog. These factors are between 1.01-1.04 and their mean value is 1.03. The reduction factors for the stars from the Fesenkov Institute catalog are much larger: 1.06-1.12 with the mean 1.06-1.07. In this case the reduction factors were taken into account for all the stars of the catalog except the brightest ones with V < 2.0 mag.

The values of the differences between spectral energy distribution data for common stars as a function of magnitude, spectral type and wavelength are presented.

[Editor's note: The need for carrying through a process of homogenization of these catalogs is discussed by Hayes (1986). The published data is spread through many papers; references to much of these data may be found in Hayes (1986), also.]

A. G. Davis Philip, D. S. Hayes and S. J. Adelman, (eds.)
New Directions in Spectrophotometry 337 - 338

REFERENCES

Hayes, D. S. 1985 in IAU Symposium No. 111, Calibration of Fundamental Stellar Quantities, D. S. Hayes, L. E. Pasinetti and A. G. D. Philip, eds., Reidel, Dordrecht, p. 225.

Hayes, D. S. 1986 in Highlights of Astronomy, Vol. 7, J.-P. Swings, ed., Reidel, Dordrecht, p. 819.

SPECTROPHOTOMETRIC DATA ARCHIVES

C. Jaschek

Stellar Data Center, Strasbourg

One critical point in spectrophotometry is the data archive. Although it seems obvious that data must be published in print, this is not enough today, because any printed spectrum is practically lost for automated retrieval. On the other hand most journals are unwilling to publish lengthly tables (flux vs. frequency or wavelength). It seems thus obvious that authors should present their data also in a computer-readable way, be it on diskettes or on tapes.

I think some discussion of archiving would be most welcome. As you may know there exists an "IAU archive of unpublished observations of variable stars" where numerical data not printed in journals can be stored. It would be worthwhile to discuss a similar data bank for spectrophotometry consisting for instance of one volunteer who collects and puts all data into a similar format. This obviously has to be done by an expert. He can then pass this data collection to any data center like NASA Goddard, CDS Strasbourg or DC Moscow for further distribution. (Since all data centers are linked by exchange agreements, the data then become accessible worldwide). This procedure avoids that, in addition to his work, the volunteer also has to copy and ship the data requested by colleagues.

Another way would be to encourage authors to send their longer data collections directly to a data center. This has the drawback that each presentation will probably be different, as well as the supporting documentation; the different collections then have to be handled by data centers who usually have no expertise in spectrophotometry. And smaller data collections probably will never find their way to data centers.

So in my opinion the first approach is definitely superior.

A. G. Davis Philip, D. S. Hayes and S. J. Adelman, (eds.)
New Directions in Spectrophotometry 339

RANDOM SAYINGS

RANDOM SAYINGS

"I wouldn't want to be a member of a club that would have me as a member."

A scanner is a failed spectrograph.

What is the effect of bleeding? It messes things up.

High Dispersion Imperialism.

Large Aperture Fever.

You get that warm fuzzy feeling that you have done something right!

Leave that to Gliese.

The net result is that it is difficult to explain the high dispersion results.

Monitoring stars is boring!

Will there be enough time to make calibration measures?

I hope all the observers take this to heart.

There is a very great cost difference between us and NASA.

The people saw that the coffee came!

If I send $1000 for one star, would it get done?

I remember padding my thesis with lots of computer output to make it thick enough to fit the word spectrophotometry on the spine.

You need to find an Angel!

INDICES

If a page number is underlined in the Name Index it indicates the name of an author of a paper. An underlined page number in the Object or Subject Index indicates that the object or subject was mentioned in the title of the paper.

NAME INDEX

OBJECT INDEX

SUBJECT INDEX

ADDRESSES OF PARTICIPANTS

Dr. Saul J. Adelman
Department of Physics
The Citadel
Charleston, SC 29409-0270
(803) 792-6943
G260031@UNIVSCVM

Dr. Thomas B. Ake III
CSC/Space Telescope Science Institute
3700 San Martin Drive
Baltimore, Maryland 21218
(301) 338-4934
AKE@STSCI.ARPA

Dr Roger A. Bell
Astronomy Program
University of Maryland
College Park, Maryland 20742
(301) 454-3005
ROGER@STARS.UMD.EDU

Dr. J. Bland
Institute for Astronomy
2680 Woodlawn Drive
Honolulu, Hawaii 96822
(808) 948-8319
BLAND@UHIFA

Dr. Ralph C. Bohlin
Space Telescope Science Institute
3700 San Martin Drive
Baltimore, Maryland 21218
(301) 338-4804
(SPAN)SCIVAX::Bohlin

Mr. Louis J. Boyd
Fairborn Observatory
629 N. 30th St.
Phoenix, Arizona 85008
(602) 275-8258/235-1897

Mr. John W. Briggs
100 Oak Rim Way, 11-C
Los Gatos, California 95030
(818) 440-1136

Dr. Carla Cacciari
Space Telescope Science Institute
3700 San Martin Drive
Baltimore, Maryland 21218
(301) 338-4700

Dr. Roy Campbell
Physics Department
Southwestern Adventist College
Keene, Texas 76059
(817) 645-3921

Mr. Tod Colegrove
Physics Department
University of Nevada, Reno
Reno, Nevada 89557
(702) 786-8872

Dr. David Crawford
Kitt Peak National Observatory
P.O. Box 26732
Tucson, Arizona 85726
(602) 325-9346

Dr. Paul B. Etzel
Department of Astronomy
San Diego State University
San Diego, California 92182-0334
(619) 265-6169
SDSU!ALNILAM!ETZEL@UCSD.BITNET

Mr. Russell M. Genet
Fairborn Observatory
1357 N. 91st Place
Mesa, Arizona 85207
(602) 986-2828

Dr. Carol A. Grady
Computer Sciences Corp./IUE Observatory
Code 684.9
NASA/Goddard Space Flight Center
Greenbelt, Maryland 20771
(301) 286-3938
(SPAN)IUE::GRADY or CHAMP::GRADY

Dr. Edwin J. Grayzeck
Physics Department
University of Nevada, Las Vegas
Las Vegas, Nevada 89154
(702) 739-3507
CI63C15@UNEV

Dr. Donald S. Hayes
Fairborn Observatory/ISO
P.O. Box 1907
Scottsdale, Arizona 85252
(602) 947-3572

Dr. Keith D. Horne
Space Telescope Science Institute
3700 San Martin Drive
Baltimore, Maryland 21218
(301) 338-4964
HORNE@SCIVAX.EDU

Dr. Hollis R. Johnson
Astronomy Department
Indiana University
Bloomington, Indiana 47405
(812) 335-4172
JOHNSONH@IUBACS

Mr. Mike Joner
Department of Physics & Astronomy
296 ESC, Brigham Young University
Provo, Utah 84602
(801) 378-4361

Dr. William C. Keel
Department Physics/Astronomy
P.O. Box 1921
University of Alabama
Tuscaloosa, Alabama 35487
(205) 348-5050
WKEEL@UALVM

Dr. Robert Kurucz
Smithsonian Astrophysical Observatory
60 Garden Street
Cambridge, Massachusetts 02138
(617) 495-7429
KURUCZ@CFAI

Dr. Lukas Labhardt
Astronomical Institute
Univ. of Basel
Venusstrasse 7
CH-4102 Binningen
SWITZERLAND
+41 61 227711
LABHARDT%URZ.UNIBAS.CH@CERNVAX.CSNET

Dr. Stephen J. Little
Department of Natural Sciences
Bentley College
Waltham, Massachusetts 02154
(617) 891-2721

Dr. Lars O. Loden
Astronomical Observatory, Uppsala
Box 515
S-751 20 Uppsala
SWEDEN
018-11 44 90

Dr. Maria L. Malagnini-Sicuranza
Dipartimento di Astronomia
Universita degli Studi di Trieste
Via G.B. Tiepolo, 11 (TS 5)
I-34131 Trieste
ITALY
39 40 768506
(EMAIL)PSI%02222403259::MALAGNINI

Dr. Karl-Heinz Mantel
Universitats-Sternwarte
Scheinerstrasse 1
D-8000 Munchen 80
WEST GERMANY
089-989021
UH10102@DMOLRZ01

Mr. Kelly McDonald
CASA
University of Colorado
Boulder, Colorado 80309
(303) 492-4050

Dr. Albert Merville
MIRA
900 Major Sherman Lane
Monterey, California 93940
(408) 659-3864

Dr. John B. Oke
Department of Astronomy
California Institute of Technology
105-24
Pasadena, California 91125
(818) 356-4007

Mr. Klaus Olesch
27659 W. Greenwood
Spring Grove, Illinois 60081
(312) 587-7862

Dr. Geraldine J. Peters
Space Sciences Center
University of Southern Californi
Los Angeles, California 90089-1
(213) 743-6962
(SPAN)CYGNUS::PETERS

Dr. A. G. Davis Philip
Van Vleck Observatory
1125 Oxford Place
Schenectady, New York 12308
(518) 374-5636
AGDP@UNION

Dr. Joachim Puls
Universitatssternwarte, Munich
Scheinerstrasse 1
D-8000 Munich 80
WEST GERMANY
089/989021
UH101AW@DMOLRZ01

Dr. Francois Querci
Observatoire Midi-Pyrenees
14, Avenue Edouard-Belin
31400 Toulouse
FRANCE
61 25 21 01
(EARN)node:FRMOP11 userid:CYRILL

Mr. William V. Schempp
PHOTOMETRICS Ltd.
2010 N. Forbes Blvd., Suite 103
Tucson, Arizona 85745
(602)623-8961

Dr. Alfred B. Schultz
Physics Department
University of Nevada, Las Vegas
Las Vegas, Nevada 89154
(702) 739-3784

Dr. Diane Pyper Smith
Department of Physics
University of Nevada, Las Vegas
Las Vegas, Nevada 89154
(702) 739-3755
CI63C19@UNEV

Dr. Benjamin J. Taylor
Dept. of Physics and Astronomy
Brigham Young University
Provo, Utah 84602
(801) 378-2233

Dr. Terry J. Teays
IUE Observatory
Code 684.9
NASA/Goddard Space Flight Center
Greenbelt, Maryland 20771
(301) 286-7537
TEAYS%IUESOC.SPAN@STAR.STANFORD.EDU

Dr. David A. Turnshek
Space Telescope Science Institute
3700 San Martin Dr.
Baltimore, Maryland 21218
(301) 338-4700

Dr. Arthur Upgren
Van Vleck Observatory
Wesleyan University
Middletown, Connecticut 06457
(203) 347-9411 X2829
AUPGREN@WESLEYAN

Dr. Wayne H. Warren, Jr.
National Space Science Data Center
Code 633
NASA/Goddard Space Flight Center
Greenbelt, Maryland 20771
(301) 286-8310
W3WHW@SCFVM

Dr. Wm. Bruce Weaver
MIRA
900 Major Sherman Lane
Monterey, California 93940
(408) 375-3220

Dr. Nathaniel M. White
Lowell Observatory
Mars Hill Road, 1400 W
Flagstaff, Arizona 86001
(602) 774-3358

Dr. Ramon L. Williamson II
Space Telescope Science Institute
3700 San Martin Drive
Baltimore, Maryland 21218
(301) 338-4541
WILLIAMSON@STSCI

Dr. Robert F. Wing
Astronomy Department
Ohio State University
Columbus, Ohio 43210
(614)292-7876
TS4718@OHSTMVSA

Other Publications in Astronomy

THE EVOLUTION OF POPULATION II STARS (1972), A. G. D. Philip, ed., Dudley Observatory Report No. 4.

MULTICOLOR PHOTOMETRY AND THE THEORETICAL HR DIAGRAM (1975), A. G. D. Philip and D. S. Hayes, eds., Dudley Observatory Report No. 9.

UBV COLOR-MAGNITUDE DIAGRAMS OF GALACTIC GLOBULAR CLUSTERS (1976), A. G. D. Philip, M. F. Cullen and R. E. White, Dudley Observatory Report No. 11.

AN ANALYSIS OF THE HAUCK-MERMILLIOD CATALOGUE OF HOMOGENEOUS FOUR-COLOR DATA (1976), A. G. D. Philip, T. M. Miller and L. J. Relyea, Dudley Observatory Report No. 12.

GALACTIC STRUCTURE IN THE DIRECTION OF THE POLAR CAPS (1977), M. F. McCarthy and A. G. D. Philip, eds., in Highlights of Astronomy, Vol 4, Reidel, Dordrecht.

IN MEMORY OF HENRY NORRIS RUSSELL (1977), A. G. D. Philip and D. H. DeVorkin, eds., Dudley Observatory Report No. 13.

IAU SYMPOSIUM NO. 80, THE HR DIAGRAM (1978), A. G. D. Philip and D. S. Hayes, eds., Reidel, Dordrecht.

IAU COLLOQUIUM NO. 47, SPECTRAL CLASSIFICATION OF THE FUTURE (1979), M. F. McCarthy, A. G. D. Philip and G. V. Coyne, eds., Vatican Observatory.

PROBLEMS OF CALIBRATION OF MULTICOLOR PHOTOMETRIC SYSTEMS (1979), A. G. D. Philip, ed., Dudley Observatory Report No. 14.

X-RAY SYMPOSIUM 1981 (1981), A. G. D. Philip, ed., L. Davis Press.

IAU COLLOQUIUM NO. 68, ASTROPHYSICAL PARAMETERS FOR GLOBULAR CLUSTERS (1981), A. G. D. Philip and D. S. Hayes, eds., L. Davis Press.

A DEEP OBJECTIVE PRISM SURVEY OF THE LARGE MAGELLANIC CLOUD FOR OB AND SUPERGIANT STARS. PART I. (1983), A. G. D. Philip and N. Sanduleak, L. Davis Press.

IAU COLOQUIUM NO. 76, THE NEARBY STARS AND THE STELLAR LUMINOSITY FUNCTION (1983), A. G. D. Philip and A. R. Upgren, eds., L. Davis Press.

IAU SYMPOSIUM NO. 111, CALIBRATION OF FUNDAMENTAL STELLAR QUANTITIES (1985), D. S. Hayes, L. Pasinetti and A. G. D. Philip, eds., Reidel, Dordrecht.

IAU COLLOQUIUM NO. 88, STELLAR RADIAL VELOCITIES (1985), A. G. D. Philip and D. W. Latham, eds., L. Davis Press.

HORIZONTAL-BRANCH AND UV-BRIGHT STARS (1985), A. G. D. Philip, ed., L. Davis Press.

SPECTROSCOPIC AND PHOTOMETRIC CLASSIFICATION OF POPULATION II STARS (1986), A. G. D. Philip, ed., L. Davis Press.

STANDARD STARS (1986), A. G. D. Philip, ed., in Highlights of Astronomy, Vol 7.

IAU COLLOQUIUM No. 95, THE SECOND CONFERENCE ON FAINT BLUE STARS (1987) A. G. D. Philip, D. S. Hayes and J. W. Liebert, eds., L. Davis Press.

IAU SYMPOSIUM NO. 126, GLOBULAR CLUSTER SYSTEMS IN GALAXIES (1988), J. E. Grindlay and A. G. D. Philip, eds., Reidel, Dordrecht.